Land and Water Issues Related to Energy Development

Land and Water Issues Related to Energy Development

Proceedings of the Fourth Annual Meeting of the International Society of Petroleum Industry Biologists

Denver, Colorado
September 22-25, 1981

Edited by

Patricia J. Rand

ANN ARBOR SCIENCE
THE BUTTERWORTH GROUP

QH
545
.E53 I57
1981

Copyright © 1982 by Ann Arbor Science Publishers
230 Collingwood, P.O. Box 1425, Ann Arbor, Michigan 48106

Library of Congress Catalog Card Number 81-70868
ISBN 0-250-40538-5

Manufactured in the United States of America
All Rights Reserved

Butterworths, Ltd., Borough Green, Sevenoaks
Kent TN15 8PH, England

8920034

PREFACE

Biologists in the energy industry are faced daily with decision-making problems their scientific training tells them require years of study to unravel, yet they often are asked for immediate answers. Mistakes or poor judgment can cost their employers millions of dollars and needless months of project delay. Wrong answers, too, may result in damage to the environment being protected. At other times, rather than too few data, there is a plethora of information from which to choose. Decision-making in the latter case thus revolves on selection of critically important items.

Since the stakes are high and so many people are involved, new ways to solve the dilemma of making the right choice appear regularly. Some survive the test of application and become incorporated procedurally; others, after a brief flurry of interest, disappear from view.

Meetings, of which this proceedings volume is an outcome, are forums in which ideas and methodologies are exchanged. Biologists associated with the energy industry, like their colleagues working elsewhere, can, and do, publish some of their research in the hundreds of specialty journals that exist. Few of these outlets, however, accept papers addressing problem-solving issues of special concern to energy industry biologists, and few are willing to publish papers in which important information is omitted because of secrecy clauses surrounding the work.

Industry biologists, trained for the most part as research scientists, are uncomfortable with this situation, for they recognize that peer review is essential to maintain and improve the quality of their work and to advance scientific knowledge. Although papers in this proceedings volume are directed primarily to biologists in the energy industry, we welcome discussion of the information by our colleagues in the scientific community at large and solicit their further cooperation in applying sound biological principles to environmental issues.

Patricia J. Rand

Patricia J. Rand is an ecologist on the corporate staff of the Atlantic Richfield Company, Los Angeles, California. As Senior Science Advisor in the Health, Safety, and Environmental Protection division, she monitors environmentally related activities and provides guidance and direction to the corporation in the general area of terrestrial ecology.

Dr. Rand received a BS in Natural Sciences and an MS in Botany and Zoology from the University of Minnesota - Minneapolis. Her PhD in Botany (Plant Ecology) and Geology is from Duke University, Durham, North Carolina.

Before joining Atlantic Richfield in 1973, she taught at the University of Minnesota, Hamline University, Duke University, the University of Arkansas-Fayetteville and the University of Nebraska-Lincoln. In addition, she has served as a Ranger-Naturalist with the U.S. National Park Service and as an environmental consultant to numerous governmental agencies and private architectural-engineering firms.

She is a Fellow of the American Association for the Advancement of Science and a member of the Ecological Society of America, the Botanical Society of America, the American Institute of Biological Sciences, the Nature Conservancy, the International Society of Petroleum Industry Biologists and Sigma Xi.

ACKNOWLEDGMENTS

The International Society of Petroleum Industry Biologists acknowledges with sincere thanks the many people who made the Fourth Annual Meeting of the group a success. We thank especially the invited speakers and the authors of contributed papers and posters. Their presentations and the discussions they stimulated make attendance at meetings worthwhile and this book of collected symposium papers interesting and valuable to read.

Invited speakers were: Nicholas C. Sonntag, Vice President of Environmental and Social Systems Analysts, Ltd. and Adaptive Environmental Assessments, Inc.; Virginia P. Carter, biologist with the U.S. Geological Survey; Robert Cook, Professor of Wildlife Management, Colorado State University, Fort Collins.

We are indebted, also, to participants in the panel discussion, "Industry and the Environment: The Role of the Biologist," who shared their insights with us. Panelists were: Ing. Jose Luis Garcia Luna, Vice President, Petroleos Mexicanos; Robert H. Nanz, Vice President, Shell Oil Co.; Charles B. Smith, President, Thunder Basin Coal Co.; Richard Hoos, Vice President, Dome Petroleum, Ltd.; Keith G. Hay, Conservation Director, American Petroleum Institute; June L. Siva, Manager, Environmental Sciences, Atlantic Richfield Co.; William J. Syratt, Environmental Officer, British Petroleum. James P. Ray, President, International Society of Petroleum Industry Biologists, was Panel Moderator.

Members of the Society gave freely of their time and skills to ensure the success of the meeting. Those in designated roles are listed here; our thanks to them and to the others who attended and helped in countless ways. The Symposium Committee was chaired by Bruce A. Cox and James Tate, Jr. Program Coordinator was Judith S. Phelps. Registration was handled by Christiane C. Garlasco. Mary Anne Sharpe was in charge of Membership. Robert L. Newell, Chairman of the Arrangements Committee, was assisted by Robert L. Dent and Tim J. Leftwich. The Scientific Program was chaired by Patricia J. Rand with the assistance of

Dee W. Chamberlain, James P. Ray and June L. Siva. Gary L. Austin, Chairman, with Dee W. Chamberlain, handled Finance Committee chores. Field trips were coordinated by John A. Lamping, Chairman, with Larry G. Kline and Gregg Squires. Mary Anne Sharpe, Kenneth E. Guziak and Patricia J. Rand chaired the scientific paper presentation sessions. Symposium opening and closing remarks were made by James Ray and Bernard Tramier.

Our deepest thanks to those individuals, not named here, without whom there would be no proceedings volume: the skillful typists who, literally, put the book together.

We are pleased to have had such associations, aid and assistance.

OVERVIEW

Papers presented at the symposium on "Land and Water Issues Related to Energy Development" are grouped in this proceedings in two main sections. They might, of course, be organized in other ways and, in fact, were presented in a different sequence at the meeting. If the categorization distorts your impression of the symposium, go directly to the papers; they contain the significant information.

Section 1 primarily addresses methods, from the general to the specific. It begins with ways to assess total ecosystems for management purposes, continues with techniques for studying individual components and ends with a method for conflict resolution.

Section 2 describes some effects of the energy industry on the environment. Expected subjects are found here, e.g., the effect of oil on birds, fish, invertebrates and, more exotically, polar bears. Some not-so-expected topics also are included in this section, e.g., the effects of ammonia on aquatic systems and the effects of brine on the native plants of two stands of boreal forest in Canada.

The leadoff paper in Section 1, by Sonntag et al., summarizes a resource evaluation method developed in British Columbia now being applied worldwide. Called "Adaptive Environmental Assessment and Management," it focuses on uncertainties in the evaluation of any system and combines computer modeling, optimization and decision analysis with the synergism of effective communication among individuals concerned in the analyses. Simulation modeling is employed to analyze impacts and to isolate data for critical decision-making.

Hartzband and Bennett also address decision-making in environmental planning and utilize mathematical formalism to model a problem and its solution. Their Multiple Attribute-Information Theoretic Decision Model (MAIT), according to the authors, can be applied to "decision problems which have well defined sets of alternatives characterized by quantifiable attributes." Manipulation of the information utilizing principal components analyses, multivariate probability density functions, nonparametric correlation methodologies, etc., then ranks alternatives and identifies the optimum one.

A mathematically less rigorous technique, but one that attempts to quantify alternatives, is described by Tate in a best-route evaluation case in Wyoming. Engineering values were combined with environmental ones in such a way that "varying percentages of environmental and engineering weight could be assessed." The method clearly demonstrated the best route to use based on both engineering and environmental criteria.

The research strategy of Truett and Douglass, described using an example from a project in the Piceance Basin of Colorado, is to limit baseline data collection to only a few species and their behavioral attributes. They conceptualize ecosystem interaction and human impacts by combining the newly collected data with information from published reports. By so focusing the investigation, they feel they are better able to identify effects and to design more efficient monitoring and mitigation systems.

Crumpacker outlines a cooperative plan presently underway to assist decision-makers in assessing impacts of energy development on wildlife species and their habitats in northwestern Colorado. The method relies heavily on the judgment of experts to make impact predictions if data are scant. A computerized data base of wildlife information will be available by 1985.

Reclamation and rehabilitation plans are of concern in most projects. Potter and Snyder point out that information gathered in comprehensive baseline studies frequently is not utilized properly, if at all, in reclamation planning. They outline a series of steps to improve its use, their goal being more ecologically sound revegetation and wildlife enhancement programs.

Ellis, too, is concerned with proper planning for reclamation, and uses as an example revegetation planning for the proposed Northern Tier Pipeline System route in the states of Washington and Montana. His method emphasizes proper identification of environmental, regulatory, political and post-reclamation land use constraints during the preconstruction phase.

Reclamation of uranium waste disposal areas in the western United States is the concern of Eaman, who focuses on soil and plant factors important to revegetation success. Easter also discusses reclamation methods. He describes a project, now in its third year, to revegetate disturbed areas in a drilling area of about 8 square miles at 9000 ft elevation in Wyoming.

A method to assess the rehabilitation potential of federal coal lease lands was developed by the Special Studies staff of the U.S. Bureau of Land Management. Incorporated in "Technical Investigations in Support of the Coal Program," it provides information needed to rank potential coal lease tracts in all major U.S. coal regions. A brief informational report of the program is included here.

The Endangered Species Act of 1973 and its amendments pose problems to industries in the United States planning development projects. Reyes-French and Shoemaker, however, believe most of the difficulties raised by the act can be obviated by adding a "threatened and endangered species" track to the engineering, environmental and permitting tracks usually operative in major development projects. The success of such integrated, early planning is demonstrated with examples from the Northern Tier Pipeline System.

Two authors use vegetation mapping as a critical tool for environmental management. Carter, one of the authors of the wetlands classification system currently used by U.S. government agencies, outlines the scheme, but cautions that understanding of the interactions of soils, vegetation and hydrology is needed for sound planning and management of the wetland units identified by the system. Degler uses detailed mapping of landforms, soils and plant community types on the arctic coastal plain in Alaska for making site selection decisions. She believes such maps are valuable planning tools and recommends their wider use.

A method of a different kind is presented by Pielou. She shows that it is possible to derive a functionally useful measure of the diversity of an aquatic community by recording the relative abundance of only the three most abundant organisms at a site. By plotting the information from many sites, a graphic picture of the "overall magnitude and the internal variability of an extensive community" emerges.

Kanter et al., while studying the biological effects of chronic low-level crude oil exposure on selected fish and shellfish, needed a way to maximize oil-water contacts. They developed an apparatus for producing solutions of water-soluble fractions of crude oil, the MBC solubilizer, which they describe in their chapter.

Tailings pond rehabilitation is the topic of MacKinnon and Retallack, who examined waters at the Syncrude Canada Ltd. oil sands plant in Alberta. They conclude that detoxification is possible relatively easily, at least on a bench scale.

Kaczynski et al. consider a mitigation measure of a different sort. These authors surveyed all of the known geothermal resource areas in seven western United States to determine the feasibility of using spent geothermal fluids to create waterfowl wetlands. From the total of 203 individual geothermal sources in the 91 geographic areas they examined, only three had the characteristics desirable for wetlands development.

Section 1 concludes with two chapters on conflict resolution. Boyce discusses the situation in Wyoming, where opposing demands for land

and wildlife resources in the coal-producing Powder River Basin have created severe management problems. Soule et al. describe a solution to a permitting problem in California, where consultations among regulatory agency and industry professionals revealed the need for more data to properly assess the situation. The involved companies collected the necessary data and, when they showed no harm would ensue to the environment, permit restrictions were lifted.

The chapters of Section 2 address specific effects of the energy industry on the environment. The section begins with what has become classic research for biologists in the petroleum industry, i.e., studies of the effects of hydrocarbons on plants, animals and their habitats. Frink describes how rehabilitation teams are organized on the East Coast of the United States to rescue and clean oiled birds. Hurst et al. discuss metabolic changes induced in polar bears by oil, whereas Wong and Engelhardt examine the effect of petroleum hydrocarbons on one of the enzyme systems of rainbow trout. In another study, Englehardt et al. assess the responses under varying conditions of *Daphnia pulex* to crude oil and some selected aromatic fractions.

The impacts of oil on total ecosystems also are of general interest. Getter reviews the literature on the effects of oil on mangrove forests in both the Old and New Worlds. He develops an "Index of Mangrove Stress" to quantify effects and monitor recovery of mangrove forests from oil spills.

Other products and activities associated with the petroleum industry may damage organisms and their habitats. Ammonia, for example, is a naturally occurring chemical whose appearance in surface waters may increase if coal conversion processes to produce synthetic fuels are accelerated. Fava et al. present an overview of the acute and chronic effects of ammonia on fish and aquatic invertebrates.

Brine solutions frequently are associated with crude oil recovery, and the potential for their accidental release is ever-present around drilling operations. Although agronomists know a great deal about the effects of saline irrigation waters and soils on the growth of agricultural plants, little is known about the effects of brine on other nonhalophytic species. Hettinger reports on the responses of native vegetation to brine solutions that he applied at two drill sites in stands of boreal forest in Alberta.

Petroleum exploration and production have the potential to affect the environment in other ways as well. Tramier, for example, reports that abandoned pits filled with drilling muds and fluids in southern France are rapidly colonized by the surrounding native plant species and thus have little detrimental effect on the landscape. The pits also can be revegetated readily with numerous introduced and native tree species.

Bennington et al. describe the effects of drilling and associated petroleum development activities on elk, bobcat and bear in Michigan. These animals were not adversely affected, and, in fact, the elk population increased during the period of the study.

The benthic inhabitants near pipeline crossings of streams in four areas of Pennsylvania, New York and Maryland were studied by Gartman, who also reports no adverse effects on numbers or variety of organisms.

Reynolds monitored the effects of winter travel of vehicles associated with oil and gas exploration activities in the National Petroleum Reserve on the North Slope in Alaska. She reports differential rates of recovery of the plant species in four different tundra types.

Cox et al., in a study still in progress, describe a hydrological investigation of a wetlands-pine forest mosaic community chosen as the site for a petroleum refinery expansion. They are determining the sources and quantities of nutrients transported from the site to an adjacent estuary as a measure of one significant effect of the project on its ecosystem.

The chapter by Mancini concludes the proceedings. He hypothesizes a worst-case spill scenario for a coal slurry pipeline and examines the possible effects on major organisms in five different types of waterways.

CONTENTS

Section 1. Environmental Assessment Methodologies

Adaptive Environmental Assessment and Management: An Integrated Approach to Resource Impact Evaluation
N. C. Sonntag, R. R. Everitt and M. J. Staley 1

A Multiple Attribute-Information Theoretic Decision Model (MAIT) for Use in Water Resource and Land Use Planning
D. J. Hartzband and C. J. Bennett 23

A Conveyance Route Selection Methodology
D. J. Tate 35

Ecosystem Analysis for Terrestrial Baseline Studies—An Example in Colorado's Piceance Basin
J. C. Truett and R. J. Douglass 57

Wildlife Conservation Strategy Associated with Energy Development in Northwest Colorado
D. W. Crumpacker 71

Improved Integration of Baseline Information and the Development of Revegetation and Wildlife Restoration Plans
G. L. Potter and B. D. Snyder 85

Revegetation Constraint Analysis for Pipeline Rights-of-Way
S. L. Ellis 95

Factors Associated with the Reclamation of Uranium Disposal Areas
T. K. Eaman 111

Land Reclamation Plan and Procedures Used on the Green Mountain (Wyoming) Uranium Exploration Project
A. R. Easter 121

Technical Investigations in Support of the Federal Coal Program
  Bureau of Land Management 139

The Endangered Species Act: Planning and the Section 7 Consultation Process
  G. A. Reyes-French and T. G. Shoemaker 141

Wetland Classification—Truth and Consequences
  V. Carter 157

Detailed Mapping for Developmental Planning
  S. A. Degler 165

Rapid Estimation of the Diversity of Polluted Communities
  E. C. Pielou 169

Apparatus for Solubilization of Crude Oil in Seawater
  R. G. Kanter, R. C. Miracle, C. J. Foley
  and M. L. Sowby 179

Preliminary Characterization and Detoxification of Tailings Pond Water at the Syncrude Canada Ltd. Oil Sands Plant
  M. D. MacKinnon and J. T. Retallack 185

Utilization of Geothermal Effluents to Create Waterfowl Wetlands
  V. W. Kaczynski, M. A. Wert and D. J. LaBar 211

Socio-political Conflicts in Wildlife Management on Coal Leases in Northeastern Wyoming
  M. S. Boyce 221

Environmental Science and the Regulatory Process: A Case Study Where Dialogue Led to Problem Solving
  D. F. Soule, M. Oguri, J. Lindstedt-Siva
  and J. P. Ray 233

Section 2. Effects on the Environment

A New Approach to Oiled Bird Rehabilitation After Oil Spills on the East Coast
  L. S. Frink 257

Metabolic and Temperature Responses of Polar Bears to Crude Oil
  R. J. Hurst, N. A. Øritsland and P. D. Watts 263

Effects of Petroleum Hydrocarbons on Branchial Adenosine Triphosphatases (ATPases) in Fresh and Salt Water Acclimated Rainbow Trout (*Salmo gairdneri*)
   M. P. Wong and F. R. Engelhardt 281

Effects of Petroleum Hydrocarbons on Trophic Factors of *Daphnia pulex*
   F. R. Engelhardt, R. G. Trucco, C. K. Wong and J. R. Strickler 297

Oil Spills and Mangroves: A Review of the Literature, Field, and Lab Studies
   C. D. Getter 303

Evaluation of Ammonia Toxicity in Site-Specific Environmental Assessments
   J. A. Fava, R. M. Kapp and W. J. Rue, Jr. 319

Vegetation Response to Brine Spill Reclamation Measures: Boreal Forest, Alberta, Canada
   L. R. Hettinger 339

Land Impact of Drilling Fluids Reserve Pits in South West France
   B. Tramier 353

The Effect of Hydrocarbon Development on Elk and Other Wildlife in Northern Lower Michigan
   J. P. Bennington, R. L. Dressler and J. M. Bridges 363

Pipeline Crossings of Streams: Benthos Recovery and Habitat Enhancement
   D. K. Gartman 385

Some Effects of Oil and Gas Exploration Activities on Tundra Vegetation in Northern Alaska
   P. C. Reynolds 403

A Hydrological Investigation of a Mosaic Ecosystem
   S. Cox, B. Hall, B. Langley, A. Winter, T. Simpson and D. Hawkins 419

Development of an Aquatic Biological Impact Assessment Methodology for Analysis of Hypothetical Coal Slurry Pipeline Spills
   E. R. Mancini 433

Appendix: The International Society of Petroleum Industry Biologists 453

Index 457

SECTION 1

ENVIRONMENTAL ASSESSMENT METHODOLOGIES

Adaptive Environmental Assessment and Management: An Integrated Approach to Resource Impact Evaluation

Nicholas C. Sonntag

Robert R. Everitt

Michael J. Staley
ESSA Environmental and
Social Systems
Analysts Ltd. and
Adaptive Environmental
Assessment Inc.

Society's recent heightened concern for man's use and abuse of the environment has spawned hundreds of public and private institutions totally committed to environmental research. However there is a gap between the research and its ability to impact environmental policy and management. This is mainly because of a failure to:

1. realize the potential of a number of available quantitative tools in environmental research; and
2. use environmental research effectively to impact environmental policy and environmental management.

Adaptive Environmental Assessment and Management (AEAM: Holling) (1), a relatively new environmental analysis methodology, specifically addresses these and other failings within environmental management. In the following pages we present the philosophy of the AEAM process and an example application.

PHILOSOPHY (BIASES) OF AEAM

Environmental systems are characterized by overwhelming uncertainty. Man's understanding of the underlying biological, social and physical processes and interactions is minimal, and will remain so for the forseeable future. The complex and pervasive nature of environmental and social issues guarantees our ignorance will always exceed our knowledge.

There are many sources of uncertainty in environmental problems. For example:

1. Often there is severe doubt about the accuracy of existing information and understanding. Data used in environmental work is often collected without consideration of the subsequent analysis. In addition deficiencies in these data are often overlooked in the rush for results and predictions.
2. Environmental systems will never be well defined. However analysis requires some bounds in space and time. These are usually artificial and inevitably lead to surprises due to phenomena outside of the scope of the analysis. The complexity of these problems within any space-time domain is often too great for any human or mechanical analysis.
3. Successful environmental analysis usually necessitates excluding some seemingly important components. Unfortunately often what is excluded dominates what is included. The dilemma is one of breadth versus depth. No matter how broad or how deep the analysis inevitably something outside will influence the results and violate the predictions. (Systems ecology can learn some lessons about predictability from meterology. Even though the physical processes of weather are well understood and the data are precise, weather on the intermediate time scale is fundamentally unpredictable. Phenomena occurring on a scale smaller than the monitoring system can cascade and rapidly dominate the weather.)

For these and other reasons we believe that environmental and social systems are fundamentally unpredictable. The problem then is if we accept this hypothesis, how can we effectively manage the environment? AEAM is an approach that attempts to deal with this problem of unpredictability (i.e.,uncertainty) and develop a workable environmental management strategy in the face of uncertainty. AEAM is by design not a fixed step by step procedural approach to environmental impact assessment. It is adaptive both in structure and intent, the objective being to minimize all the obvious and not so obvious biases that are so often the "albatross" accompanying a procedural approach. AEAM is in a sense a synergistic phenomenon that is more than the sum of its parts. AEAM combines the power of the mathematical and statistical sciences (i.e.,modelling, optimization, statistics, decision analysis, etc.) with effective modes of communication (i.e.,workshops, audio-visual production, graphic demonstrations) and imbeds it all in a philosophy of critical evaluation recognizing that the uncertain and unexpected lie in the future of any chosen direction.

In brief, AEAM's key features are as follows:

1. Ecological and environmental knowledge is incorporated with economic and social concerns at the beginning of a strategic analysis rather than at the end of a design process.
2. Since linked resource/social systems are dynamic rather than static and linear, techniques of simulation modelling, qualitative modelling, and policy design and evaluation are chosen to reflect these features.
3. Scientists, managers, and policy people are involved and interact from the beginning and throughout the process of synthesis, analysis, and design so that social learning becomes as much of a product as does problem solving.
4. Direction, design, and understanding are in the hands of those from the region who analyze, select and endure policies rather than in the hands of a separate group of analysts who lack the knowledge of needs, the responsibility, and the accountability.
5. Although prediction can be improved, the uncertain and unexpected lie in the future of every design. Hence policies are designed both to explore opportunities and pitfalls as well as to fulfill immediate social needs.

The major quantitative tool utilized within AEAM is computer simulation modelling. Simulation modelling is not new to resource management. It has been used extensively by systems ecologists for the last decade and has greatly enlarged the ecologists' "bag of tricks" for investigating the dynamic and stochastic properties of ecosystems (i.e., Hall and Day, 1977 (2); Watt, 1968 (3); Kremer and Nixon, 1978 (4); May, 1976 (5); Clark et al., (1979 (6)). What was learned precipitated a qualitative shift in the questions being asked about ecosystem behaviour although system complexity, and temporal and spatial considerations limited the contribution simulation could make to understanding "complete ecosystems". However, despite the wide use of simulation in ecology, it has failed to become an effective tool in environmental research and environmental management. Does this mean we should abandon simulation modelling as a useful tool in environmental science? No, instead we must change how resource models are perceived and used.

First and foremost we must never promote models as a panacea for prediction of the fate of society and its environment. Second, we must recognize that decisions are always made under uncertainty, ignorance, and unpredictability. Realizing this we must capitalize on the aspects of the modelling process that helps us focus ideas, integrate concepts, and guide decision making. Our experience has shown that modelling can help in these areas. It is our belief that better decisions can be made with a clear picture of both knowledge and ignorance, and a broad appreciation of the consequences of action.

No discussion of modelling philosophy would be complete without some mention of validation since most simulation modelling exercises spend large amounts of resources on validation. This activity may be useful in systems with few uncertainties and high predictability. However, in our opinion, there is little to be gained from validating a model of complex environmental systems.

In AEAM, validation is approached from the opposite direction: invalidation. In truth no bio-physical model is valid since it is by necessity simple in comparison to the real world. Further, a model is an hypothesis describing how the modellers think the world behaves and, if one wishes to remain true to the scientific method, the objective of evaluation (i.e., invalidation) should be to disprove the hypothesis (i.e., the model). If one fails, this doesn't necessarily mean the hypothesis is true (i.e., it could mean the right test has not yet been developed). In any case the more tests the model survives, the greater one's confidence in its predictions. However one must be careful not to believe the model, but rather use its output to suggest ideas and issues that require action (i.e., monitoring, research analysis, etc.). One cannot a priori, identify the limits of a model's predictive power or robustness, no matter how much effort goes into testing the model. There is invariably a real world process not included that will eventually cause a divergence between model and observation.

In utilizing models for resource management it is important to de-emphasize the quantitative nature of the model output and concentrate more on the qualitative aspects. We are not so interested in future projected numbers but more in qualitatively how the system responds given various perturbations imposed (i.e., pollutants, drought, low nutrients, floods, immigration, etc.). In other words we want to use the model to help:

1. develop an understanding of how robust the system is to stress;

2. determine if there are management actions that can partially or completely mitigate an adverse impact; and
3. come to some agreement on the quantity, quality, and kinds of data we need to both improve our understanding of the system and help us monitor the "health" of the system.

MODELLING WORKSHOPS: THE MODEL BUILDING PROCESS WITHIN AEAM

Simulation modelling within the AEAM process is facilitated in a unique environment, a simulation modelling workshop. The modelling workshop process was developed as part of AEAM in the early 1970's by Drs. Holling and Walters and associates at the Institute of Animal Resource Ecology, University of British Columbia and the Institute of Applied Systems Analysis in Vienna, Austria. Over the last 10 years, more than 60 workshops have been conducted covering a wide range of problems from pest management (Walters and Peterman, 1974) (7) to hydroelectric development (Walters, 1974) (8) to fisheries management (Walters et al., 1978) (9).

What Goes on in a Workshop?

A modelling workshop is a 3-5 day meeting of a group of scientists, planners, and managers involved in the design and execution of an environmental study. No papers are presented, there is no keynote speaker, there are simply 3 to 5 days of focussed activity on the problems at hand. The development of a computer simulation model of the physical, biological, and social aspects of the problem serves as the focus for the workshop.

The organizational structure of the workshop that has been found most useful is shown in a typical agenda for a 5 day workshop.

Day 1 morning   - introductory lectures
      afternoon - problem definition; looking outward

Day 2 morning   - complete looking outward; break into subgroups
      afternoon - subgroup meetings

Day 3 morning   - subgroup meetings
      afternoon - subgroup meetings

Day 4 morning   - overall model test runs
      afternoon - scenario development and gaming

Day 5 morning – gaming
afternoon – intergration and synthesis

Beginning on the morning of the first day introductory lectures briefly sketch the workshop goals and discuss the technique of computer simulation. At this stage the roles of the workshop facilitators and participants are clarified.

The role of the facilitators is to translate participant input into quantitative relationships that can be programmed into the simulation model. The facilitators can be viewed as information translators for it is the participants who conceptualize the model.

Participants in the workshop do <u>not</u> need any knowledge of computers or modelling to contribute to the workshop and to gain from its results. The model that evolves from a workshop is as much a product of the ideas and concerns of people unfamiliar with modelling as it is a product of those familiar with simulation techniques.

The problem identification exercise that occurs on the afternoon of the first day is of critical importance. This exercise, sometimes called bounding, defines what will and will not be considered by the workshop. At the end of this stage the general problem has been made explicit and has been broken down into manageable components or subsystems.

Looking outward, the next stage of the workshop, focusses attention on those variables that are required by a specific subsystem X from all other systems. In looking outward the standard question of analysis is recast. Instead of asking "what do you need to know to describe subsystem X?", the question of "to predict how subsystem X will behave, what do you need to know about all other subsystems" is asked. This question demands a more dynamic view and forces one to "look outward" at the inputs into a particular subsystem. At completion of the looking outward exercise each subsystem has a list of required inputs and outputs.

The workshop now breaks into subgroups, one subgroup for each subsystem. Each subgroup is in charge of developing a submodel for eventual inclusion in the overall system simulation model. One facilitator works within each subgroup and acts as the submodel programmer. Each subgroup has two basic charges: generation of output variables required by other submodels and generation of indicator variables of special interest to the subgroup.

At the conclusion of the subgroup meetings the facilitator is charged with programming each of the submodels on the computer. The final integration of all submodels into a working simulation model occurs during the fourth day. Often the model is not performing satisfactorily at the beginning of the day and most of the facilitators are occupied making corrections. During this period the participants are directed to specify a spectrum of interesting scenarios representative of the options available for project development. It is useful to propose extreme cases to test model behaviour and sensitivity.

On the fifth day the workshop turns to the task of evaluation and integration. The morning is spent gaming with the model using the scenarios generated previously. Clear and comprehensive presentation of model predictions is facilitated using SIMCON (Hilborn, 1973) (10), an interactive simulation language with graphical display capabilities. SIMCON's interactive nature permits rapid comprison between alternative scenarios.

The remainder of the day is directed towards synthesis of and reflection on the preceding four day's activities.

Why Use the Workshop?

The obvious objective of a modelling workshop is to build and run a computer simulation model of the biophysical system of interest. However, the resultant model is not an end in itself. Usually its predictions are not very precise and it often lacks obvious features of the actual system. Rather, the model is a focus for communication, promoting objectivity, and honesty. Building the model forces the participants to formalize their understanding of the system components and interactions. This facilitates easier evaluation of the importance of interactions to the system and the workshop objectives. Often favoured factors turn out to be irrelevant for prediction, therefore requiring less future effort both in model development and possible data-gathering programs.

As with many modelling studies, a workshop generated model confers the ability to test different management policies without risking the realities of policy failure. However, that is where the similarity usually ends. Since a workshop model is designed and built by all the participants its structure and resultant dynamics are "transparent" to the user. The model is comprehensible. This inspires trust and increases insight thereby promoting generalization of the model projections. It also facilitates easier evaluation of those factors left out of the model.

The modelling workshop style prevents the building of a sophisticated state of the art simulation model. The workshop model is invariably simple in structure and inefficient in operation. Further, by striving to simplify the problem the need for sophisticated tools such as complex implicit (or explicit) finite difference methods for approximating differential equations are usually avoided. Although the precision of the model's predictions may suffer, such simplification does not deter from the objectives of participant involvement, communication and understanding. People are the key components in the modelling workshop not the model. This does not have to be the case after the workshop. Often the model structure established serves as an excellent guide for future modelling efforts when more consideration can be given to the "art" of simulation modelling.

As stated earlier, by definition any simulation model is "wrong" since it must be a simplification of reality. However, complex models do not necessarily make better predictions. As the number of variables multiplies, so does the number of assumptions about how they are related and the chances of making a critically wrong assumption rise rapidly. Therefore, parsimony is the underlying workshop modelling theme. Interactions and relations should make sense when interpreted in terms of physics and biology. Logical consistency and clarity are stressed in the building of the workshop model and go far in maintaining its "transparency" to the participants.

Orchestration of Workshops Within AEAM

The incorporation of the modelling workshops in the AEAM process is designed to be iterative. This can best be demonstrated by the following hypothetical plan for a one year assessment. We stress hypothetical since each application will be different, however, the steps described here can be moulded to meet specific requirements.

I. January, a resource development problem is identified. A program manager is given the charge to produce a report evaluating the consequences of a development within 1 year. The manager (having chosen the AEAM methodology) next selects a small team of specialists (core group) who have specific responsibility and/or experience with the problem at hand. This team then meets with the workshop facilitators for a one or two day scoping session. A number of tasks must be accomplished at this time. The workshop objectives, management options and conflicting interests must be identified. A first cut is made at determining the spatial and temporal extent of the workshop model and the relevant variables to be included. Emerging from the scoping meeting is a list of participants, an understanding of the general form of the model and an assignment of responsibilities.

Although a number of decisions have now been made, it is tactically important for them to remain invisible during the first workshop and abandoned if appropriate. In the workshop these decisions will be made again by incorporating the expertise of all the participants.

II. February, the first workshop is held. This workshop is attended by the core group, facilitators and the subject specialists. Also, high level decision makers and managers are involved as much as possible. They may only attend the first day, or even the first hour, but their presence is essential to ensure what is done is relevant to their needs.

The structure of the first workshop focusses on problem definition, variable selection, and model design. All going well a rough model of the system will be running by the end of the workshop. In any case all the subject matter specialists must have a firm picture of the information and data needs to address the problem at hand.

After the workshop the core group and facilitators work on improving the model and the specialists develop their research plans for a coming field season.

III. April, a second 2-3 day workshop is held. By this time errors in model structure and coding have been corrected and the specialists have formulated their research plans and obtained as much information as possible from the literature. All new information is incorporated in the model and then used to explore suggested functional and/or management alternatives. Finally, each specialist's field work plans are reviewed to assure the data are **relevan**t and adequate.

Emerging from this meeting is a set of research plans for the specialists and a set of management options to be tested by the core group. The core group and facilitators then begin a process of improving the model through analysis, gaming, and technical meetings with some of the specialists as needed.

IV. September, a third and final workshop is held. The model is updated with the newly acquired data and insights gained in the past months. Then the participants, including some policy people, are given as much time as possible to game with the model in an interactive environment.

V. In the remaining months everybody's task is communication. Management options are explored with the model; information packages are produced; reports are prepared; demonstrations of the model are given to high level administrators and other interest groups. The purpose now is to affect decision making and all the creativity of the team should be directed towards that end.

This schedule is fairly ambitious both in timing and effort. It requires a strong commitment, both individual and institutional. It is, of course, an idealized structure and problem specific. Often the workshops will be spread out over a number of years and more than three workshops will be held. For example, a project requiring more than one field season could have a workshop between field programs with the intent of iteratively aiding in experimental design and contributing to current management.

In the above idealized schedule the AEAM methodology was initiated at the beginning of a study. Although certainly it is the recommended strategy, the AEAM process, and specifically modelling workshops, can be successfully implemented at any point in a study, right up to the end. The integration and coordination aspects of modelling workshops provide a useful vehicle for communication among those responsible for preparing environmental overviews and assessments. This ensures they are pertinent, credible, and address the questions being asked. Further, the resultant model (after some refinement) provides a very effective device for summarizing and presenting the results of a study to the policy makers, administrators and/or funding agencies.

Modelling Workshop Example: Bering Sea Crab Model

Background

To meet the energy needs of the United States during the remainder of the decade and throughout the next, expeditious development of Outer Continental Shelf (OCS) fossil fuel deposits has been determined to be in the national interest. The Alaskan OCS oil and gas deposits are potentially the largest undeveloped source of petroleum during a time of critical need. In each OCS area for which development is proposed, extensive environmental studies must be conducted before such development is allowed. If these studies show that development of specific areas will result in unacceptable environmental risks, those areas will not be leased or mitigating measures will be designed to alleviate these risks. As manager of the Outer Continental Shelf Leasing Program, the Bureau of Land Management (BLM) of the Department of the Interior (DOI) initiated the Outer Continental Shelf Environmental Assessment Program (OCSEAP) to conduct studies to ensure that the Alaskan marine environment is not damaged. Study programs for a number of OCS lease areas adjacent to Alaska were planned and should eventually be conducted under the interagency agreement for BLM by the Alaska Field Office of the Office of Marine Pollution Assessment (OMPA) of the National Oceanic and Atmospheric Administration (NOAA), U.S. Department of Commerce.

Because of the importance of the Bering Sea crab populations and the lack of a clear understanding of their dynamics, OCSEAP/BLM planned to sponsor a study in the 1981-1985 period, the initial objective of which would be to develop an understanding of how the crab populations are supported by critical physical and biological processes. A second phase of the project would determine the effects of oil on the important processes and, in turn, on the crab populations themselves. It was decided that the first step in addressing these objectives should be the development of a conceptual model of the southeastern Bering Sea ecosystem, which would then be used to identify hypotheses for testing in a field program. To this end, OCSEAP sponsored an AEAM modelling workshop, which was held at the Asilomar Conference Center in Pacific Grove, California, from July 7-11, 1980 (Sonntag et al., 1980)(11).

Biophysical System

The southeastern Bering Sea has received more scientific study than any other portion of the Bering Sea, (probably as a result of the lucrative commercial fisheries). OCSEAP has supported reconnaissance-level studies of circulation, biology, contaminant levels, and geology. The NSF-supported PROBES program (Processes and Resources of the Bering Sea) explored the cause of the area's high productivity, primarily through studies of the walleye pollock.

Results of these studies demonstrated the presence of three distinct hydrographic domains, based on vertical structure, temperature, and salinity (Figure 1). Inshore of the 50m isobath, the inner domain is vertically homogeneous and separated from the adjacent middle domain by a narrow (approximately 10 km) front. Between the 50m and 100m isobaths, the middle domain tends toward a strongly stratified two-layered structure and is separated from the adjacent outer domain by a weak front. Between the 100m isobath and the shelf break (approximately 170m depth), the outer domain has surface and bottom pronounced fine structure, as oceanic water intrudes shoreward from the weak haline front over

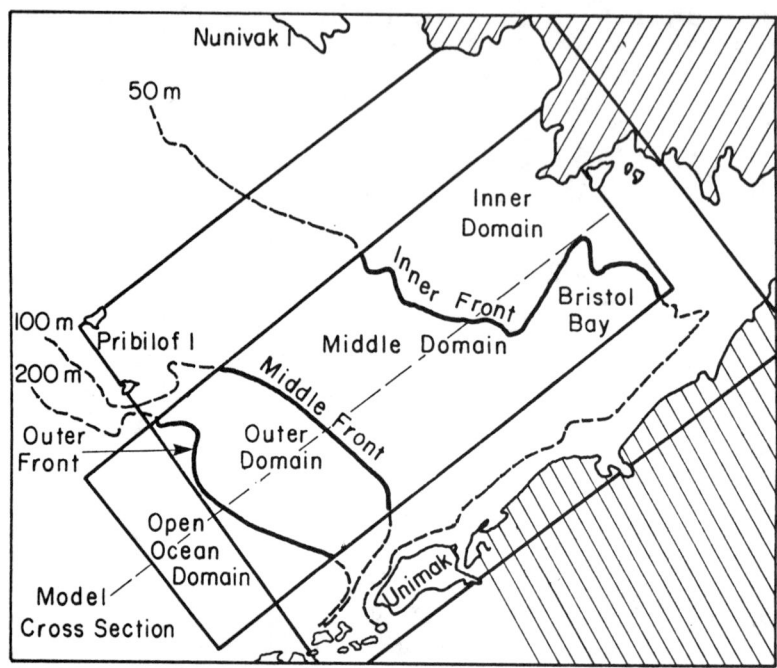

Figure 1. Map of southeastern Bering Sea showing the three distinct hydrographic domains and their associated fronts.

the slope, and shelf water (middle domain) passes seaward across the 100m isobath. These domains and their bordering fronts tend to persist through winter, although the absence of positive buoyancy often makes the middle shelf vertically homogeneous.

Important commercial species include the Tanner crab, located primarily in the middle domain and the inner front, the king crab, which migrate between the inner domain and the outer domain, and the walleye pollock, located primarily in the outer domain. The outer domain also contains dense populations of marine birds and animals, including the once nearly extinct northern fur seal. Large numbers of pink salmon migrate up the northern side of the Aleutian Chain and the Alaska Peninsula shoreward of the 50m isobath and into Bristol Bay, where they are harvested commercially.

Evidence indicates that the pelagic food web is only effectively coupled to phytoplankton production in the outer domain, and hence the major yield of pelagic animals, including pollock, is confined to this zone. In the middle domain the large phytoplankton (i.e., diatoms) are not effectively grazed, and most sink to the seabed where they nourish a benthic food web. This Bering shelf production scheme suggests that only in the region between the 100m isobath and the shelf break does the system yield pelagic animals, whereas on the remainder of the shelf, by far the major portion of its area, the yield would be in benthic-related animals. Evidence is also being gathered which suggests that the distribution of seabirds and the migratory routes of marine mammals are controlled by this system of fronts and domains.

Simulation Model

In bounding the model the "actions" were predominately related to the activities of oil and gas development (i.e., oil spills, blowouts on the bottom, chronic leakages, deposition of heavy metals and smothering from drilling muds) and the "indicators" were the biomass of selected pelagic and benthic populations, (i.e.,Tanner and king crabs, clams, crab predators, crab food). The spatial resolution of the model was based primarily on the oceanographic work of the PROBES program. The resultant structure (Figure 2) had six water columns designed to represent the division of the southeastern Bering Sea by the fronts and divisions of depth in multiples of 50m. The time horizon for model projections was 30 years with a resolution of one month.

The system was disaggregated into four subsystems: physical, pelagic, benthic, and crabs. For each of these subsystems a conceptual model was developed by the participants and a computer code prepared. These submodels explicitly accounted for the between submodel interactions developed during the looking outward exercise (Table I) thereby facilitating the operation of an integrated representation of the ecosystem.

The physical submodel was concerned with the lateral and horizontal transport of crab larvae, clam larvae, and detritus, and generating temperatures for each model spatial unit. It also determined spill-related concentrations of oil and dissolved toxicants in the water column and sea bottom.

The pelagic submodel represented daily primary production, detrial fallout associated with the primary production, standing stock of two zooplankton size classes, and fecal pellet production. Because of the model's focus on the benthic community, a detailed model of the population dynamics of the plankton community was deemed unnecessary.

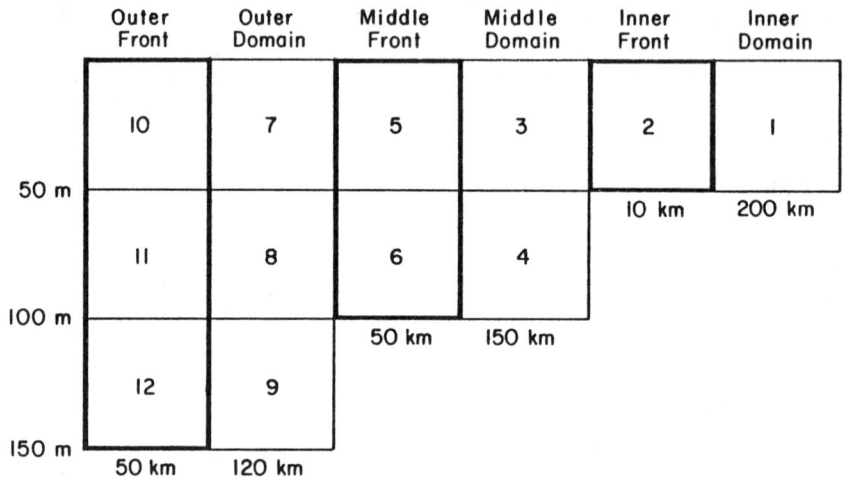

Figure 2. Schematic representation of model's spatial structure showing each of the spatial units explicitly included in the model. This vertical cross-section corresponds to the line indicated in Figure 1.

The benthic submodel was concerned primarily with the dynamics of the benthic organisms that comprise food for crabs. The biomass of food was a result of growth and recruitment for the various organisms, minus the effects of natural and oil-induced mortality, and predation by both crabs and their competitors.

The crab submodel was concerned with the dynamics of the Bering Sea crab population (i.e., Tanner and king) under both oil free and oil contaminated conditions. The populations were represented with an explicit age structure; growth and mortality were a function of available food, temperature, predator numbers, oil concentrations, and commercial harvest.

For illustrative purposes two example runs of the integrated model are shown in Figures 3 and 4. Figure 3 is a run of the model in the "natural" state (i.e., no oil or contaminants in the water). The overriding feature is the

Table I. Looking outward matrix developed by workshop participants.

FROM \ TO	Pelagic	Physical	Crab	Benthic
Pelagic		fecal pellets ($g\ C/m^2$)	small zooplankton (<1mm) ($g\ C/m^3$) large zooplankton (>1mm) ($g\ C/m^3$)	detritus fallout ($g\ C/m^2$) (planktonic and marcophytic)
Physical	temperature (^0C) ice cover (%)		temperature ^0C (monthly mean) new larval densities ($\#/m^3$) oil (ppm)	oil (g/m^2) diss. toxicants (ppm) temperature (^0C) lateral detritus movement ($g\ C/m^3$) new clam larvae ($\#/m^3$)
Crab		old larva densities ($\#/m^3$)		food benthos removed ($g\ C/m^2$)
Benthic		detritus to be moved ($g\ C/m^2$) old clam larvae ($\#/m^3$)	food benthos ($g\ C/m^3$) (clams and polychetes)	

tight coupling between benthic food and crab biomass. The cycles were a combined result of the variation in detrital flux, water temperatures, and predation within the year.

Figure 4 is an extreme scenario in which a chronic release from drilling activities resulted in the death of all crab larvae in spatial unit 4 for 6 years. The 6 year kill did have a noticeable effect on the crab biomass and resulted in a decreased harvest for about 6 years starting 7 years after the event. However, as with all other scenarios, the population was eventually re-established at previous population levels.

The basic conclusion drawn from the model runs was that even under the worst case scenarios the simulated crab population was very resilient and able to recover readily. The scenarios, however, only considered each impact separately and the model did not (nor did it have the capability to) consider any possible synergistic effects that could result from an oil-related activity.

The major feature of the crab population that determined its resilience in this context was its age structure and the

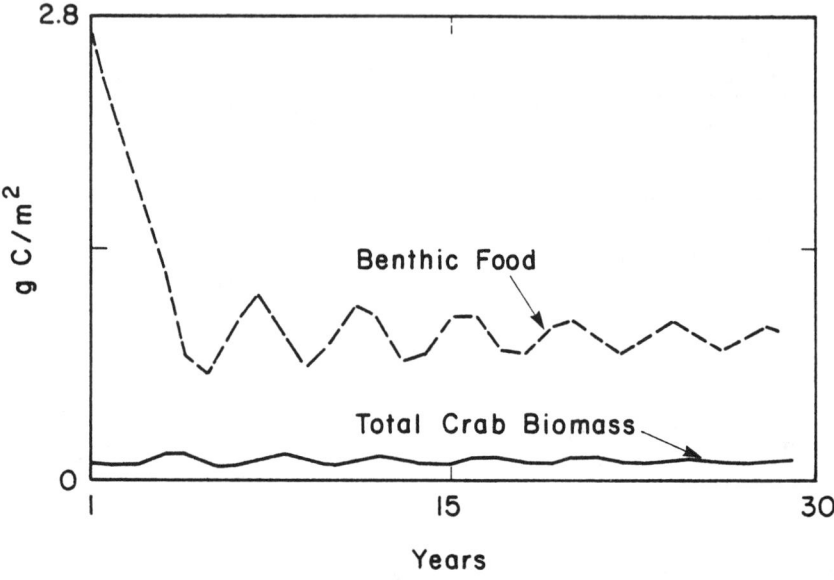

Figure 3. Standard scenario. Total crab biomass and total benthic food biomass in spatial unit 4.

fact that the crabs reach sexual maturity at 5 years but are not harvested until year 7. Even if one or more cohorts are severely depleted the remaining adults restore the population quickly. However, the age structure can also be the source of some problems. If, for example, the impact of man's activity is very age specific (i.e., larval stage only) then analysis of crab catch data would not flag the impact for a number of years after the initiation of the event. If the impact is long term or chronic and continues throughout the period of our ignorance, the results could be socially, economically and/or biologically catastrophic. Therefore, it is essential to carefully consider the variables to be monitored to ensure that critical events are not overlooked before it is too late to correct them.

Conclusions of Workshop

The objectives of the workshop were met with encouraging success. A conceptual and numerical simulation model integrating the physical and biological processes in the Bering

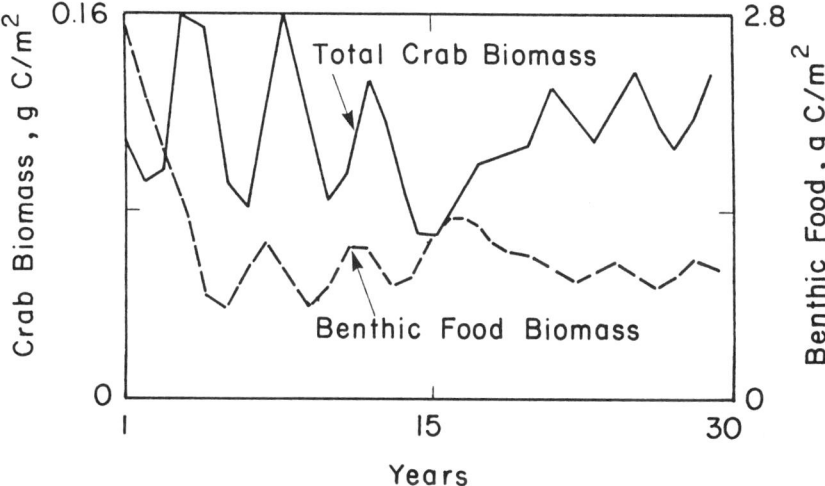

Figure 4. Egg mortality scenario. Total crab biomass and benthic food biomass in spatial unit 4 with a 6 year kill of crab eggs between year 10 and 16.

Sea considered relevant to crab population ecology was successfully developed. During the development and analysis of this model the other workshop objectives (identification of hypotheses, interdisciplinary communication) were also productively addressed. Table II is a summary of the conceptual gaps and information needs identified by the participants in the workshop.

Considering the small amount of information available on crab and benthic ecology in the Bering Sea the submodels developed were conceptually exciting and were good initial attempts to capture the dynamics of the whole system.

These submodels can be considered as preliminary hypotheses of how the subsystems operate. In the process of their development a number of information gaps provided the basis for the identification of research needs and hypotheses for testing in ensuing field programs. The identification of these needs in the interdisciplinary context of the workshop proved insightful for a number of participants. Hope-

Table II. Summary of conceptual and information needs agreed upon by participants at the workshop.

	Conceptual	Information
Physical	vertical transport of oil	adsorption of oil to particles; ingestion by zooplankton; oil solubility
		transport data for: (a) North Aleutian shelf; (b) Unimak Pass
Pelagic	cropping of detritus by bacteria and amphipods; what is the loss of detritus as it moves down through the water column?	timing and transport of detrital rain (rates and seasonality) sources of detritus and relative proportions
Benthic Food	relationship of benthic productivity to detrital rain and the effects of oil on this process	growth rates of food benthos; productivity
	population dynamics of benthic food	identification of shallow water benthos (distribution and abundance versus fronts)
	relationship of benthic community metabolism to secondary production	rates of accumulation and decay of oil in sediments, including physical resuspension and bioturbation
		extent of walrus foraging area (crab competitor)
		benthic food population parameters: (a) recruitment; (b) mortality; (c) benthic community metabolism; (d) competition and predation
		toxicity of oil on benthos
Crabs	larval survival: (a) predation of larvae, (b) relation between oil and larval survival, (c) cannabalism	temporal and spatial pattern of larvae in water column
	relationship of temperature to egg development and survival	larval and post larval population parameters: (a) feeding rates on zooplankton and benthic food, (b) recruitment, (c) mortality, (d) competition and predation
	post larval survival: (a) predation, (b) cannabalism	winter distribution of juvenile and adult crabs
	fishing fleet dynamics and harvesting pressure	toxicity of oil in water column and in sediments on the various crab life stages

fully this improved insight will encourage the collection of information which is more useful to an understanding of the system as a whole and thus to an assessment of potential impacts.

As with any ecosystem the biophysical dynamics of the Bering Sea are complex and will probably defy complete understanding for the forseeable future. What is critical is to keep in mind the objectives of the exercise, namely to identify the possible effects of oil drilling on the crab populations and the resultant commercial harvest. These effects can be direct or indirect, therefore requiring a more complete conceptual understanding of the dynamics of the crab population. Any base line data work must be underpinned with a well structured process-oriented look at the system. Not only are population estimates important but also rates of growth, mortality, reproduction, mobility and a number of other dynamic processes. Only by improving understanding of these processes will it be possible to adequately address the issues of impacts in both the short and long term.

Finally, it became clear during the workshop that the pelagic, benthic, and crab communities experience a great deal of natural variation due to an array of biophysical causes. It is critical that the impacts of man's activity be compared to natural variation (i.e., a time series) rather than an average (i.e., one number). If it is not possible to tell the difference in degree of variation between the natural and the natural-plus-oil-activities situations, then perhaps the impacts of the oil activity are acceptable. The situation to worry about is when as a consequence of man's activity the degree of variation shows a change, be it an increase or decrease.

WHAT HAVE WE LEARNED?

The AEAM process represents the combined learning of a number of international scientists and practitioners and is in a state of "dynamic equilibrium", continually adapting in light of new experience. But the term "adaptive" stresses a more important lesson, the need for research and management to be open to change and to be able to adapt in both style and content when new information becomes available. While AEAM itself is in a continual state of change, its two underlying themes, expect the unexpected, and learn to plan and plan to learn, never change.

The many applications of AEAM have taught that the modelling process is not only for the modellers; more will be gained if the users help build the model. This need for the users to identify with the model is well known; AEAM is a proven process satisfying this need.

Application of AEAM with an interdisciplinary team has many benefits. These benefits, which have arisen in almost all applications, may be considered a set of necessary conditions for successful application of AEAM in environmental research and environmental management.

An environmental analysis must:

1. focus discussion on the issues surrounding the environmental problem;
2. promote communication amongst the many disciplines represented in environmental problems;
3. make explicit the assumptions underlying the current understanding of the problem;
4. identify information needs;
5. provide a framework to evaluate the relevancy of existing information; and
6. provide a framework to evaluate different management actions and guide policy design.

In the first four conditions, it might be more appropriate to use the phrase "the process of executing the environmental analysis" instead of the "environmental analysis" because it is the process that ensures the success of the application. The sixth condition reflects a bias that contends that ultimately environmental research must be used to enhance public policy.

In public policy there is a growing awareness of the need to integrate environmental concerns with economic and social dimensions. Though admirable and well meaning in intent, current legislation designed to protect the environment has severe limitations. In an attempt to encompass all relevant issues Environmental Impact Statements (EIS) have become encyclopedic and lacking in rigorous analysis. The application of AEAM has taught the valuable albeit painful lesson that most of the information contained in EIS's is useless from a decision making or policy perspective. This lesson emerged repeatedly when decision makers, included as participants in workshops, were asked "What kind of information would you like the simulation model to provide to aid you in making your decisions?" From their perspective most of the information in an EIS is unnecessary, the remainder insufficient.

AEAM through its use of simulation in the workshop setting provides a mechanism for undertaking the analysis of impacts rather than being a way to massage baseline data. It is through thoughtful analysis and effective communication of environmental information that environmental researchers will reach decision makers. This is perhaps the most important lesson that AEAM has to teach.

CONCLUSIONS

Adaptive environmental assessment and management is in our opinion the most sensible, usable and diverse methodology available to orchestrate and coordinate environmental research and management. It is not perfect, but it works.

LITERATURE CITED

1. Holling, C.S. (editor). 1978. Adaptive Environmental Assessment and Management. Wiley Inter-Science, Chichester, 377 pp.

2. Hall, C.A.S., Day, J.W., Jr. (editors). 1977. Ecosystem Modelling in Theory and Practice: An Introduction with Case Histories. John Wiley & Sons, New York, 684 pp.

3. Watt, K.E.F. 1968. Ecology and Resource Management. McGraw Hill, New York, 90 pp.

4. Kremer, J.N., Nixon, S.W. 1978. A Coastal Marine Ecosystem: Simulation and Analysis. Springer-Verlag, Berlin, 217 pp.

5. May, R.M. (editor). 1976. Theoretical Ecology: Principles and Applications. W.B. Saunders, Philadelphia, 317 pp.

6. Clark, W.C., Jones, D.D., Holling, C.S. 1979. Lessons for ecological policy design: a case study of ecosystem management. Ecological Modelling 7: 1-53.

7. Walters, C.J., Peterman, R.M. 1974. A systems approach to the dynamics of spruce budworm in New Brunswick. Proc. 23rd Mts. Entomol. Soc. Can. Quaset. Entomol. 10: 177-186.

8. Walters, C.J. 1974. An interdisciplinary approach to development of watershed simulation models. Technol. Forecast. Soc. Change 6: 299-323.

9. Walters, C.J., Hilborn, R., Peterman, R.M., Staley, M.J. 1978. Model for examining early ocean limitation of Pacific salmon production. J. Fish. Res. Board Can. 35(10): 1303-1315.

10. Hilborn, R. 1973. A control system for FORTRAN simulation programming. Simulation 20: 172-175

11. Sonntag, N.C., Everitt, R.R., Marmorek, D., Hilborn, R. 1980. Report of the Bering Sea Ecological Processes Study Modelling Workshop. NOAA, Office of Pollution Assessment, Seattle. 96 pp.

A MULTIPLE ATTRIBUTE-INFORMATION
THEORETIC DECISION MODEL (MAIT) FOR USE
IN WATER RESOURCE AND LAND USE PLANNING

D. J. Hartzband
   Normandeau Associates, Bedford, NH

C. J. Bennett
   Normandeau Associates, Denver, CO

INTRODUCTION

    Many current problems in environmental planning require the development and evaluation of a complex set of alternatives in order to identify a "best" alternative which meets certain a *priori* needs. Examples of such problems would include the evaluation and selection of an optimum plant or mine site from a series of potential sites with respect to economic, environmental and sociological criteria, or the identification of an optimum water supply strategy from a set of alternative strategies with respect to economic, environmental, technical water usage, and public opinion criteria. These decision problems are simple to solve if few alternatives are available and if the criteria or attributes distinguishing the alternatives are distinct and non-overlapping. Unfortunately, few real life problems are so obliging; in fact, even if a problem has few alternatives, they are usually represented by a large number of criteria many of which are colinear (statistically indistinguishable), overlapping, or contradictory. In such cases, it is often helpful to have a mathematical formalism which can model the decision problem and allow a solution to be found. This paper presents a multiple attribute-information theoretic decision model for use in environmental planning. This model is demonstrated with data and recommendations from problems in the literature and from recent studies.

    There are a number of ways to quantitatively approach decision models. A very conservative procedure would be to rely on hypothesis testing methods with associated multiple comparison tests such as analysis of variance (Scheffe, 1959). There are many problems with this approach related to the violation of basic assumptions by the data concommitant with decision problems. The most intractable problem, however, is that such a procedure does not really answer the decision maker's question, namely which alternative is best. Gupta and Huang (1981) present a method which addresses the decision maker's question but which requires the evaluation of multivariate probability density functions. This is not always easy or even possible. Many other methods have been developed including other formal probability models (Keeney and Raiffa, 1976), multidimensional scaling decision analysis (Nijkamp, 1979) and multiple objective or design models which are generally linear programs with literally hundreds of constraints (Cohon, 1978).

Hwang and Yoon (1981) have divided multiple criteria decision problems into multiple objective and multiple attribute problems with their respective models. Multiple objective decision models (MODM) are not associated with problems where the alternatives are predetermined. MODM's are used to design a best alternative given a well defined set of constraints referring to quantifiable objectives. On the other hand, multiple attribute decision models (MADM) are used to select an optimum alternative from a small, well defined set of alternatives which are characterized by a series of quantifiable attributes. These attributes can be related to economic, environmental, engineering or sociological factors.

METHODS

The multiple attribute-information theoretic decision model (MAIT) presented in this paper is developed to be applied to multiple attribute problems in the sense of Hwang and Yoon, that is those decision problems which have well defined sets of alternatives characterized by quantifiable attributes. The quantified values for these attributes are expressed in an attribute (rows) by alternatives (columns) matrix. The attributes are then ranked among alternatives and a rank decision matrix is formed. At this time attributes can be either maximized or minimized by adjusting the direction of the ranks, i.e. if cost is to be minimized the highest cost is given the lowest rank and other costs ranked accordingly. Attributes to be maximized are generally given the highest ranks (see example in next section). The rank decision matrix is then projected mathematically into a multidimensional space by means of reciprocal averaging ordination (RA). RA is a variation of principal components analysis (Noy-Meir, 1970; Mardia et al., 1980) which cross calibrates alternatives with respect to attributes (and vice versa) to provide an accurate representation of the rank decision matrix in reduced dimension. It accomplishes this by transforming the elements, $x_1, \ldots, x_p$, of the rank decision matrix into a new set of elements, $y_1, \ldots, y_p$; such that: a) each y is a linear combination of each x, b) the x's are cross calibrated against each other to produce the y's, c) y_1 has the greatest variance of all possible combinations and scales the y's along the first axis of the solution, d) y_2 has the next greatest variance and scales the y's along the second axis of the solution, and so on. The reduced dimensionality is provided because it is common for the first few axes of the solution to explain the majority of the variance (>90%) of the rank decision matrix and thus serve as an adequate representation of that matrix. The view of the original matrix developed by this multivariate method is very rich both because it depicts the decision problem in reduced dimension and because the eigenvalue matrix of the solution is a linear combination of the multivariate covariance matrix associated with the alternatives of the decision problem and is therefore a representation of the multivariate probability density function relating the alternatives.

This eigenvalue matrix is then used to identify the optimum alternative as follows. The eigenvalues for each alternative are scaled by the percent variance accounted for on each axis and summed to give a measure of the importance of each alternative in accounting for the variance in the rank decision matrix. These sums are then scaled by a measure of the quantitative information content of each alternative first proposed by Shannon (Shannon and Weaver, 1949). This information content is a function of the entropy or amount of uncertainty represented by the multivariate probability density function of the rank decision matrix and as such can be shown to be related to the amount of uncertainty in the solution (Brillouin, 1962). These results are then ranked and an optimum alternative identified. This solution takes into account the attributes of all alternatives simultaneously and scales the alternatives with respect to the amount of variance each explains in the rank decision matrix and the amount of uncertainty each accounts for in the solution.

This solution can be greatly enhanced if the attributes most important in structuring the reciprocal averaging ordination can be identified. A previously reported technique (Hartzband and Boesch, 1979; Hartzband and LaPenn, 1979) called effective parameter analysis (EFPA), is a partial nonparametric correlation methodology which allows comparisons to be made between the rank structure of the attributes and the location of the alternatives along each appropriate ordination axis. This method was first suggested by Gaugh and Whittaker (1972) as a means of comparing ordination techniques.

EXAMPLES

Two examples will be used to illustrate the multiple attribute-information theoretic decision model (MAIT). The first is taken from the literature and has previously been analyzed by a number of methods without obtaining a unique or appealing solution. This is the case study in long range water supply planning in central Europe first described by David and Duckstein (1976) and subsequently analyzed in Duckstein and Opricovic (1980). The second is taken from an environmental and socio-economic study done, in part by Normandeau Associates, Inc. for the New England District Army Corps of Engineers, to evaluate a series of alternatives for port development in Lynn, Massachusetts.

David and Duckstein (1976) presented a decision problem based on a set of five alternatives for long range water supply planning in the central Tisza river basin in Hungary. Table I defines the five alternatives and the 12 attributes which are ranked to characterize these alternatives. David and Duckstein used the ELECTRE (Benayoun & Sussman, 1966; Nijkamp, 1975) model to analyze this problem while Keeney and Wood (1977) used multiattribute utility theory. Duckstein and Opricovic (1980) applied compromise programming to determine a solution. All of these methods produced either a nonunique or alternating solution, or a solution which had to be "negotiated" among several alternatives. The alternatives identified in these solutions were 1 and 2. Table II presents the result of the MAIT model for this problem along with the intermediate analytic products. The MAIT model identifies alternative 1 as the optimum alternative based on the eigensolution representation of all attributes scaled for variance explained and information content. This solution is appealing in that it is one of the two alternatives previously identified by other analytic methods.

Table III defines the alternatives and attributes of the second example; planning for harbor improvement in Lynn, MA. There are eight alternatives ranging from taking no action except maintaining existing navigational channels to dredging the entire harbor with construction of a breakwater and development of recreational facilities. There were several other alternatives proposed including the development of a deep-water port facility along with a marine industrial park

TABLE I. DEFINITION OF ALTERNATIVES AND ATTRIBUTES
FOR LONG RANGE WATER SUPPLY PLANNING PROBLEM
FROM DAVID AND DUCKSTEIN, 1976.

ALTERNATIVES:
1. Tisza and Danube water resources both used, multi-purpose canal system developed.
2. Tisza River pumped reservoir system.
3. Tisza River flat land reservoirs.
4. Tisza River mountain reservoir system.
5. Tisza River and groundwater resource for regional management.

ATTRIBUTES:
1. Total annual cost   (minimize)
2. Probability of water shortage   (minimize)
3. Water quality   (maximize)
4. Energy (reuse factor)   (maximize)
5. Recreation   (maximize)
6. Flood protection   (minimize)
7. Land and forest use   (minimize)
8. Manpower impact   (maximize)
9. Environmental architecture   (maximize)
10. International cooperation   (maximize)
11. Development possibility   (maximize)
12. Sensitivity   (maximize)

---

and/or the development of a container port facility. These alternatives were not considered by the Corps of Engineers as part of their planning process and so are not included in this analysis. Table IV presents the results of the MAIT model for this problem along with the intermediate analytic products. The MAIT model identifies alternative 3 as the optimum alternative again based on the eigensolution representation of all attributes scaled by variance explained and information content. This alternative was proposed by the Lynn Economic Development and Industrial Commission.

EFPA was used to determine which attributes were most important in structuring the solution to this problem. Two groups of attributes were highly correlated with the alternative structure on Axis 1 which accounted for 84% of the variance in the rank decision matrix. These were cost and environmental impact (most important) and recreational benefit

TABLE II. MAIT MODEL FOR LONG RANGE WATER SUPPLY PLANNING PROBLEM FROM DAVID AND DUCKSTEIN, 1976.

A. Variance proportioned by axis
  1. 0.661
  2. 0.177
  3. 0.149
  4. 0.012
  5. 0.000   Only four axes used in solution.

B. Eigenvalues

AXES

1	2	3	4
100.00	92.85	100.00	35.92
52.16	74.09	0.00	100.00
42.48	1.88	44.07	0.00
0.00	99.99	75.06	33.97
32.70	0.00	88.71	91.22

C. Solution matrix

1	2	3	4	SUM	H' (INFORMATION CONTENT)
66.10	16.43	14.90	0.43	97.87	1.26
34.48	13.11	0.00	1.20	48.79	0.99
28.08	0.33	6.57	0.00	34.98	0.77
0.00	17.70	11.18	0.41	29.29	1.06
21.61	0.00	13.22	1.09	35.92	1.12

D. Decision matrix

Alternative	Decision Coefficient	Rank
1.	77.67	1
2.	49.28	2
3.	45.43	3
4.	27.63	5
5.	32.07	4

. . Alternative 1 is optimum choice.

TABLE III. DEFINITION OF ALTERNATIVES AND ATTRIBUTES FOR PORT DEVELOPMENT PLANNING PROBLEM FROM NAI, 1980.

ALTERNATIVES:
1. No action, maintain existing channels.
2. Dredge navigational channels and turning basin.
3. Construct breakwater for harbor protection, dredge navigational channels.
4. Circular access channel, no breakwater.
5. Circular access channel with breakwater.
6. Same as #3 but with dredged mooring areas for recreational craft.
7. Same as #5 but with dredged mooring areas for recreational craft.
8. Dredge entire harbor with extended breakwater for recreational development.

ATTRIBUTES:
1. Total project costs
2. Annual maintenance costs
3. Net income gain from commercial fishing
4. Benefit/cost ratio
5. Excess net benefit
6. Environmental impact
7. Recreational benefit
8. Public opinion
9. Land enhancement from spoil disposal

All attributes maximized except 1, 2 and 6.

and land enhancement. Axis 2, which accounted for 14% of the informational variance, was entirely structured by public opinion. The knowledge of which attributes are most important in structuring the solution allows the decision maker to determine how appealing the solution is. If it is structured with respect to attributes which are considered of secondary importance, the decision maker may want to weight specific attributes and repeat the MAIT process. This is not the case in the port development problem described here in that cost, environmental impact and public opinion appear to be the most important attributes in identifying an optimum alternative.

TABLE IV. MAIT MODEL FOR PORT DEVELOPMENT PLANNING PROBLEM FROM NAI, 1980.

A. Variance proportioned by axis
   1. 0.838
   2. 0.135
   3. 0.015
   4. 0.008   Only three axes used in solution

B. Eigenvalues

	AXES	
1	2	3
100.00	71.32	68.21
75.39	35.92	67.52
49.28	0.00	79.23
79.11	67.48	21.74
30.91	40.11	66.70
36.01	7.38	62.31
14.06	43.95	0.00
0.00	100.00	100.00

C. Solution matrix

1	2	3	SUM	H' (INFORMATION CONTENT)
83.30	9.63	1.02	94.45	0.56
63.18	4.85	1.01	69.04	0.48
41.30	0.00	1.19	42.49	0.18
66.29	9.11	0.33	75.73	0.57
25.90	5.41	1.00	32.31	0.84
30.18	1.00	0.93	32.11	0.39
11.78	5.93	0.00	17.71	0.92
0.00	13.50	1.50	15.00	0.47

TABLE IV (Continued)

D. Decision matrix

Alternative	Decision Coefficient	Rank
1.	168.66	2
2.	143.83	3
3.	236.06	1
4.	132.86	4
5.	38.46	6
6.	82.33	5
7.	19.25	8
8.	31.91	7

∴ Alternative 3 is the optimum choice.

DISCUSSION AND SUMMARY

In most cases the multiple attribute-information theoretic decision model (MAIT) presented in this paper provides an unambiguous solution to the problem of identifying a "best" alternative from a given set of alternatives. As with any model, this solution can not be better than the information contained in the rank decision matrix. That is, if the attributes chosen to characterize each alternative are adequate descriptors; and if the quantitative values of the attributes accurately reflect the relationships among alternatives then the MAIT will provide a straightforward and appealing solution. It is not always possible to formulate a rank decision matrix that accurately represents the problem under consideration, and once this matrix is formed not all decision makers will agree that it does represent the problem. For example, the Lynn, MA decision problem left out two potentially attractive alternatives because no Federal participation would have been needed to accomplish them. In addition, there was some disagreement among economic consultants with respect to the figures developed for excess net economic benefit and benefit/cost ratios for this project. Thus the decision maker must determine the validity of the results of the MAIT model in terms of the quality and accuracy of the available information.

No method for solving multiattribute decision problems is universally applicable or valid. The MAIT model in combination with EFPA, however, provides several advantages in that it: 1) projects the decision problem into a multidimensional mathematical space in order to evaluate all attributes simultaneously, 2) scales the solution with respect to both the variance inherent in the rank decision matrix and its information content, and 3) generally identifies a unique optimum alternative. It thus allows the decision maker to consider recommendations with respect to complex environmental planning problems based on a quantitative and repeatable procedure which is both appealing and interpretable.

LITERATURE CITED

Benayoun, R. B. and N. Sussman, "Manual de Reference du Programme Electre", Note de Synthese et Formation, No. 25, Direction Scientifique SEMA, Paris, 1966.

Brillouin, L., Science and Information Theory, Academic Press, New York, U.S.A., 1962.

Cohon, J. L., Multiobjective Programming and Planning, Academic Press, New York, U.S.A., 1978.

David, L. and L. Duckstein, Multicriterion ranking of alternative long-range water resource systems, Water Resources Bulletin, 12(4), 731-754, 1976.

Duckstein, L. and S. Opricovic, Multiobjective optimization in river basin development, Water Resources Research, 16(1), 14-20, 1980.

Gaugh, H. and R. Whittaker, Comparison of ordination techniques, Ecology, 53(5), 868-875, 1972.

Gupta, S. S. and D.-Y. Huang, Multiple Statistical Decision Theory: Recent Developments, Springer-Verlag, Berlin, F.G.R., 1981.

Hartzband, D. J. and D. F. Boesch, Benthic ecological studies: meiobenthos, Middle Atlantic outer continental shelf. Special Scientific Report Number 195, Virginia Institute of Marine Science, Gloucester Point, VA, U.S.A., 1979.

Hartzband, D. J. and R. LaPenn, New methods for quantitative analysis of large ecological data sets, 1979 Benthic Ecology Meetings, University of New Hampshire, Durham, NH, U.S.A., 1979.

Hwang, C.-L. and K. Yoon, Multiple Attribute Decision Making: Methods and Applications, Springer-Verlag, Berlin, F.G.R., 1981.

Keeney, R. L. and H. Raiffa, Decisions with Multiple Objectives: Preferences and Value Tradeoffs, Wiley, New York, U.S.A., 1976.

Keeney, R. L. and E. F. Wood, An illustrative example of the use of multiattribute utility theory for water resources planning, Water Resources Research, 13(4), 705-712, 1977.

Mardia, K. V., J. T. Kent and J. M. Bibby, Multivariate Analysis, Academic Press, New York, U.S.A., 1980.

Nijkamp, P., Reflections on gravity and entropy models, Regional Science and Urban Economics, 5(2), 203-255, 1975.

Nijkamp, P., Multidimensional Spatial Data and Decision Analysis, Wiley, New York, U.S.A., 1979.

Noy-Meir, I., Component analysis of semi-arid vegetation in south-eastern Australia, Ph.D. Dissertation, Australian National University, 1970.

Scheffe, H., The Analysis of Variance, Wiley, New York, U.S.A., 1959.

Shannon, C. and W. Weaver, The Mathematical Theory of Communication, University of Illinois Press, Urbana, IL, U.S.A., 1949.

A CONVEYANCE ROUTE SELECTION
METHODOLOGY

D.J. Tate
International Environmental
Consultants Inc.

A proposed coal mine in Wyoming needed a conveyance route to move its product to the railroad. The development and evaluation of alternative routes was triggered by the need for a right-of-way permit, as well as by initially anticipated conflicts.

An environmental evaluation graded the impacts of each alternative route on the basis of a one to ten scale for several disciplines: wildlife; vegetation; reclamation; meteorology; surface hydrology; recreation; and scenic resources. Appropriate criteria were quantified and variously weighted within each discipline to arrive at a score for that discipline.

The composite environmental score, a weighted mean of the separate discipline scores, was graphically compared with a client-derived engineering score in such a way that varying percentages of environmental and engineering weight could be assessed.

INTRODUCTION

A proposed coal mine in Wyoming, located near the Haystack Mountains and about 18 miles due north of a railroad mainline, needed a conveyance route to move its product to the railroad. An original route involving a haul road, conveyor belt and railroad spur line was proposed. An assessment of this route as well as the development and evaluation of alternative routes was triggered by the need for a right-of-way permit across federal lands, as well as by initially anticipated conflicts with big game migration routes and wintering areas and also with prime raptor nesting sites.

Three railroad spur alternatives (1, 2 and 3), and four haul road alternatives (1A1, 1A2, 1A3 and 1B) were considered (Figure 1). All four haul roads led from the mine to Spur 1. This spur separated into

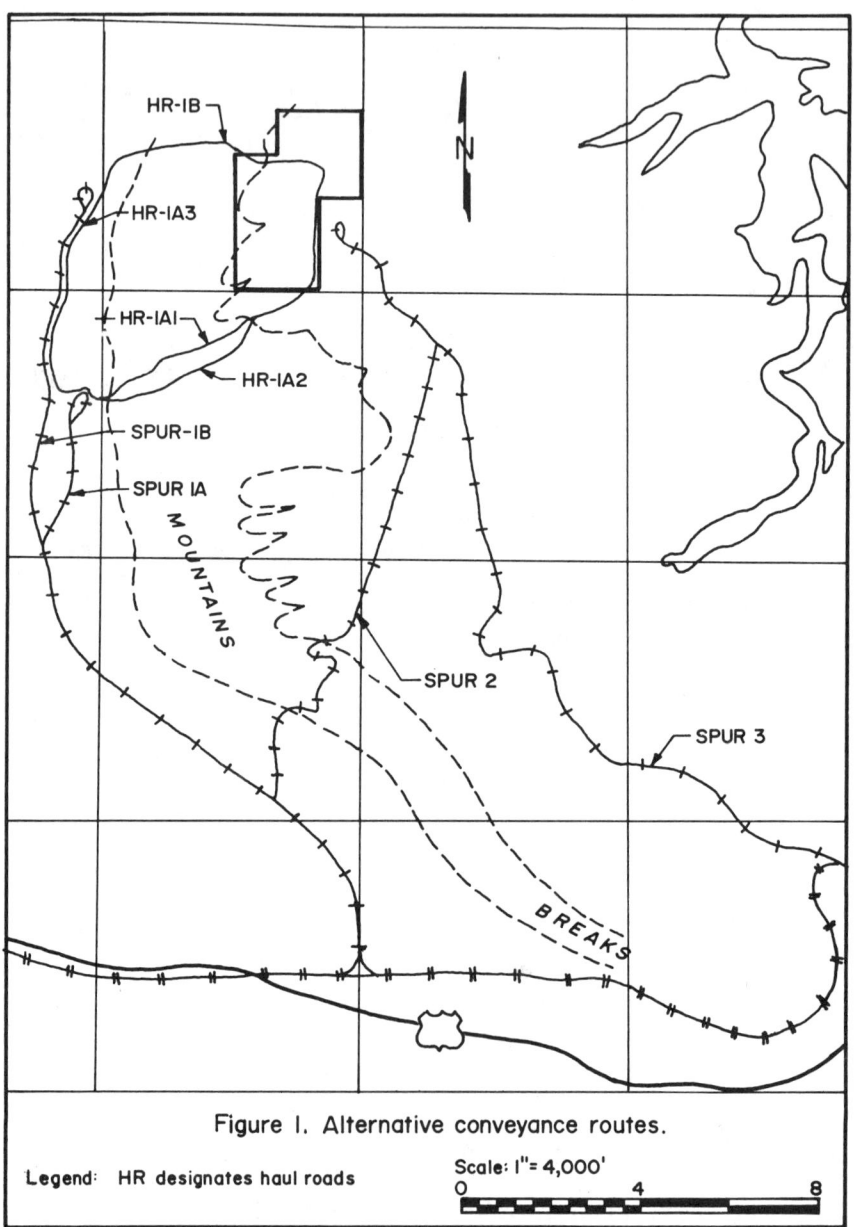

Figure 1. Alternative conveyance routes.

Legend: HR designates haul roads

Scale: 1" = 4,000'

two alternatives at the north end termed 1A and 1B; Haul Road 1A2 connected with Spur 1A via a long conveyor system.

Railroad Spur 1A was 17.7 miles long and associated with a 7.8 mile haul road as Alternative 1A1, with a 6.7 mile haul road and a 6,600 foot conveyor as Alternative 1A2, and with a 12.3 mile haul road as Alternative 1A3. Railroad Spur 1B was 21.9 miles long and associated with a 7.2 mile haul road. Railroad Spurs 2 and 3 were 22.9 and 24.6 miles long, respectively, and began less than half a mile from the proposed pit. IEC was retained by a client to prepare an environmental evaluation of these route alternatives and synthesize this with an engineering evaluation based on an analysis by the client's staff.

ENVIRONMENTAL ANALYSIS

IEC assembled a team of four professionals to evaluate the effect of the various routes on the environment. Their evaluation was based on existing field data and their previous field experience with the project. For this analysis the environmental impacts were graded from 1-10, with minimal impact scored at 1 and highly detrimental impact scored at 10. The IEC environmental analysis considered impacts in the following disciplines: wildlife; vegetation and reclamation; meteorology; surface hydrology; recreation and scenic resources. Other disciplines, even though they were important to the overall project, were not analyzed in this study, because they were not affected differently by various alternatives.

The determination of relative environmental values for the six disciplines analyzed was an iterative process that began with a detailed discussion by all team members to determine initial weighting among disciplines. Each discipline was then analyzed and a summary score developed for each route. Initial weightings were then reevaluated by team members to arrive at the following discipline weights: wildlife, 40; vegetation/reclamation, 20; air quality, 10; surface water, 10; recreation, 10; and scenic values, 10, for a total of 100 percent. These weights were based on a consensus among team members of the relative impacts of haul roads on different aspects of the environment.

All considered wildlife values of major importance in this study. The area is heavily used by deer, Pronghorn, grouse and raptors, and certain alternatives would penetrate habitats of all of these animals.

Vegetation and reclamation were combined since reclamation is based on knowledge of existing vegetation. In this dry climate, averaging less than 12 inches of precipitation annually, revegetation is difficult and these disciplines together were considered second in importance to wildlife.

Finally, anticipated impacts in four disciplines were considered moderate and about equal. Air emissions were expected to be small and/or controllable; hence, in this instance air quality was considered less important than the previous two disciplines. Surface hydrology impacts were also considered moderate. Only one permanent stream could be affected, and this primarily by Alternative 2. Impacts to the surface hydrology of other alternatives would be small. Recreation would certainly be affected differently by the various alternatives, and impacts to fishing, hunting, picnicking and boating would vary. However, recreational use of areas to be affected is only moderate. Scenic resources would similarly be affected adversely and differently by each alternative. Yet, the area is not considered to have extremely high quality scenery, and the impacts would be only moderate.

Other environmental parameters would be strongly impacted by the project, but not be differentially impacted by the various alternatives. Thus, socioeconomic impacts are of major concern for this project, but all alternatives apparently affect such concerns as housing, transportation, etc. to the same degree. Similarly, ground water impacts are of major import for this project, yet all alternatives would be surface transportation routes having minimal effect on ground water; thus, this parameter was not analyzed.

Wildlife

A variety of wildlife would be impacted by the route alternatives. Often the impacts would be species specific, yet in a study of this type only the most obvious species or species groups can be considered. For analysis, Mule Deer, Pronghorn, Sage Grouse, raptors, other birds, other mammals and aquatic organisms were considered separately. For each wildlife group, such factors were considered as the areal impact of each alternative route, the time span of the impact, the sensitivity of the species or group, the ease with which the wildlife group can recolonize after completion of reclamation efforts, and the indirect effect of traffic and other impediments to the movement of the wildlife group onto traditionally used portions of its range. The length of each route passing through the habitat of each wildlife group was tabulated. Since average disturbance widths for railroad construction in this area are about 30 feet, while average haul road disturbance widths are about 120 feet, haul road lengths were multiplied by four to account for their greater areal extent. If the human intrusion facilitated by the haul road was considered particularly detrimental to a wildlife group, the haul road length was multiplied by an additional 2.5, resulting in an overall multiplier of ten. Habitats used differentially by a species were weighted to reflect their importance to the wildlife group as discussed individually below. The direct scores are presented in Table I for each wildlife group.

TABLE I: Direct wildlife impact scores on habitat change, traffic use and human interference.

Alternatives		Pronghorn	Mule Deer	Sage Grouse	Raptors	Other Birds	Other Mammals	Aquatics
RR Spur	Haul Road							
1A	1A1	109.0	376.0	114.0	95.0	49.0	49.0	–
	1A2	98.0	365.0	110.0	78.0	45.0	45.0	–
	1A3	154.0	561.0	300.0	118.0	67.0	67.0	–
1B	1B	107.0	514.0	278.0	70.0	51.0	51.0	–
2		60.0	58.0	34.0	12.0	23.0	23.0	2.0
3		28.0	25.0	94.0	11.0	25.0	25.0	1.0

To score the six route alternatives for Pronghorn, the length of each route alternative crossing both crucial winter range and yearlong range, as determined by Wyoming Game and Fish Department (WGFD) range, was measured. Based on unpublished WGFD data for the South Ferris - Seminoe Mountain Area, Pronghorn densities in modereate winters may be 2.1 times greater than in summer. However, severe winters are known to reduce the winter range available in the region by two-thirds (Taylor 1975). Under these conditions, Pronghorn winter densities may be 6.2 times summer densities. On this basis, impact on crucial winter range was considered 6.2 times greater than on yearlong range. Haul roads were considered ten times more detrimental than railroads because of their greater width and frequency of use and because of the human access they facilitate. Thus, the Pronghorn score for a given alternative is the sum of the yearlong range sum (YR sum) and 6.2 times the critical winter range sum (CWR Sum). The YR Sum and CWR Sum equal ten times the length of haul road plus the length of railroad crossing yearlong and critical winter ranges, respectively.

Mule Deer scores were also based on the length of route alternatives crossing crucial winter range and yearlong range. Neither the Bureau of Land Management (BLM) nor the WGFD have recently surveyed the numbers of deer summering or wintering in the Haystack Mountain area. Ongoing research on radio-collared deer by Joseph Springer of the WGFD and Lorin Ward of the U. S. Forest Service (USFS) indicates that about 1300 deer move into the Haystacks from the south and as many as 270 move in from the north. Thus, the winter deer population in the Haystacks increases by as many as 1570 animals. Fall and winter big game surveys of the Haystack Mountain area by IEC indicate that winter populations of Mule Deer are at least eight times summer populations. This multiplier is also consistent with casual comments by people familiar with the wildlife in the Haystack Mountain area. Therefore, in scoring the six route alternatives for Mule Deer, impact on crucial winter range was considered eight times greater than impact on yearlong range. As discussed for Pronghorn, haul roads were considered ten times more detrimental than railroads.

For Sage Grouse, impact of haul road and railroad lengths on habitat and on known leks was considered. WGFD grouse maps divide habitat into that used yearlong (spring, summer, winter, fall), for breeding only (spring, summer, fall or spring and summer), or for wintering (winter and spring). The length of route passing through each of these habitat use areas was tabulated. Haul road length was multiplied by four based on the greater areal disturbance of haul roads, but assuming human disruption of habitat use would be minimal. These habitat areas were weighted as follows: yearlong, 40 percent; breeding, 30 percent; and wintering, 30 percent. In addition, the length of haul road or railroad within two miles of leks known active in 1978 or 1979 was tabulated for each alternative.

These haul road lengths were multiplied by ten to compensate for their greater areal disturbance and the impact human intrusion would have on displaying birds. Also multiplied by ten were any lengths of route passing within one-quarter mile of a lek. The lek scores were doubled because of the long term traditional nature of their use and added to the habitat area scores to arrive at the total score for this species.

For raptors, impact was based on the length of haul roads and railroads passing within one mile of prime cliff nesting habitat. Haul road lengths were multiplied by ten, because of their greater areal extent and the impact of human intrusion. Further, Haul Roads 1A1 and 1A2 were considered to have 1.5 times the impact of the other two haul roads, because 1A1 and 1A2 put people right in the midst of prime nesting habitat.

Impact on birds and mammals not considered above was assessed solely on the basis of areal disturbance of habitat. The total length of each railroad plus four times the haul road length was used as a measure of areal disturbance of each alternative. This approach underestimates the probable impact on these species; this is warranted because of their more numerous offspring and shorter generation length.

Impacts on aquatics were based on the length of North Platte River affected. Only Alternatives 2 and 3 would impact the river. The impact of Alternative 2 was assumed four times greater because of its probable greater cut and fill requirements. Impacts on the river were assumed to last only 1.5 years out of a 20 year mine life and scores were proportionally reduced before being recorded in Table I.

Finally, the scores of various wildlife groups were weighted on the basis of their relative sensitivity, the relative ease with which they can be expected to recolonize the area and the relative impact the routes would have on their access to other appropriate habitat. The weighting used is presented in Table II.

In Table III are the weighted scores for each wildlife group derived by multiplying the Table I values by the composite weighting for the appropriate species in Table II. These figures are totaled and reduced to a 1-10 final wildlife score for each alternative. As can be seen, Alternatives 2 and 3 have the least impact on wildlife and rank about the same. This is primarily because the greater impact which Alternative 2 has on Pronghorn and Mule Deer is counterbalanced by its lesser impact on Sage Grouse as compared to Alternative 3.

All routes associated with Railroad Spurs 1A and 1B have greater environmental impacts than Alternative Routes 2 and 3. This is

TABLE II: Indirect wildlife weights.

	Pronghorn	Mule Deer	Sage Grouse	Raptors	Other Birds	Other Mammals	Aquatics
Sensitivity	4.0	7.0	5.0	10.0	2.0	2.0	3.0
Ease of recolonization	3.0	3.0	7.0	8.0	2.0	2.0	2.0
Ease of other habitat access	7.0	5.0	3.0	2.0	1.0	2.0	1.0
Composite weighting on a relative scale	2.8	3.0	3.0	4.0	1.0	1.2	1.2

TABLE III: Weighted impact scores totaled across groups for each route and adjusted to a 1-10 scale.

Alternatives		Pronghorn	Mule Deer	Sage Grouse	Raptors	Other Birds	Other Mammals	Aquatics	Total	1-10 Score
RR Spur	Haul Road									
1A	1A1	305.0	1,128.0	342.0	380.0	49.0	59.0	–	2,263.0	6.2
	1A2	274.0	1,095.0	330.0	312.0	45.0	54.0	–	2,110.0	5.8
	1A3	431.0	1,683.0	900.0	472.0	67.0	80.0	–	3,633.0	10.0
1B	1B	300.0	1,542.0	834.0	280.0	51.0	61.0	–	3,086.0	8.4
2		168.0	174.0	102.0	48.0	23.0	28.0	3.0	546.0	1.5
3		78.0	75.0	282.0	44.0	25.0	30.0	1.0	535.0	1.5

because the haul roads associated with Spurs 1A and 1B are considered to have considerably greater detrimental impact on wildlife. If alternative routes associated with haul roads are arranged from best to worst for Mule Deer, Sage Grouse, other birds and other mammals; the order is 1A2, 1A1, 1B and 1A3. This correlates directly with increasing areal disturbance of habitat by these routes. For Pronghorn, the order of haul road associated routes, from best to worst, is 1A2, 1B, 1A1 and 1A3. Alternative 1B is slightly better for Pronghorn than Alternative 1A1, because even though the length of railroad passing through year round habitat is longer, the length of the associated haul road passing through year round range is shorter. For raptors, Alternative 1B is considered better than either Alternatives 1A1 or 1A2, because the latter two alternatives transect cliff nesting habitat and thereby have an even greater potential for disturbing it.

Vegetation/Reclamation

Vegetation and reclamation were considered together due to their close relationship. Both areal disturbance and productivity estimates were utilized in the evaluation of vegetation. Soils were not included in the evaluation because specific data regarding topsoil depth, fertility and reclamation potential in this area were absent. To obtain the areal disturbance scores, the linear distances of alternative haul roads were multiplied by four, while the length of the associated railroad was multiplied by one. These areal disturbance components are given in Table IV.

TABLE IV: Vegetation impacts.

Alternatives		Distances			
RR Spur	Road	Haul RR Spur (1)	Haul Road (4)	Total	Score
		x2	x2	x3	
1A	1A1	17.7	1.0	6.8	125.0	6.2
	1A2	17.7	0.0	6.7	115.8	5.8
	1A3	17.7	6.6	5.7	156.6	7.8
1B	1B	21.9	1.5	5.7	124.2	6.2
2		22.9	-	-	45.8	2.3
3		24.6	-	-	49.2	2.5

The productivity of different vegetation types was used to weight the areal disturbance scores. The major vegetation types affected by the six alternatives would include grassland, sagebrush, greasewood and saltbush, along with various admixtures. An initial vegetation analysis of the proposed mine site and surrounding area by IEC provided maps and coverage data for these types. While productivity values vary among these types, all are primarily upland, used for livestock and wildlife grazing, and their productivity is often quite similar. Since the railroad spur alternatives all pass through several of these types, they were rated equal. The productivity values of the grassland, shrub and shrub-grassland mixtures in the Haystack Mountains were considerably higher, and these areas were considered 1.5 times as valuable. This figure was based on both limited sampling and general observation.

This weighting of upland vegetation types is supported by the Soil Conservation Service (SCS) range site productivity values in the Technician Guide (SCS 1978), Section II-E for this region, the High Plains Southeast, which falls within a 10-14" annual average precipitation zone. The SCS considers saline upland range sites dominated by Gardner's saltbush to support 0.25 AUM's/Acre, while clayey and loamy sites dominated by mid-grasses support some 0.4 AUM's/Acre. Utilizing these data as general support, a 3:2 ratio for the Haystack Mountains versus the railroad spur area appears reasonable, and this ratio has been used to modify the areal values. Thus, we have multiplied haul road areal disturbance in the Haystack Mountains (1A2 and parts of 1B, 1A1 and 1A3) by three and all other areas by two. The final scores reflect these weights (Table IV).

For reclamation, costs and difficulties for each mile of alternative were assumed to be approximately the same regardless of soil, terrain or vegetation differences. Therefore, the reclamation impacts were based on approximate areal disturbance scores that were not weighted (Table V).

TABLE V: Reclamation impacts.

Alternatives		Distances			
RR Spur	Haul Road	RR Spur	Haul Road	Total	Score
		x1	x4		
1A	1A1	17.7	7.8	48.9	4.9
	1A2	17.7	6.7	44.5	4.5
	1A3	17.7	12.3	66.9	6.7
1B	1B	21.9	7.2	50.7	5.1
2		22.9	-	22.9	2.3
3		24.6	-	24.6	2.5

Air Quality

There are normally two major air pollution problems associated with mine development and operation: particulates and diesel engine emissions. Particulates include blowing coal dust and road dust from heavy vehicular traffic. Diesel engine emissions include sulfur dioxide, oxides of nitrogen, carbon monoxide and carbon dioxide from train or truck engines. Based on IEC experience with both factors in Wyoming coal mines, chemical emissions would be relative minor with use of appropriate emission controls while particulate emissions would be by far the most important. The rating values reflect this.

All haul roads associated with Railroad Spurs 1A and 1B would be extensive, gravel and traversed by large trucks. Even with the generous and frequent use of surficants, such systems usually result in sizeable particulate emissions. This would be accentuated by the windiness of this part of Wyoming, particularly in the high parts of the Haystack Mountains, which all roads would cross. In contrast, Alternatives 2 and 3 would involve haul roads less than half a mile long with unit train loading facilities adjacent to the mine site. While the railroad distances would be somewhat longer, there would be a minimum of particulates and diesel engine emissions. Thus, these routes were ranked considerably better environmentally.

Table VI gives the scores for these alternatives. Alternatives 2 and 3 have the best score, owing to the absence of haul roads and truck use. Alternative 3 is more exposed to wind than Alternative 2 and is one point higher. Alternatives associated with Railroad Spurs 1A and 1B score much higher because of their haul roads. All alternatives associated with Railroad Spur 1A are slightly worse environmentally, owing to the winds characteristic of the higher elevations their haul roads cross.

TABLE VI: Air quality impacts.

RR Spur	Alternatives Haul Road	Scores
1A	1A1	8.0
	1A2	8.0
	1A3	8.0
1B	1B	7.0
2		3.0
3		4.0

Alternatives 1A1, 1A2 and 1A3 have the same score. While the conveyor (1A2) would avoid a road down the hill, anticipated high winds at the transfer point would probably result in the same particulate emissions as 1A1. Alternative 1A3, while located on more gentle slopes, would be much longer and consequently rates the same score.

Surface Water Hydrology

Analysis of the impacts of route alternatives on surface water hydrology focused on the following four features; stream crossings; streams paralleling the route; difficulty of restoration of hydrologic functions; and erosion potential. Scores were assigned as discussed below.

For each alternative route, four factors were used to analyze stream crossings. First, the number of streams crossed, regardless of size, was tabulated. Second, a standard measure of stream size, the Horton Stream Order, was weighted to the size of the stream. Third, the total drainage area was summarized for each stream. Fourth, the length of all streams was tabulated. Each factor was scored on a 1-10 basis, and each of the four factors was considered equal.

These measures of impact on streams parallel to the various routes were next analyzed. First, the length of each stream was measured. Second, the drainage area for each stream was calculated. Finally, the potential for flooding of the route was subjectively evaluated. Again, each factor was scored on a 1-10 basis, and each of the three factors was considered equal.

The two final factors, difficulty of restoration of hydrological features and erosion potential, were evaluated subjectively. Each was scored on a 1-10 basis, and each was considered equal.

The results, presented in Table VII, reveal that while Alternative 2 appears to be the the most detrimental environmentally, there is no extreme difference among alternatives. The higher score for Alterntive 2 reflects the potential problems caused primarily by proximity to the North Platte River. The remaining five alternatives have similar scores. These close scores reflect the arid nature of the area and the similarity among routes in this discipline.

Recreation

The conveyance route location could have a significant impact on certain site-specific recreational activities in the project area, including hunting, fishing, sight-seeing, camping and boating. Analysis of impacts was based on these recreational activities. For each alternative, up to four points were assigned to hunting, up to

TABLE VII: Surface hydrology impacts.

| Alternative | Stream Crossings | | | Parallel Streams | | | Other Features | | Score |
	Number	Size	Basin Area	Length	Length	Basin Area	Flood	Restoration Difficulty	Erosion Potential	Sum/10
1A1	3.0	1.0	3.0	2.0	3.0	2.0	6.0	3.0	3.0	2.6
1A2	2.0	1.0	1.0	1.0	2.0	2.0	6.0	3.0	3.0	2.1
1A3	4.0	2.0	5.0	5.0	1.0	1.0	6.0	3.0	3.0	3.0
1B	4.0	2.0	5.0	5.0	1.0	1.0	6.0	3.0	3.0	3.0
2	1.0	3.0	2.0	4.0	4.0	10.0	6.0	9.0	8.0	4.7
3	2.0	4.0	4.0	3.0	5.0	4.0	3.0	3.0	3.0	3.1

three to fishing and up to three to all other types of recreation. Thus, an alternative with prime hunting, fishing and other recreational opportunities would score ten and have the greatest adverse impact on recreation.

Prime Mule Deer hunting areas, used both because of game availability and the isolated nature of the terrain, are in the Haystack Mountains. Thus, all haul roads, associated with Railroad Spurs 1A and 1B, would cause serious adverse impacts to hunting in this area because of frequent and heavy truck use. Railroad Spurs 1A, 1B and 3 would cross good Pronghorn hunting areas that are well used; however, their impact would be lessened by the infrequent train traffic. Alternative 2 would have a much smaller impact on hunting, as it would parallel a highway for much of its route.

Fishing impacts would be confined to Railroad Spurs 2 and 3 that would cross or parallel the river, as none of the smaller streams in the area support sport fishing. Since Alternative 3 would bridge the river, its impacts would be minor. Alternative 2, however, would have major impacts on both river and streambank fishing sites, and this would be a major adverse impact.

Impacts on sight-seeing, camping, picnicking, boating and canoeing would be minor for Alternative 3, but major for Alternative 2, especially along the river. Impacts would also be major for all alternatives involving haul roads due to their disturbances in the Haystack Mountains. Table VIII indicates that all scores are similar. Alternatives 1A1 and 1A2 have the highest scores; while Alternatives 1A3, 1B and 3 have the lowest scores. Alternative 2 is intermediate.

TABLE VIII: Recreational impacts.

RR Spur	Alternatives Haul Road	Hunting	Fishing	Other	Score
1A	1A1	4.0	0.0	3.0	8.0
	1A2	4.0	0.0	3.0	8.0
	1A3	3.0	0.0	2.0	6.0
1B	1B	3.0	0.0	2.0	6.0
2		1.0	3.0	3.0	7.0
3		2.0	1.0	2.0	6.0

Scenic Resources

Although subjective, the scenic impact of road and railroad construction and operation on the existing semi-natural and natural landscape is important and should be evaluated. Evaluation was based primarily on the numbers of people who would see the coal transportation operation; on the intensity of the impact on the quality of scenery; and the sensitivity of the viewer.

The numerous viewers from the road to the reservoir could see primarily the effects of the transportation operation. Furthermore, many of these viewers would be recreation-oriented, using the river or reservoir, and highly conscious of scenic impacts. For this reason, the reservoir road view was considered the major portion of the scenic score value. Thus, using U.S. Forest Service (USFS) terminology, the Sensitivity Level is high (Level 1), whereas views from less used roads and areas are Level 2 or 3. Therefore, because Alternative 2 would parallel the reservoir road and probably cross it several times, it was given a very high adverse score (9) in this category. All other alternatives were scored at 2 (Table IX); Railroad Spurs 1 and 3 would cross the reservoir road only once in similar terrain (Class III); scenic impacts from these crossings would be similar and minor.

TABLE IX: Scenic resources impacts.

Alternatives					
RR Spur	Haul Road	RR Spur	Bridge	Haul Road	Score
1A	1A1	2.0	-	2.0	4.0
	1A2	2.0	-	1.0	3.0
	1A3	2.0	-	1.0	3.0
1B	1B	2.0	-	1.0	3.0
2		9.0	-	-	9.0
3		2.0	2.0	-	4.0

Alternative 3 would also bridge the river, having a major scenic impact on boaters and fishermen in that region. Two adverse points were added for this reason.

Because of the minimal number of viewers, haul road impacts were given a single adverse point; Alternative 1A1 was given an additional point due to the major disturbance that would be created by descending the steep terrain to the loadout facilities and the attendant blasting necessary for the descent.

The BLM Visual Resources classification for this area, as shown on Map 10 in Appendix A of the South Central Wyoming Coal Draft Environmental Statement (BLM 1978), was also examined. The BLM considered the dugway where Alternative 2 crosses the Haystack Mountains parallel to the river, as Class II, and the remainder of the Haystack area as Class III. Class II is an area of more valuable scenery where management activities should not take place. Class III is less scenic, and management activity "may be evident" but should remain "subordinate". Thus, according to this classification, changes in the dugway would probably not be compatible with its scenic nature, while construction and operation in other areas would probably be suitable.

ENGINEERING ANALYSIS

An engineering evaluation was done by the client to examine the same six alternative routes. Costs and engineering difficulty were assessed in detail for five of the routes. The sixth route, Alternative 2, which goes through the dugway, was also examined, but because of its difficult engineering, no cost data were developed.

Costs for the various routes were broken down into development costs, depreciable capital costs and yearly operating costs. Only those costs unique to the haulage system were compared. From this information an economic evaluation was made for a 20-year mining period.

As a result of this study, engineering scores were assigned to the five feasible routes as shown in Table X. It can be seen that the operating cost of Alternative 3 would be the lowest of all the alternatives. Alternative 3 would also have the best net-present-value comparison.

CONCLUSIONS

The final engineering scores have just been presented in Table X. Table XI summarizes the environmental scores, developed during the environmental analysis of the six alternatives. These values are combined in Figure 2. In this figure, the horizontal axes indicate the percent environmental and percent engineering weight at any given point on a vertical line between the axes. When moving from the far left to the far right, the engineering influence decreases from 100 percent to zero, while the environmental influence increases simultaneously from 0 to 100 percent. The midpoint, at which both values are equal (50:50), is indicated by the vertical line in the center. Such a line intersects scores composed of half environmental and half engineering values and is a reasonable ratio for evaluating the alternatives.

TABLE X: Engineering summary scores.

Alternatives		Engineering & Construction	Operational	Economical	Combined Total (x)	Final Score* (10-x)
RR Spur	Haul Road					
1A	1A1	4.50	4.25	3.75	12.50/3 = 4.17	5.8
	1A2	4.50	5.00	2.80	12.30/3 = 4.10	5.9
	1A3	6.00	4.25	3.10	13.35/3 = 4.45	5.6
1B	1B	6.00	6.00	3.50	15.50/3 = 5.17	4.8
2	Not Rated					
3		7.50	10.00	8.70	26.20/3 = 8.73	1.3

* The combined engineering totals (x) have been subtracted from ten to give a final score where higher numbers indicate greater cost, as has been done for the environmental scoring.

TABLE XI: Environmental summary scores.

Discipline	Weight	1A1	1A2	1A3	1B	2	3
Wildlife	(40)	24.8	23.2	40.0	33.6	6.0	6.0
Vegetation	(10)	6.2	5.8	7.8	6.2	2.3	2.5
Reclamation	(10)	4.9	4.5	6.7	5.1	2.3	2.5
Air Quality	(10)	8.0	8.0	8.0	7.0	3.0	4.0
Surface Hydrology	(10)	2.6	2.1	3.0	3.0	4.7	3.1
Recreation	(10)	8.0	8.0	6.0	6.0	7.0	6.0
Scenics	(10)	4.0	3.0	3.0	3.0	9.0	4.0
TOTAL		58.5	54.6	74.5	63.9	34.3	28.1
ROUNDED SCORE		5.9	5.5	7.5	6.4	3.4	2.8

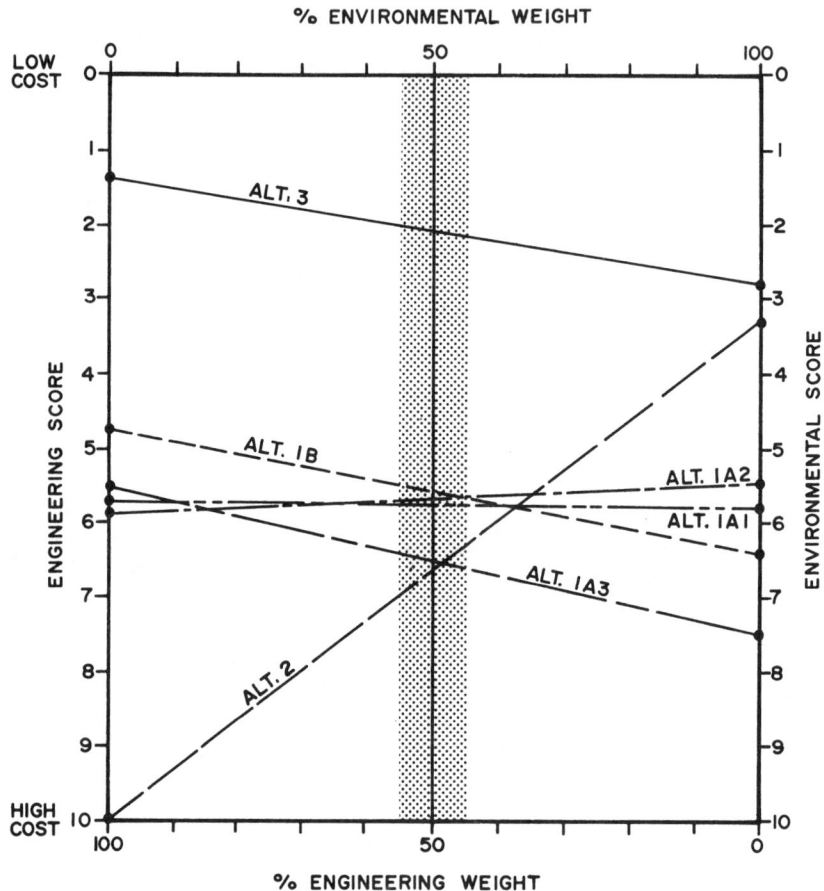

Figure 2. Comparison of environmental and engineering scores.

When analyzing the alternative routes using this 50:50 ratio, the following conclusions are evident:

1. Alternative 3 would clearly be the lowest cost engineering alternative and the least detrimental environmental alternative. Thus, it ranked best in both categories and was considered the first choice.

2. All alternatives using Railroad Spur 1 coupled with a haul road system had similar values and all ranked well below Alternative 3 in both environmental and engineering values. Operationally they would be more costly, they would be more complex in terms of manpower, and they would also have greater environmental impacts. They formed a second group that ranked well below the first choice.

3. While Alternative 2 ranked second environmentally, and while it ranked only slightly below 1A3 at the 50:50 ratio, it was so poor from an engineering standpoint that it was not considered feasible. It was rated the poorest choice of all alternatives.

Thus, when both the engineering and environmental factors were considered equally, Alternative 3 was almost twice as desirable as the other alternatives. When an engineering score of 10 was arbitrarily assigned to Alternative 2, its poor ranking was evident. All other alternatives were clustered in between.

All ratings assume that, to minimize impacts on any route, best engineering practices would be used in construction, and slight variations in routing would be possible to avoid particularly valuable sites such as grouse leks and unique riparian habitat. This must be borne in mind especially when viewing the low scores given to Alternative 3.

This selection methodology allowed site-specific quantification of route evaluation and a visual means of ascertaining the relative impacts of each route based on various weighting of environmental and engineering criteria. As a result of these analyses, a single score for each of five feasible routes could be obtained. These scores, using equal environmental and engineering weighting, varied from 2.1 for Alternative 3 to 6.6 for Alternative 1A3 and clearly demonstrated the route that should be used. Once initiated, this methodology facilitated communication between environmental and engineering personnel and provided an objective means of cooperative evaluation.

ACKNOWLEDGMENTS

This paper is based on a report prepared for a client by a multidisciplinary IEC Study Team. Various sections of this original report were contributed by various task leaders: wildlife, Dr. D.J. Tate; vegetation/reclamation, recreation, scenic resources, Dr. P. Kilburn; air quality, Mr. L. Crow; and surface hydrology, Mr. P. Rechard.

REFERENCES CITED

BLM (Bureau of Land Management). 1978. Draft environmental statement; Proposed development of coal resources in south-central Wyoming. U.S. Department of the Interior. Interrupted pagination.

SCS (Soil Conservation Service). 1978. Technician guide to range sites and range condition with initial stocking rate. U.S. Department of Agriculture, Soil Conservation Service, Casper, Wyoming. n.p.

Taylor, E. 1975. Pronghorn carrying capacity of Wyoming's Red Desert. Wyoming Game and Fish Department, Wildlife Technical Report No. 3. 65 p.

ECOSYSTEM ANALYSIS FOR TERRESTRIAL
BASELINE STUDIES -- AN EXAMPLE IN
COLORADO'S PICEANCE BASIN

Joe C. Truett
   LGL Ecological Research
      Associates
   Grand Junction, CO 81502

Richard J. Douglass
   LGL Ecological Research
      Associates
   Grand Junction, CO 81502

ABSTRACT

   A biological baseline study being conducted prior to a mining operation in the Piceance Basin of Northwest Colorado is going beyond conventional descriptions of biota to focus on predicting impacts, suggesting mitigative measures, and designing an impact monitoring plan. The research strategy requires limiting the scope of study to a few species and species attributes, using a conceptual model of ecosystem interactions and projected human activities to identify probable points of impact, and making extensive use of available literature to interpret the new data collected. Some conclusions to date are that (1) mule deer are sensitive to changes in availability and productivity of selected shrubs (mainly antelope bitterbrush) in winter, (2) songbirds respond drastically and predictably to structural changes in vegetation, and (3) most small mammals (and perhaps their predators) respond to structural changes in vegetation, but in ways that are sometimes unclear, and different among species. In terms of monitoring impacts, conventional schemes of annually or seasonally repeated measurements will usually be unable to sort development-caused changes from normal fluctuation, except where effects are already predictable.

INTRODUCTION

Multi Mineral Corporation (MMC), a subsidiary of Charter Company, is preparing to develop a sodium mine in the Piceance Basin, Colorado, about 40 km (25 mi) airline distance southwest of Meeker. MMC is required by the U.S. Geological Survey Conservation Division (USGS), to conduct pre-development environmental studies to accompany lease development. Requirements for these studies are the same as those for shale oil developments in the area.

LGL Ecological Research Associates (LGL) is under contract to MMC to conduct biological baseline studies in partial fulfillment of these requirements. These baseline studies focus on vertebrate wildlife and vegetation. (No aquatic habitats exist on the lease area.)

MMC has approved a program that differs in important ways from conventional baseline studies. Determining the distribution and abundance of animals and the productivity of vegetation were significant program objectives, but this paper focuses on the less conventional schemes of ecosystem analysis--those that are enabling us to predict impacts and design a workable monitoring program.

The objectives and design of the studies were founded on three major assumptions:

Assumption 1: The ultimate purpose is not simply to describe the biota; it is to find out how the biota will respond to development.

Assumption 2: It is futile to try to assess how the biota will respond without limiting studies to a few of the many ecosystem components because there are far too many components to address.

Assumption 3: To determine how selected components will respond to development, one must (1) acquire new data that are different from those provided by conventional baseline studies and (2) interpret new and existing data much beyond the level of interpretation by conventional studies.

OBJECTIVES

The general objectives of this study were to analyze lease area biota sufficiently to provide the following:
(1) A description of the abundance and distribution of selected vertebrates;
(2) A description of the annual net production of selected plants;
(3) An analysis of the important ecological interdependencies such that indirect effects of the proposed development actions on selected ecosystem attributes can be predicted, and,

(4) A plan by which MMC can measure (monitor) the effects of development and mitigative actions as they occur.

The last two objectives are the important ones for purposes of this paper; similar objectives are seldom emphasized in baseline studies. But without them neither this study nor others could accurately predict impacts or develop viable plans for detecting or mitigating them.

STRATEGY AND RATIONALE

The strategy we are using to address these objectives includes (1) focusing on selected ecosystem attributes (species or species qualities), (2) developing a conceptual model showing how these species function in the ecosystem, (3) designing field research to clarify important relationships among these species and between them and their habitats, and (4) interpreting both new and existing data to determine how development actions might affect the species directly or affect the factors that control their well-being.

Focusing the Study

As is true in most impact analysis programs, the MMC lease area has far too many biological components for a baseline study to effectively address. We are focusing the study on those components that USGS and MMC consider important from their points of view. In most cases these components are animal species, either of direct interest to society at large or thought by USGS to be useful as monitors of impact that would indirectly affect society's interests. Components being studied are (1) songbirds, (2) birds of prey (raptors), (3) sage grouse (Centrocercus urophasianus), (4) small mammals (rodents and rabbits), (5) carnivores, (6) mule deer (Odocoileus hemionus), (7) feral horses, and (8) vegetation production.

We are further restricting the field portion of the study by addressing only the portions of these components that are amenable to measure by efficient techniques. Thus we are focusing on songbirds detectable by Emlen's census technique (1), raptors detectable by sight, small mammals susceptible to trapping with Sherman live-traps or detectable by roadside census, and carnivores detectable by scent-station survey methods (2) or counts of tracks in fresh snow.

Developing a Conceptual Model

We have developed a conceptual model of the important ecological interrelationships that suggests which data should be collected and how they should be analyzed. Components of the model are (1) the species identified by MMC and USGS that are amenable to study by existing methods, (2) the factors that may regulate populations of these species and (3) the development actions that may impinge on the regulating factors or directly on the species. This model was developed on the basis of a literature synthesis at project initiation; it is continually being revised as new data become available during the course of the project.

The model is diagrammatic (Fig. 1); with additional inputs of money and time it could be computerized. The important point is that it is strongly focused on the components of interest (therefore relatively simple) and easily modified. Its primary purpose is to show the important relationships among (1) the components of interest, (2) their regulating factors, and (3) industry actions that affect one or both of these. The study design and data analyses are strongly influenced by the need to validate the model.

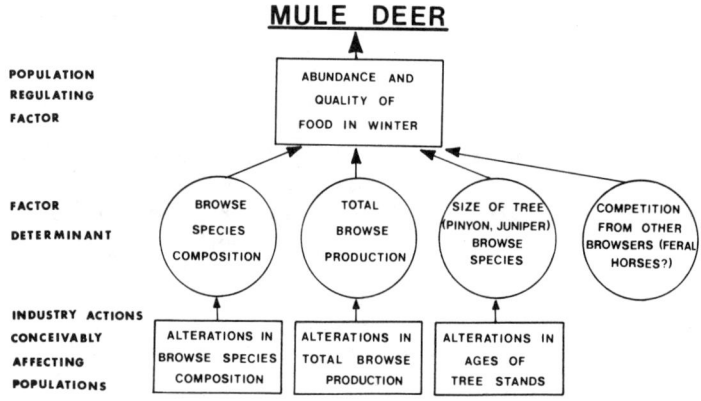

Figure 1. Example of diagrammatic modeling used to focus research studies on the Multi Mineral lease, Piceance Basin, Colorado.

Clarifying Ecosystem Function

As suggested above, field research is designed to reveal important functional relationships as well as describe the distribution and abundance of organisms. We are trying to find which factors regulate populations of sage grouse, feral horses, mule deer, and selected songbirds, raptors, small mammals, carnivores, and plant production. This information will suggest which development actions are likely to indirectly affect these entities.

Interpreting the Data

Data interpretation is one of the most important aspects

of this study. It requires extensive examination of existing literature as well as new data, and a careful application of known ecological principles to both new and existing data. The state of the art in impact assessment does not permit a useful level of impact prediction without extensive synthesis because most of the information required for valid predictions cannot be collected as new data within the funding and time constraints of single studies.

FIELD METHODS

Songbirds

We study songbird distribution and abundance during the breeding season by using the Emlen transect method (1). Several permanently marked transects were established per habitat type; each is monitored 3 times during the breeding season. Data are recorded by transect segment. From the data we estimate bird density (based on the number of singing males detected) by species and by habitat type. Because it is known that bird occurrence and abundance are normally regulated by vegetation structure (3,4,5), we plan to measure structural attributes of the vegetation within each transect segment to determine what traits of the vegetation influence bird presence and abundance.

Small Mammals (Rodents)

Small mammals are sampled on live-trap grids each month for 3 consecutive days. Each grid contains 100 traps; there are 2 grids per habitat type. Live-trapping enables us to calculate (by species) (1) density, (2) biomass, (3) productivity, and (4) biomass lost (to predation, etc.). It gives us a good idea of the magnitude of predator population that can be supported by rodent prey. Measurements made of vegetation structure in each grid enables us to correlate rodent occurrence with habitat structure--small mammal abundance in some areas is known to be associated with habitat structure (6,7,8,9,10).

Mule Deer

Mule deer are surveyed monthly from a small fixed-wing airplane such that the observer(s) view the entire lease area in successive parallel flight strips. In this way we estimate monthly deer densities and habitat affiliations. Additionally, deer pellet groups are counted on belt transects (several per habitat type). The dispersion of the pellet groups in relation to the dispersion of potential food plants is being evaluated to help determine whether deer are attracted more to some plant species than to others. (Because the study is in important deer winter range, on which food supply probably regulates populations, we want to find which plants are important.)

Other Components

Raptors, sage grouse, carnivores, feral horses, and vegetation production are also being addressed. Nests of cliff-nesting raptors and locations of raptors seen during aerial surveys are mapped. Scent-station surveys and track counts in snow are conducted to acquire indices to carnivore (mainly coyote, Canis latrans) abundance. Sagebrush habitats are searched for evidence of breeding sage grouse in April. Shrubs and grasses are clipped and weighed in September to estimate annual net production of the common species.

RESULTS

Some field data have been analyzed, but results of the completed study will not be available for some time. Both literature syntheses and new data contribute to the following discussions, which emphasize how the attributes we focused on are regulated and how mine development might alter them.

Songbirds

As expected, songbird distribution and abundance seem to be regulated by vegetation structure. Vertical components of plant structure tend to regulate bird species composition; horizontal components tend to regulate density. Some birds seem to respond to selected plant species (e.g., Brewer's sparrows, Spizella breweri, are faithful to sagebrush) and some respond to structural attributes provided by more than one species (e.g., cavity nesters use both pinyon and juniper trees; green-tailed towhees, Chlorura chlorura, use brush habitats irrespective of the species composition). It is probable that, where habitats are optimum, ceiling densities are established by intraspecific aggression (territoriality).

In terms of important ecological interrelationships, the physical structure of the habitats strongly control the bird populations, but no other significant ecosystem components depend on birds. The biomass of birds available to predators is insignificant in comparison to alternate sources of food (e.g., rodents), even to traditional bird-eaters such as sharp-shinned (Accipiter striatus) and Cooper's (A. cooperii) hawks. Bird foods probably fluctuate independently of bird populations. As suggested by Peterson (11), the birds could entirely disappear with probably few significant impacts to other parts of the ecosystem.

Judging from development scenarios currently projected for the area and the known responses of birds to man's activities, we conclude that the greatest impacts to birds will probably be a consequence of altered habitat structure. Because our data will eventually show which structural features regulate which birds, we will be able to predict the consequences of a given change in habitat structure. O'Meara et al. (12) found that chaining of pinyon-juniper woodlands eliminated the cavity-nesting forest birds and introduced the brush and grassland species (Fig. 2). We are finding almost identical parallels in this study, and predict that bird responses will continue to be the same in the future. Responses to other kinds of changes in structure (removal of sagebrush, increases in grass cover, etc.) should be equally predictable.

Small Mammals (Rodents)

Environmental factors regulating rodent distribution and abundance are less obvious than those which regulate birds. Although some species are predictably restricted to specific habitats (sagebrush voles, Lagurus curtatus, to sage habitats; pinyon mice, Peromyscus truei, and Colorado chipmunks, Eutamius quadrivittatus, to pinyon-juniper habitats), the most abundant ones--deer mice (Peromyscus maniculatus) and least chipmunks (Eutamius minimus)--are ubiquitous. Some literature (7) suggests that associations between rodent and habitat features are strong (though often not obvious) and can be detected by multivariate analysis. We are performing such analyses but have no results as yet.

Population levels of deer mice and least chipmunks are probably not directly controlled by habitat features such as food or den sites, but by social interactions that may vary in intensity among habitats (13). Our data show these species to be common in all habitats.

In contrast to birds, rodents are important to other components of the ecosystem. They can significantly affect vegetation structure and productivity (14) and may provide most of the food for carnivores and raptors on the MMC lease.

Approximately 10,000 kg of rodents were lost (most probably to predation) on the 15,000-acre study area this year. This amount would support approximately 37 coyotes year-round, but is seasonally variable (Fig. 3). In winter the biomass lost dropped to about 250 kg per month (enough to support 11 coyotes). During the breeding season for rodents the amount lost increased to about 2200 kg per month (enough to support 100 coyotes). Because the rodent food for coyotes is not equally available throughout the year, and other predators compete for rodents, our data suggest that during at least some parts of the year availability of small mammals as prey may affect predator populations. Food habits studies of

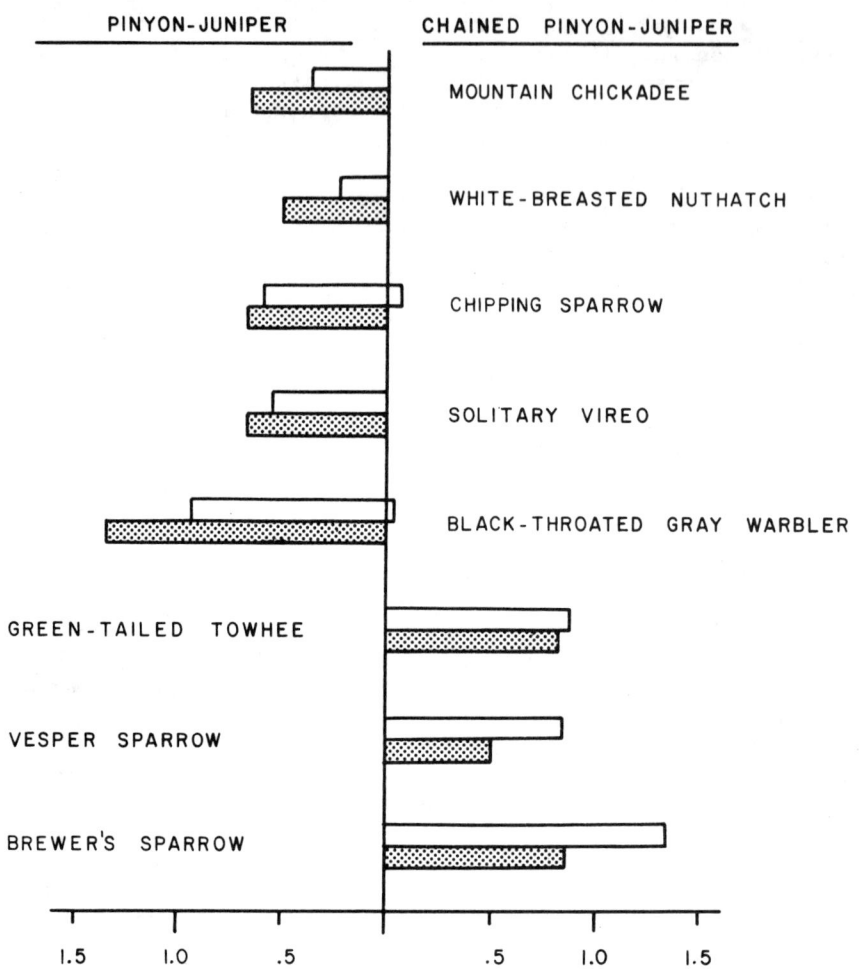

Figure 2. Densities (breeding birds/ha) of common birds in unmodified and modified pinyon-juniper habitats in Piceance Basin, Colorado, as found by 2 studies conducted in different places and years (clear bar = This Study; dotted bar = O'Meara et al. {12}). The similarities between studies and the differences between habitats suggest bird responses to be predictable and drastic.

coyotes (15) and raptors (16,17) show that rodents comprise large proportions of the diets of these predators.

Gross habitat changes caused by development will affect some small mammals, but not all, in predictable ways. Habitat changes will have the greatest impact to rodents that are restricted to specific habitats. The impact to ubiquitous

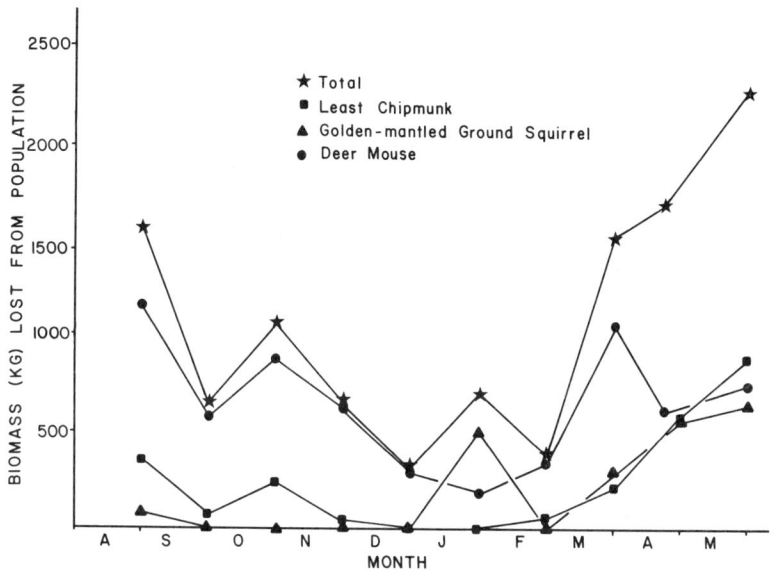

Figure 3. Estimated monthly losses of small mammals (biomass) from the 15,000-acre Multi Mineral lease area in Piceance Basin, Colorado.

species is not as predictable. O'Meara et al. (12) found that chaining of pinyon-juniper woodlands reduced species diversity. Chaining of the MMC lease increased species diversity. The most abundant species in both the O'Meara study and our study increased in density with chaining.

Subtle habitat changes may have a greater effect than gross changes on some rodents. Stoecker (Stoecker-Keammerer Associates, pers. comm. 1981) found a significant reduction in deer mouse and least chipmunk densities when an upland sage habitat in Northwest Colorado (Cb Oil Shale Tract) was sprinkle irrigated. Some impacts may not be evident as changes in rodent densities, but as changes in population dynamics. It may be possible to detect subtle impacts as they occur by monitoring socially-regulated population processes such as recruitment, survival, and reproductive activity.

Mule Deer

The availability of selected food plants probably regulates mule deer numbers in Northwest Colorado. On the study area, deer are almost non-existent between May and October, but they are abundant there in winter. Existing data strongly suggest that deer in the region surrounding the study area are regulated in abundance by food availability on winter range. Our data suggest that availability of one plant species—antelope bitterbrush (Purshia tridentata)—may be the most critical component regulating deer populations on the study area and adjacent areas. We are finding that deer concentrate near, and consistently hedge, bitterbrush plants in preference to the other browse species.

Deer have important ecosystem functions, both as consumers and as prey. It is clear that their browsing modifies the plant community structure and productivity on the study area. They also may be important food for coyotes, and perhaps other predators and scavengers, especially in severe winters.

Information, mostly from other studies, suggests that deer may be sensitive to direct disturbance (road traffic, facility operation, etc.). However, it is likely that the most important factor affecting their welfare will be the continued availability of nutritious food. We are trying to clarify which aspects of food availability (time of year available, plant species and age, etc.) are most crucial in influencing the welfare of deer. Expected effects of development actions on these factors can then be translated into how they will affect mule deer.

Other Components

Cliff-nesting raptors (golden eagles, Aquila chrysaetos; red-tailed hawks, Buteo jamaicensis) are almost certainly regulated by the scarcity of suitable nest sites on the area. Although there are many low cliffs and bluffs, few provide nest sites inaccessible to potential predators. To date we have found no active raptor nests on cliffs, but a few inactive ones. Golden eagles and red-tailed hawks commonly hunt on the lease area.

Sage grouse had not been reported to breed on the area, and we found no breeding sites (leks). Some grouse were found in winter, substantiating the published notion that the study area might support wintering grouse. The potential of the sagebrush habitats as grouse breeding areas remains uncertain; most nearby breeding areas are at higher elevations.

Coyotes appear to occur in high densities relative to those in surrounding areas. Although there is heavy trapping pressure in and near the area in winter, which influences how many coyotes survive to spring, we suspect that prey availability helps regulate their density and productivity. The losses suffered by the rodent portion of the prey base are unexpectedly low, suggesting that they may be relatively secure from predators. We do not know what factors affect prey vulnerability.

Feral horses are few compared to what the range could support, mainly because of Federal horse-removal programs. They therefore probably have little effect on other components of the ecosystem.

Vegetation production measured to date falls within the ranges expected for the area. It is expected to vary greatly among years and plant species, because the moisture regime varies greatly and so do the grazing pressures by livestock.

CONCLUSIONS AND IMPLICATIONS

Preliminary results suggest that the approach we are using will provide much new information that can be used to help direct the course of environmental research and reclamation and mitigation practices. Examples of tentative conclusions and their expected utility follow.

(1) The major impacts of development to songbirds will probably be caused by changes in habitat (mostly vegetation) structure. In many cases enough is already known about habitat requirements of the common species to allow us to predict the impacts to birds with almost as much precision as we could later document the impacts with monitoring programs. Impacts suffered by birds would usually not ramify through the ecosystem--songbirds may be an ecological "dead-end" in terms of their ability to influence other systems components.

The implications of this are clear. Environmental studies that try to find what birds currently exist in common habitats, or to predict or monitor response of birds to habitat changes, may be redundant--the information may already be available.

(2) In rodents, the levels and kinds of impacts will differ among species. Populations of some rodents, like those of birds, will predictably respond to gross habitat changes by changes in numbers. But the expected responses of the species that are most abundant in this study (and the most important to other ecosystem components) are less evident. Existing literature suggests that these species may respond to subtle environmental changes, but not necessarily by exhibiting great changes in population levels. The changes may come, for example, as alterations in production or variability in susceptibility to predators. The implications of this are that monitoring programs that measure only abundance may not detect important changes in population dynamics of species, changes that influence predators and perhaps other components of the ecosystem.

(3) The continuing production of selected browse species may be a key to maintenance of mule deer populations. We suspect the well-being of the deer population on and near the MMC leases to be closely tied to the availability of bitterbrush in winter. The deer themselves probably affect the well-being of predators--mountain lions and possibly coyotes. Implications are that proper management of bitterbrush, or reclamation with bitterbrush or equivalent browse, may greatly in-

fluence whether impacts to deer, and possibly to their predators, are positive or negative.

(4) Evidence from our study and from the literature suggests that raptor populations on the lease area are probably regulated by specific structural attributes of the habitat such as high cliffs, large trees with stick nests established by other raptors, and cavities in trees. Disturbances near, or destruction of, traditional nest sites is the only adverse consequence of development that we foresee.

(5) Circumstantial evidence suggests that numbers of coyotes may be regulated by the food base, and that in summer rodents form the preponderance of this food. Implications of this to assessment of impacts are not yet clear.

(6) The normal annual, seasonal, and spatial variability we are finding in most ecosystem components will preclude easy documentation of impacts. A monitoring program that will sort development-caused change from normally-expected change will require an extremely rigorous sampling design, drastically different from that of most existing programs in Northwest Colorado and elsewhere.

ACKNOWLEDGEMENTS

We thank Multi Mineral Corporation for the funding that made this project possible. James Meredith and Susan Summers of Multi Mineral deserve special credit for approving a research approach we feel is providing the kinds of information needed for a high level of environmental protection.

LITERATURE CITED

1. Emlen, J.T. 1977. Estimating breeding season bird densities from transect counts. Auk 94:455-468.
2. Linhart, S.B., and F.F. Knowlton. 1975. Determining the relative abundance of coyotes by scent station lines. Wildl. Soc. Bull. 3:119-124.
3. Odum, E.P. 1945. The concept of the biome as applied to the distribution of North American birds. Wilson Bull. 57:191-201.
4. MacArthur, R.H. 1964. Environmental factors affecting bird species diversity. Amer. Naturalist 98(903): 387-397.
5. Cody, M.L. 1968. On the methods of resource division in grassland bird communities. Amer. Naturalist 102(924): 107-147.
6. Douglass, R.J. 1976. Spatial interactions and microhabitat selections of two sympatric voles, Microtus montanus, and M. pennsylvanicus. Ecology 57(2):346-352.
7. Duesser, R.D., and H.H. Shugart. 1979. Niche pattern in a forest-floor small-mammal fauna. Ecology 60(1):108-118.

8. Feldhamer, G.A. 1979. Vegetative and edaphic factors affecting abundance and distribution of small mammals in southeast Oregon. The Great Basin Naturalist 39: 207-218.
9. Geir, A.R., and L.B. Best. 1980. Habitat selection by small mammals of riparian communities: evaluating effects of habitat alterations. J. Wildl. Manage. 44: 16-24.
10. Barry, R.E., and E.N. Franq. 1980. Orientation to landmarks within preferred habitat by Peromyscus leucopus. J. Mammal. 61:292-303.
11. Peterson, S.R. 1980. The role of birds in western communities. Pages 13-26 In: Workshop Proceedings: Management of Western Forests and Grasslands for Nongame Birds. USDA For. Ser. Gen. Tech. Rep. INT-86.
12. O'Meara, T.E., J.B. Haufler, L.H. Stelter, and J.G. Nagy. 1981. Nongame wildlife responses to chaining of pinyon-juniper woodlands. J. Wildl. Manage. 45(2): 381-389.
13. Fairbairn, D.J. 1978. Dispersal of deer mice, Peromyscus maniculatus. Oecologia 32:171-193.
14. Grant, W.E., and N.R. French. 1979. Evaluation of the role of small mammals in grassland ecosystems. Eco. Modeling 8:15-37.
15. Johnson, M.K. and R.M. Hansen. 1979. Coyote food habits on the Idaho National engineering laboratory. J. Wildl. Manage. 43:951-956.
16. Phelan, F.J.S., and R. Robertson. 1978. Predatory responses of a raptor guild to changes in prey density. Canadian J. Zool. 56(12):2565-2572.
17. Fitch, H.S., F. Swenson, and D.F. Tillotson. 1946. Behavior and food habits of the red-tailed hawk. Condor 48:205-237.

WILDLIFE CONSERVATION STRATEGY ASSOCIATED WITH ENERGY DEVELOPMENT IN NORTHWEST COLORADO

David W. Crumpacker
Department of Environmental, Population and Organismic Biology, and Center for Research on Judgment and Policy, University of Colorado at Boulder

ABSTRACT: A consortium has been formed to develop a strategy that will aid public and private decision makers in protecting terrestrial and aquatic wildlife during the period of intense energy development that may occur in Northwest Colorado. Members include Colorado State University, the University of Northern Colorado, the University of Colorado, and the Colorado Division of Wildlife. The Northwest Colorado Wildlife Consortium will develop methods for predicting and mitigating cumulative, site-specific and regionwide, impacts on wildlife associated with habitat loss and degradation, and with increased use of habitats by people. The impact prediction process will utilize existing, regional, automated data processing and geographic information systems, as well as useful analytical methods already developed for prediction of wildlife impacts resulting from specific human activities. Expert opinion will be used to make final, comprehensive predictions in terms useful to decision makers. The Consortium methodology will be capable of producing rapid impact predictions and mitigation recommendations. It will be designed for use by private companies, local governments, and state and federal agencies.

INTRODUCTION

   The potential for a vast amount of energy development exists in Northwest Colorado. Resources include oil shale, coal, gas, oil, uranium, and water (hydroelectric). Primary impacts emanating from energy development sites and secondary impacts associated with the necessary buildup of socioeconomic support systems could seriously affect the area's terrestrial and aquatic wildlife. A regionwide conservation strategy is needed to ensure that a quality wildlife resource will be provided during and after the period(s) of energy development that may occur.

   Large amounts and diverse types of environmental information have been gathered in Northwest Colorado over the past 20 years. Much of it is pertinent to wildlife populations and habitats. However, this information does not presently exist in a form that can be effectively used for predicting and mitigating the cumulative wildlife impacts that are expected to occur in association with energy development. The information has been produced by numerous public and private organizations which have different intended uses for it, depending on their own missions and goals. Much of it is not available in easily obtainable, centrally located places such as public research libraries. It is of varying quality and often has not been subjected to outside peer evaluation. Much of the information tends to be site- and time-specific, with little reference to the broader, regional setting and to future years. Some serious gaps also occur in the existing wildlife data bases for Northwest Colorado. These concern such things as life histories and population dynamics of wildlife other than big game, selected small game, sport fishes, and threatened and endangered species.

   There is a great need to identify, compile, compare, evaluate, simplify, and summarize existing information on the wildlife resource in Northwest Colorado; to supplement it with information from other areas to the extent that this is feasible and desirable; and to present it in an easily accessible, centrally located form. An information bank of this type would form the basis for construction of a comprehensive impact prediction and mitigation system.

THE NORTHWEST COLORADO WILDLIFE CONSORTIUM

   Production of a wildlife conservation strategy for a region as extensive as Northwest Colorado is a large and complex task. Furthermore, time is short. For maximum effectiveness, an impact prediction and mitigation system should be devised and placed in operation during the preliminary planning stages of energy development. These considerations led the Colorado Division of Wildlife and three Colorado universities to pool their resources and organize the Northwest Colorado Wildlife Consortium.

Goals

The Consortium was formed June 9, 1981 to devise a procedure that can be used to aid in the orderly development of natural resources in Northwest Colorado by providing adequate protection for wildlife. The Consortium's activities will be accomplished in two phases as follows:

Phase I - Compilation and evaluation of existing: (1) information on wildlife populations and habitats in Northwest Colorado and (2) methods for predicting, assessing and mitigating cumulative, regionwide, wildlife impacts.

Phase II - Development of procedures for: (1) predicting cumulative, regionwide, wildlife impacts from energy development and (2) recommending mitigations based on land use and management decisions involving reclamation, manipulation of existing wildlife habitats and populations, etc.

If adequate, timely funding can be obtained, the Consortium plans to finish Phase I by September 30, 1982 and Phase II no later than 1985.

Although the completion of Phases I and II could be viewed as an interesting academic exercise, the intent is a very practical one. Success will be measured by the extent to which the Consortium process is accepted and used by the following groups:

(1) Local governments in making land use decisions.

(2) Energy companies in devising efficient and economical operation, reclamation and mitigation plans.

(3) State and federal agencies in issuing mining and other energy development permits.

If the Consortium process is accepted by these groups, it may provide a model for dealing with cumulative, regional, wildlife impacts from various types of natural resource utilization in other parts of Colorado and the United States. Development of the process will also undoubtedly result in the identification of important information gaps which should be filled by future research conducted by appropriate agencies, universities and other groups. This type of information can then be fed back into the Consortium process in later years to improve its effectiveness.

Members

The three other members of the Consortium, in addition to the Colorado Division of Wildlife, are Colorado State University, the University of Northern Colorado, and the University of Colorado. Two delegates from each of the four member organizations comprise a governing board that is responsible for designing studies related to Phases I and II, raising funds, choosing investigators, administering the studies, and disseminating results. Investigators will be chosen primarily from university and college faculties throughout Colorado. Hence, the Consortium will draw from a professional talent pool which is broader than its own membership.

The Consortium also intends to cooperate closely with Region 6 of the U.S. Fish and Wildlife Service, which has a group of scientists and technicians that are working to accomplish similar goals to those of the Consortium, but on a multi-state basis.

DEVELOPMENT OF PHASE I

The Phase I compilation of existing information will emphasize 14 subject matter areas as follows: big game; small game furbearers and other mammals; small game waterfowl and upland game birds; game and nongame fishes; nongame mammals; nongame birds; reptiles and amphibians; molluscs and crustaceans; terrestrial ecosystems; aquatic ecosystems; prediction, assessment and mitigation of wildlife impacts; built environment and prediction of socioeconomic impacts; reclamation and rehabilitation; and legal/political.

All of these Phase I investigations will be coordinated and bounded by the common need to develop information that will be pertinent to the problems of predicting and mitigating Northwest Colorado wildlife impacts in Phase II.

Information on Impact Prediction and Mitigation

A knowledge of the present built environment and methods of predicting future effects of energy development on the built environment and on the type and amount of future human populations in the region will be critical to the process of predicting cumulative wildlife impacts. The locations, sizes and kinds of energy development operations will provide the primary "driving forces" that will generate wildlife impacts. The supporting industries, businesses, population centers, recreational areas, and transportation systems that will also develop are examples of socioeconomic factors that will provide secondary driving forces in the wildlife impact generating system.

Prediction of wildlife impacts is a newly developing technology. No satisfactory, comprehensive models exist which can be used to relate a wide range of human activities to cumulative wildlife impacts. Partial wildlife impact prediction models are now being built in several agencies such as the U.S. Fish and Wildlife Service, the U.S. Bureau of Land Management, the U.S. Forest Service, the U.S. Geological Survey, and State Divisions of Wildlife and Departments of Game and Fish. These models begin by associating wildlife species and their respective needs for food, water, breeding, and cover with habitat characteristics [e.g., see 1, 2, 3]. Terrestrial habitats may be characterized by factors such as dominant plant species, successional stage, horizontal habitat structure (edge, patchiness, ecotones), and vertical habitat structure (number of vegetative layers, amount of closure in each layer). Aquatic habitats are more conveniently described in terms of their physicochemical parameters (bank stability, instream cover, flow, temperature, dissolved oxygen, substrate, major ions, etc.) Association of wildlife needs with habitat characteristics provides the conceptual basis for predicting wildlife impacts. For example, removal of the tree canopy from a mature pinyon-juniper woodland or lowering the flow of a particular stream reach will have predictable effects on various wildlife species.

An important aspect of Phase I will be to survey and evaluate existing automated data processing systems to determine which ones have the greatest potential for predicting wildlife impacts in Northwest Colorado. Special emphasis will be placed on operational, computerized, geographic information systems that can be used to determine areas of potential conflict between wildlife and other resources, to choose siting alternatives for energy facilities and socioeconomic support systems, and to prioritize different areas with respect to their wildlife values [e.g., see 4, 5]. Surveys will be made of additional automated components that relate specific types of human activities to specific types of wildlife impacts. Examples of these are the Fish and Wildlife Service's Instream Flow Methodology which predicts impacts on fish populations from changes in stream flows due to impoundments and diversions [6], and Forest Service models which relate timber management actions to impacts on terrestrial wildlife [7].

Attention will also be given in Phase I to procedures which place less emphasis on formal analytical methods and depend more on the use of expert opinion [e.g., see 8, 9, 10]. This is especially important because no formal analytical methods presently exist for making all of the complex impact predictions needed to address the accumulation of wildlife impacts over time and space. Hence the Consortium process will have to rely to an important degree on the use of expert judgments to produce the types of wildlife impact predictions that will be most useful to policy and decision makers.

From the wildlife viewpoint, reclamation is a process whereby disturbed land or water is returned to a condition that provides suitable habitat for species which were originally present. Rehabilitation is a process by which disturbed land or water is returned to a situation that conforms with a prior land use plan and which leads to a stable ecological condition. The desired land use may or may not emphasize wildlife. Reclamation and rehabilitation methods suitable for Northwest Colorado will be evaluated for two reasons. First of all, they represent important tools for helping to mitigate energy-related wildlife impacts. Secondly, any meaningful wildlife impact prediction made with the aid of the Consortium process developed in Phase II will have to be based on the assumption of certain amounts of reclamation and rehabilitation. The more accurate existing information is concerning the effectiveness of, e.g., an assumed reclamation method, the more accurate the impact prediction will be.

Two types of legal/political information will be considered in Phase I. Laws, regulations and policies designed to conserve and protect wildlife during energy development in Northwest Colorado will be one area of emphasis. They provide much of the impetus for planning and mitigation, both by government agencies and private industry. The other area involves laws, regulations and policies related to wildlife management. These are administered by the Colorado Division of Wildlife and, in the case of federal threatened and endangered species, by the U.S. Fish and Wildlife Service. Regulation of fish and game harvests, reintroduction of depleted species, habitat improvement projects, and monitoring of wildlife populations are examples of legally mandated approaches that can provide important mitigations of potential wildlife impacts.

Baseline Biological Information

The Phase I survey and evaluation of the existing natural environment in Northwest Colorado will utilize an ecosystems approach. Information will be assembled on the major vegetation and aquatic types which form the basis for the wildlife habitats of the region. Consideration will also be given to special habitat features such as riparian zones, cliffs, caves, migration corridors, and traditional reproduction areas. Concurrent surveys will be made of the various categories of wildlife mentioned earlier. Wildlife species that are valuable indicators of the general health of each major terrestrial and aquatic ecosystem type must be identified, since it will not be feasible to treat all species in equal detail. In addition, special emphasis will be placed on game animals and sport fish, state and federal threatened and endangered species, and Colorado resident species of special significance.

Once the species of interest have been identified, data of special ecological significance must be compiled and evaluated for each of them. Again, the survey must be selective. Emphasis will be placed on aspects of species life histories and population dynamics which are essential to wildlife planning processes; e.g., geographical distribution, degree of dispersion, population sizes and trends, species status, special habitat requirements not necessarily identified by vegetation types (breeding grounds, nesting sites, migration corridors, etc.), detailed breeding and feeding niches, guild associations, vagility, and **home ranges**.

The baseline information collected in Phase I can be used in Phase II to construct lists of wildlife species requiring one or more terrestrial/aquatic ecosystem types at one or more life cycle stages. This will provide a general basis for predicting the impacts of habitat modification or loss on wildlife in Northwest Colorado.

Information concerning the known impacts on wildlife of past habitat modifications and other activities must also be collected and evaluated in Phase I. This will include the direct and indirect effects of phenomena such as air, water and soil pollution, lowered primary and secondary productivity, altered plant and animal species composition, roads, noise, sightings of people, and hunting and fishing pressures.

DEVELOPMENT OF PHASE II

The process developed by the Consortium for predicting cumulative wildlife impacts will provide a basis for effective wildlife planning by government agencies and energy firms. It will also provide the rationale for recommendation of both site-specific and regionwide mitigation measures. The most promising impact prediction methods identified in Phase I will be used whenever possible in the Phase II methodology. The Consortium will develop a limited number of new methods, as required. The remaining gaps in the prediction process, for which no explicit, analytical methods are yet available, will be filled in by the use of expert opinion. Methods identified in Phase I for eliciting, analyzing, evaluating, and quantifying expert opinions will be an important part of the Phase II process.

A Tripartite System

It seems likely that three interconnected levels will be included in the Consortium's impact prediction methodology: (1) an automated data processing system that will store information on the built and natural environments, provide rapid retrieval of various data summaries and relationships, and be continually updated by one or more established agencies such as the Colorado Division of Wildlife and the U.S. Fish and Wildlife Service; (2) a group of computer-based, analytical methods that can utilize this data bank to make various types of impact predictions; (3) an interdisciplinary panel of experts, including wildlife biologists and plant ecologists familiar with Northwest Colorado and the underlying components of the system, who will make the final impact predictions and translate them into terms that will be of maximum value to decision makers. As new impact prediction methods become available and old methods are modified by information obtained from observations, experiments, and project monitoring, improvements can be made in the Consortium's system. The process will be designed to yield rapid impact predictions and mitigation recommendations of a site-specific or region-wide type, at various projected times in the future.

A Suggested Scenario Approach

The Consortium has neither the time nor money to develop and validate a comprehensive, detailed, regional, land use model that predicts effects of different human activities on wildlife in an area as large and complex as Northwest Colorado. Working "from the bottom up" in this fashion might be intellectually satisfying, but is clearly impractical. A fully operative system is needed by 1985. An attractive alternative which will be considered is to construct a wide range of future development scenarios and estimate the types of wildlife populations and habitats that will be associated with each of them at various times in the future. The difference between each such estimate and the existing, baseline situation will constitute a prediction of the cumulative wildlife impact associated with that particular scenario. This "from-the-top-down" approach has the potential of relatively rapid and relatively inexpensive accomplishment. An example of how this might be done is given below.

1. Develop a realistic range of Northwest Colorado scenarios for 2000 AD and later that will include various combinations of:

    a. Future land use desires;

    b. Predicted levels of energy development;

    c. Predicted types of socioeconomic support systems.

2. Predict the quantity and quality of wildlife populations and habitats for each scenario by:

    a. Assuming reclamation and other mitigations and land use decisions consistent with the desires and restrictions imposed by each scenario;

    b. Use of the best available predictive methods, including:

        (1) Explicit, analytical models;

        (2) Intuitive judgments of experts.

3. Analyze relationships between changes in amount of energy/socioeconomic systems development and predicted cumulative wildlife impacts, in hopes of finding function forms that can be used to predict impacts associated with any scenario that lies within the range given in 1.

4. If no useful function forms are found, place extra emphasis on developing methods to speed up the entire process of producing new impact predictions "from scratch" by the methods described in 2.

An example of a generalized version of one scenario is given below:

Scenario X

Future land use desire	- less quantity but higher quality of wildlife populations and habitats than there is now
Predicted level of energy development	- moderately extensive and very rapid
Predicted type of socio-economic support system	- dispersed, relatively unplanned residences, recreation areas, and transportation systems

In order to make a wildlife impact prediction that is consistent with the future land use desire of this scenario, it would be logical to assume that the majority of mined land would not be reclaimed for wildlife habitat, nor would much undisturbed land be dedicated to wildlife habitat as another type of mitigation. Instead, some lesser amount of mined and/or undisturbed land would be dedicated to high quality wildlife habitat, presumably in conjunction with improved management and possibly with more restricted access for hunting and fishing.

Any number of scenarios could be constructed by sampling one level at a time from a large number of levels of future land use, energy development, and socioeconomic support system development. However, only a few diverse levels for each of the three categories would be needed to generate, by random combination, a realistically wide range of scenarios.

Each scenario would obviously need to be defined in more detail than that provided in the discussion of Scenario X. For example, the socioeconomic support system associated with a particular scenario would need to specify such things as the predicted increase in human population size in the region, the number and types of new recreation areas to be built, the increase in vehicle miles traveled, etc. At this point there would still be several alternative ways in which the scenario could be implemented, depending on the location and size of cities, towns, roads, etc. Each alternative would be associated with a specific wildlife impact prediction. Thus it would be possible to calculate an impact variance which would represent the scatter of impacts around the mean impact for a particular scenario. The impacts associated with a scenario may exhibit relatively little variation, regardless of how the scenario is implemented. Other scenarios may cause greatly different impacts depending on how they are put into effect. It is possible that the variance of predicted impacts within some scenarios might exceed the variance between the scenarios. It follows that a poorly implemented, moderate energy development scenario might create a more severe wildlife impact than a properly instituted, heavy energy development scenario. Information of this sort would be critical from the wildlife standpoint and the system must be capable of developing it rapidly and presenting it to planners and decision makers in understandable terms. Computerized assistance in the development of impact variances would clearly be of great value.

Some Other Considerations

Measurement Scales

Construction of function forms that relate levels of development to cumulative wildlife impacts requires the use of scales to measure these effects. A search should be made for scalar variables that integrate as much information as possible, because of the simplification this would provide in the analyses. An example of a dependent variable of this sort, which provides some measure of cumulative wildlife impacts, is species diversity. Impacts measured by the latter variable incorporate the concept of numbers of species as well as that of numbers of individuals within species. However, the variable is somewhat difficult to interpret from the ecological standpoint [11], is not easy to translate into tangible terms for decision makers, and provides no indication about which species are most affected by a certain type of development. An example of an independent variable that may integrate a large amount of information on socioeconomic support system development is the per capita energy consumption in Northwest Colorado. Higher values of this variable should be related to increased use of the area's natural systems for hunting, fishing, backpacking, jeeping, snowmobiling, picknicking, etc. As the human population density increases, these secondary aspects of energy development could impact wildlife more heavily than the primary ones. This would be especially likely if advanced technologies were used in energy development operations to control pollution and to reclaim disturbed lands, and if concerted efforts were made to avoid high-value wildlife areas during the siting of energy facilities.

Effects of Fragmented Habitats

The primary and secondary effects of energy development in Northwest Colorado may create extensive areas of fragmented, as well as degraded, habitats. It is possible, then, that some of the simplest concepts in the field of conservation biology [12] may be useful for both prediction and mitigation of wildlife impacts. As habitats became smaller and more isolated from similar habitat types, species populations may decrease below some minimal critical size and become locally extinct [13, 14]. The first species to be lost would be those with large areal needs, low mobility, and poor capabilities for crossing areas of unfriendly habitat. Once a species population of this sort has been lost from a habitat isolate, it will be difficult for it to be reestablished by immigration. It may be possible to achieve important mitigation of local species extinctions resulting from habitat fragmentation by choosing development alternatives which avoid extreme reduction in size and isolation of habitats and which leave maximum amounts of connecting corridors between habitat isolates. Species-specific information from Phase I on amount of habitat needed per individual, general mobility, ability to cross barriers of unfriendly habitat, and minimum number of individuals needed to maintain a viable population can be used in Phase II to predict the amount of local species extinction which will occur due to the habitat fragmentation expected with various scenarios.

It is unlikely that the full range of successional stages typical of a particular plant community (vegetation) type can be maintained by natural processes in small fragments of the community. In order to mitigate the loss of wildlife adapted to specific seral stages of a plant community, each stage may need to be managed as a separate entity. For example, if there were 10 major vegetation types, each with three well defined seral stages, 30 plant community/ successional types might have to maintained and managed separately.

CONCLUSION

Methods for effective conduct of the information gathering and evaluation activities (Phase I) of the Northwest Colorado Wildlife Consortium's program have been described. However, the means by which impact predictions and mitigations will be accomplished (Phase II) are only crudely perceived at present. Since no completely analytical, automated method for relating an array of energy and socioeconomic development activities to wildlife impacts can be devised under existing circumstances, expert opinion will have to be used to finalize the impact predictions in a form understandable to decision makers.

The most difficult problem of all will be to produce credible predictions of points on the development scale(s) at which the cumulative effects on wildlife may be expected to exceed critical threshold levels and lead to rapid deterioration in both quantity and quality of the wildlife resource. The most important types of mitigation will be those which prevent the wildlife resource from decreasing below these critical levels.

LITERATURE CITED

1. Mason, William T., Jr., Cushwa, Charles T., Slaski, Lawrence J. and Gladwin, Douglas N. 1979. A procedure for describing fish and wildlife — Coding instructions for Pennsylvania. FWS/OBS — 79/19, December, 1979. 21 pp. Eastern Energy and Land Use Team, U.S. Fish and Wildlife Service, Kearneysville, West Virginia.
2. Patton, David R. Run wild II: A storage and retrival system for wildlife data. Transactions of the 44th North American Wildlife and Natural Resources Conference, pp. 425-430. Wildlife Management Institute, Washington, D.C.
3. Kingery, Hugh E., and Graul, Walter D., eds. 1978. Colorado bird distribution latilong study. Nongame Section, Colorado Division of Wildlife, Denver, Colorado. 58 pp. + appendices.
4. Porter, L. Ruggles, Towns, Grady W., Carlson, LeRoy W., Hamill, John F., and Fresquez, Victoria F. 1979. Promising methodologies for fish and wildlife planning and impact assessments. U.S. Fish and Wildlife Service, Region 6, Environment, Denver, Colorado. 28 pp.
5. Tessar, Paul A., and Caron, Loyola M. 1980. A legislator's guide to natural resource information systems. National Conference of State Legislatures, Denver, Colorado. 59 pp.
6. Stalnaker, Clair B. 1979. The use of habitat structure preferenda for establishing flow regimes necessary for maintenance of fish habitat. In Ward, James V., and Stanford, Jack A., eds., The ecology of regulated streams, pp. 321-337. Plenum Press. New York, New York. 398 pp.

7. Thomas, Jack Ward, tech. ed. 1979 Wildlife habitats in managed forests — The Blue Mountains of Oregon and Washington. Agriculture Handbook Number 553. 512 pp. Pacific Northwest Forest and Range Experiment Station, Portland, Oregon.
8. Holling, C. S., ed. 1978. Adaptive environmental assessment and management. John Wiley and Sons, New York, New York. 377 pp.
9. Hammond, K.R., Stewart, T.R., Brehmer, B., and Steinmann, D. O. 1975. Social judgment theory. In Kaplan, M.F., and Schwartz, S., eds., Human judgment and decision processes, pp. 271-308. Academic Press, New York, New York. 325 pp.
10. Edwards, W. 1977. How to use multiattribute utility measurement for social decision making. Institute of Electrical and Electronic Engineers Transactions. Systems, Man, and Cybernetics SMC-7(5): 326-340.
11. Hurlbert, S. H. 1961. The non-concept of species diversity: A critique and alternative parameters. Ecology 52: 577-586.
12. Soulé, Michael, E., and Wilcox, Bruce A., eds. 1980. Conservation biology: An evolutionary-ecological perspective. Sinauer Associates, Inc., Sunderland, Massachusetts. 395 pp.
13. Franklin, Ian Robert, 1980. Evolutionary change in small populations. In Soulé, Michael E., and Wilcox, Bruce A., eds., Conservation biology: An evolutionary-ecological perspective, pp. 135-149. Sinauer Associates, Inc., Sunderland, Massachusetts. 395 pp.
14. Soulé, Michael E. 1980. Thresholds for survival: maintaining fitness and evolutionary potential. In Soulé, Michael E., and Wilcox, Bruce A., eds., pp. 151-169. Sinauer Associates, Inc., Sunderland, Massachusetts. 395 pp.

IMPROVED INTEGRATION OF
BASELINE INFORMATION AND THE
DEVELOPMENT OF REVEGETATION
AND WILDLIFE RESTORATION PLANS

G. L. Potter, PhD
    Engineering-Science, Inc.
    (Denver, Colorado)

B. D. Snyder, MS
    Engineering-Science, Inc.
    (Denver, Colorado)

Abstract: The gathering of environmental baseline data on proposed mine sites is an important element of the premining activities. The intent of information collection on fish, wildlife, vegetation and soil characteristics is to provide a foundation for developing successful strategies to accomplish the ultimate post-mining site restoration. The initial terrestrial and aquatic baseline studies and subsequent long-term monitoring efforts should contribute to the development of mining operations that cause minimal long-term adverse environmental effects to the local biotic resources. Frequently, however, mine revegetation and wildlife restoration strategies are seemingly developed without an awareness or independent of the results of the baseline investigations. Review of a wide range of mining baseline studies and companion restoration plans has identified several important aspects of wildlife and vegetation ecology that are not adequately integrated into the development of the site reclamation plans. These aspects seem to be inadequately addressed regardless of the amount of field effort and expense spent on the baseline and monitoring studies. These phenomena are typically demonstrated in several ways. These include the following: (1) poor correlations between the vegetation studies and the habitat utilization rates of important wildlife species; (2) failure to recognize and incorporate the important habitat concepts of plant interspersion, plant succession, diversity, and edge into design of the revegetation plans; (3) a tendency not to use the extensive scientific literature available on the habitat requirements of key wildlife and plant species; and (4) tendencies to generalize or over-simplify the ultimate consequences of mining operations on the biotic resources. This paper discusses these and other topics in detail and suggests methods to alleviate these conditions. A systematic procedure is presented that assists the land manager to integrate baseline information into specific tasks that will facilitate establishment of a sound ecologically valid restoration program and wildlife enhancement plan.

PURPOSE OF BASELINE DATA

The implementation of environmental baseline studies is intended to serve two major purposes. First, environmental studies provide a method for inventorying and evaluating the natural resources present on a mining site. Second, the intensive, and often extensive, data collection effort provides the specific information base for the later development of site-specific resource restoration plans. Both of these purposes should be of major interest to those concerned with the planning, permitting, and operation of large surface mining operations.

Site Inventory and Characterization

Federal, state, and local natural resource management agencies are generally sensitive to any proposed major land use change that occurs within their jurisdiction. Extensive surface coal mining operations are included. Generally, few personnel within agency review staffs are intimately or even casually familiar with the biological resources present on any particular proposed mine site. These individuals rely on the baseline ecological data provided in the mining application to evaluate the site resources and make a determination of the acceptability of the proposed project. The primary concerns of the review staff and the agency consist of five very simple and basic interests; namely, species, community, and habitat:

- Presence,
- Abundance,
- Distribution,
- Susceptibility to project-induced changes, and
- Significance of any alterations.

Some state regulatory agencies have achieved fairly high levels of sophistication in assuring that their minimum baseline data requirements are satisfied in a consistent manner. For example, the Wyoming Department of Environmental Quality, Land Quality Division, has published a series of comprehensive guidelines and minimum requirements for conducting, among other things, vegetation and wildlife baseline studies for surface coal mine operations. Recommended survey and sampling procedures are outlined in an ordered fashion. Minimum statistical requirements and designs are identified. Tabular formats are provided to help in data organization and presentation. In other states, baseline data requirements are less well defined and may change from project to project, depending on circumstances.

Most mine plan applications adequately and consistently address resource presence, abundance, and distribution. There are many well-tested standardized techniques available for obtaining and analyzing this type of information. Consequently, the approaches are well known and their application is wide spread. However, this is not the case with the issues of resource susceptibility and impact significance. Resolution of these issues requires the more difficult and ill-defined process of integrating field numeric and observational data with literature data, education, past experience, and deductive reasoning to arrive at site-specific conclusions. Answering these two questions constitutes the most important information responsibility of baseline studies. Unfortunately, they are also the issues least adequately addressed. As a result of extensive field studies, the actual investigators should be the most qualified and the most able to comprehensively describe and evaluate the important ecological features of the site. Concurrently, the investigators should be able to clearly define the effects of the proposed surface mine on these features.

Development of Restoration Plans

The second important purpose of the baseline investigations is to provide an adequate data base for the development of subsequent vegetation and wildlife restoration plans. In most cases the intended purpose of these plans is to re-establish similar plant and wildlife communities on mined sites following completion of coal extraction phases. This is possible in theory, but difficult to accomplish in practice. However, more technical aspects of restoring selected critical components of biotic communities are known and have been successfully applied in actual practice.

More are becoming available with increased testing. The key element to these successes is the accumulation of an accurate and relevant baseline profile. The successful completion of these efforts requires accurate measurements of numerous site-specific environmental parameters that determine the quality and abundance of the local premining biotic populations. In many cases, these premining levels become goals for post-mining restoration The baseline data constitute the technical nucleus for determining the environmental features that control vegetation dynamics and ultimately the wildlife use of the mining area. These data dictate the types of techniques and the magnitude of the restoration efforts required to restore key or critical environmental components. If the data are too general or of insufficient quantity, many dollars and manhours may be expended unsuccessfully with little or no benefit.

Common Deficiencies in Data Application

The goals of baseline studies are to: 1) identify and evaluate the significant ecological features of a permit area; 2) determine the environmental factors regulating these features; and 3) provide the technical information base to facilitate their protection or subsequent restoration. Therefore, this information should be reflected in 1) the design of the vegetation and wildlife restoration plans; 2) assessment of project-related impacts; and 3) a comprehensive summary evaluation of the permit area's outstanding ecological features. However, these goals are typically not realized because of several deficiencies.

Restoration Plan Design

Mine plan applications frequently have few correlations between the descriptions of significant plant and wildlife resources and the companion revegetation, wildlife habitat restoration, wildlife protection, and grazing management plans included in the operational plan. This represents the single greatest deficiency in the use of baseline data. This trend seems to appear consistently regardless of the amount of time expended on the baseline investigations and the application preparation. Baseline studies should not be conducted simply to fulfill regulatory requirements. More consideration should be given to the subsequent use of this information in planning restoration programs that are specifically designed for a particular permit area and its unique characteristics. Restoration and management plans and the quantitative results of vegetation, wildlife, soils, and hydrologic data acquisition efforts are seemingly generated completely independently of each other. A synthesis of these data into a truly integrated ecological context rarely occurs. It is the synthesis and integration of these data into subsequent restoration plans that represent the critical linkage between baseline studies and restoration plans.

For example, frequently vegetation and wildlife baseline sections provide extensive details about the plant species composition, habitat utilization rates, and community structural details for high-value wildlife areas. In western states, these often include riparian zones and critical winter ranges. The value and utilization rates of these particular areas are documented by quantitative studies and descriptions. Habitat characteristics are detailed by the vegetation baseline analysis. Accompanying restoration plans need to reflect the importance of these areas by putting more emphasis into reestablishing these units during the restoration phases. Final revegetation plans need to show the integration of these available site-specific data on vegetation characteristics, wildlife use, and habitat distribution and composition. Often the revegetation plan for such areas proposes reseeding these areas with general range seed mixtures of highly adaptable and tolerant grasses and forbs, instead of plant types more characteristic of these sites. Greater emphasis is needed to formulate plans that account for critical habitat parameters such as plant structural diversity, shrub height, canopy coverage, shrub density, plant interspersion, and species composition. The importance of these parameters to key wildlife species has been documented repeatedly and is often reconfirmed by the initial pre-mining baseline studies. However, proposed habitat restoration plans often fail to reflect the importance of this information. Consequently, reviewers are left to speculate about the intentions of the revegetation or habitat restoration plan. A critical evaluation of a detailed pre-mining habitat description compared with a too generalized revegetation plan may give the wrong impression that major long-term plant community changes and concurrent shifts in wildlife composition will persist if habitat restoration efforts are implemented as stated.

Impact Assessment

A second major area where baseline data are not used to the maximum potential is in the assessment of project-induced impacts on wildlife and wildlife habitat. In some cases the discrepancies between the comprehensive and advanced methods employed to acquire population density data, habitat utilization values, and seasonal distribution patterns contrast markedly with analytical conclusions, that it is impossible to accurately assess the environmental impacts that will occur to local wildlife populations. For example, baseline investigations often report the occurrence of mule deer migration corridors, sage grouse leks, raptor nest sites, and populations of endangered species within a mine area, all of which have a potential for significant adverse wildlife impacts. The companion impact assessments, however, may minimize the significance of these issues without having adequately evaluated the consequences of the activity on these resources. The baseline data collected on population densities, habitat utilization rates, and seasonal distributions for key species should allow a fairly accurate estimate of certain quantitative changes. For example, quantitative estimates of percent reduction in key wildlife populations should be computed from data compiled showing the number of animals wintering on portions of their range to be removed by mining, the amount and quality of habitat to be lost, and the fraction of the total winter range that the mining area includes. Also, it is possible to estimate the approximate length of time that will elapse before the disturbed site can potentially support wildlife populations roughly equivalent to premining levels. This can be done by using available information on natural recovery and successional rates of various ecosystem types and comparing that information with actual baseline measurements.

These types of relevant evaluations are needed more often in impact analyses, since the basic data have been collected already, sometimes with great effort, and are readily available for use. The length of time an effect will persist is an especially important evaluation criterion. The failure to use baseline data for such purposes represents another critical application that is not routinely established between environmental studies and project evaluations.

Habitat Characterization

A third common deficiency is the tendency not to thoroughly develop a summary of relative overall habitat values for important wildlife species. Except for the investigators, other biologists have usually not spent time on a proposed mining site and rely on the experiences of the field investigators to identify and evaluate important wildlife sites or zones. They are dependent on the baseline study to integrate all the quantitative field data collected during the study and summarize the final evaluations in a concise, yet relevant manner. Too often baseline studies simply provide extensive tabular numeric data accompanied by simple descriptive text. The implications and significance of these data on wildlife are not addressed. However, it is precisely these aspects that are of greatest interest and constitute the main purposes of conducting the study. Delays, misunderstandings, and unncessary confusion often result as attempts are made to assess the significance of the quantitative measurements. The accuracy and efficiency of this integration and summarization process can be enhanced significantly if the baseline report judiciously uses the available scientific literature to evaluate the site compared to other similar biological sites.

SYSTEMATIC APPROACH

The following approach may help future investigators more thoroughly integrate quantitative vegetation and wildlife baseline data into products that satisfy the intended purposes of baseline investigations. Seven basic steps comprise the approach which culminates in submission of the mining permit application.

Review Mining Regulations

Although this is a very simple, elementary step, it seems that many applicants fail to recognize that many guidelines and minimum requirements for collecting baseline data are provided by Federal and appropriate state regulations. A thorough review of the regulations will indicate the kinds, scope, and intensity of data that will be required. The application of this information, once collected, will also become apparent during the review. Some regulations are necessarily generalized to permit site-specific flexibility. Such situations require personal contacts with appropriate agency representatives for clarification.

Contact Appropriate Agency Representatives

In addition to obtaining clarification on general requirements, this step allows the applicant to accurately confirm specific concerns of the regulatory agencies. Special-interest issues may be identified. The magnitude and level of sampling programs can be discussed and perhaps modified, if needed, before intensive field efforts begin. Sources of existing agency data may be identified and substitution for original field work may be authorized. At this point the applicant can determine what types of ecological evaluations and discussions the various reviewers will be most interested in. This provides the applicant an opportunity to find out exactly how the subsequent data will have to be summarized and what the most important topics of concern will be. Most importantly, the applicant can determine preliminarily what level(s) of vegetation and habitat restoration will be required. These requirements will dictate certain baseline data demands which will have to be planned for and budgeted.

Schedule Pre-Application Conference

This step may not be required depending on project-specific circumstances. If, after contacting the appropriate agency representatives, the applicant encounters conflicting or duplicative data collection requirements from two or more closely related agencies, then a preliminary conference with all interested parties to resolve the issues is warranted. This procedure often streamlines the data collection program.

Conduct Baseline Sampling Program

This step needs little explanation. All basic data collection efforts are conducted during this period. Special attention is given to collecting quantitative data that will both satisfy the standard regulatory requirements and that will also provide an adequate data base to resolve the special-interest issues identified by previous agency interviews. The accumulation of information should be guided by the five basic intentions of requiring baseline investigations previously discussed. In addition, the formulation of acceptable revegetation and habitat restoration plans will depend extensively on the adequacy of the baseline information.

Develop Baseline Report

This report presents extensive quantitative and qualitative information in both tabular and narrative forms. The narrative information should summarize the relative important values of the plant communities, wildlife populations, and other environmental features. Areas and habitats of importance should be clearly identified and discussed. Any special-interest issues should be thoroughly addressed. The baseline data should also be used to document or support value judgments. Numeric data should be (1) integrated into the summary statement of the site's value; (2) used in the identification of special value areas; and (3) the basis for the realistic assessment of environmental impacts. The goal of the baseline report should be to (1) clearly identify whether significant biological resources occur on the permit area and (2) determine what effect the project will have on these resources.

Develop Resource Restoration and Protection Plans

The various resource restoration plans are formulated concurrent with development of the baseline environmental reports. These can include separate plans for revegetation, wildlife habitat restoration, wildlife protection and mitigation, grazing management, and monitoring. Much planning is required to develop an approach that accomplishes its goals. Extensive preplanning is needed to understand and classify site conditions that might be encountered. Success is conditioned by a host of factors that are completely unique to each site. The integration of the diverse baseline data previously collected is critical to this step. It is unwise to assume that one strategy or choice of plants for revegetation will satisfy the diverse habitat requirements for all wildlife species at any given time. Yet this approach is taken in many restoration plans. Instead, greater consideration should be given to the habitat requirements of those wildlife species that are considered important or key species. These key species can be determined independently by the applicant. A safer approach would involve participation by the agency during the initial planning phase. Because of the many complicating factors that can affect wildlife response to restoration attempts, it is imperative that the site-specific ecological data collected during the baseline studies be used to formulate the restoration plans specifically for the key species. Otherwise, the most promising general-program may not produce the desired effect of increased wildlife diversity and productivity. Plan characteristics, such as rooting habit, season of growth, seedling vigor and competitive ability, growth rate, persistence, palatability, and tolerance to grazing must be considered. The important concepts of succession, edge, diversity, and interspersion must be included in the design. Administrative decisions should be made whether or not to increase the carrying capacities of undisturbed areas peripheral to the mine site in order to increase wildlife diversity and productivity thus compensating for habitat losses in the mine area.

File Application

The completed ecological baseline sections with accompanying vegetation and wildlife restoration plans are submitted as part of the final mine and reclamation plan for review and approval.

SUMMARY

The intent of pre-mining vegetation and wildlife baseline studies is to provide natural resource managers with accurate, objective, and integrated data on the ecological resources of the proposed mining area. Generally, five broad types of information are required about species, community, and habitat characteristics. These include determinations of presence, abundance, distribution, susceptibility to project-induced changes, and estimates of significance of any alterations. The goals of baseline studies are to: (1) develop a comprehensive summary evaluation of the permit area's outstanding ecological resources; (2) conduct an accurate assessment of project-related impacts; and (3) design vegetation and wildlife restoration plans. However, these goals are not often consistently achieved because the critical and complex linkages between obtaining quantitative baseline data and formulating integrated restoration plans and resource evaluations are frequently not realized. Reasons for this situation may include:(1) an unclear understanding of why baseline studies are required; (2) difficulties with synthesizing diverse ecological information into a holistic perspective; and, (3) a lack of understanding of how baseline data should be used in planning and resource evaluation processes. Consequently, several deficiencies routinely occur in the formulation of revegetation and wildlife restoration plans. These include: (1) poor agreement between the evaluations of significant plant and wildlife resources and the restoration plans intended to re-establish these resources after mining, and (2) a tendency to generalize or over-simplify the ultimate environmental consequences of mining on the ecological resources.

REVEGETATION CONSTRAINT ANALYSIS
FOR PIPELINE RIGHTS-OF-WAY

Scott L. Ellis
 Environmental Research &
 Technology, Inc.,
 Fort Collins, Colorado

SUMMARY

This presentation outlines the procedures that can be used to define environmental and regulatory constraints from which revegetation techniques and seed mixtures can be selected for specific on-site conditions. A stepwise method of incorporating a literature review, ecological data, environmental and political constraints, and post-reclamation land use objectives into the revegetation planning process is presented. The importance of various types of environmental data and their relationships is shown.

INTRODUCTION

A large crude oil pipeline that crosses several states presents a unique challenge for the revegetation planner. The project affects public and private landowners whose revegetation requirements for pipeline construction vary from state to state, and from county to county. A long pipeline of 500 to 1000 miles may cross a series of natural communities ranging from forest receiving 100 cm or more of annual rainfall to semi-arid to arid rangelands that receive less than 25 cm per year. Pipeline construction proceeds at varying rates, and completion of different segments may not coincide with the best season to reestablish vegetation.

The purpose of this paper is to 1) identify the major types of constraints or limitations on the revegetation methods used and species selected; 2) how these constraints are analyzed and incorporated into revegetation planning; 3) how constraints are accounted for throughout the life of the pipeline project.

PIPELINE REVEGETATION GOALS AND CONSTRAINTS

Several short-term and long-term revegetation objectives should be considered in revegetation planning. Short-term objectives are related to procedures that benefit soil stabilization and seedling establishment immediately following the construction phase, and include:

- Protect exposed soil using short-term wind and water erosion stabilization measures such as mulches and cover crops where necessary;

- Provide the best quality soil available for the seedbed;

- Prepare the seedbed for maximum germination and establishment of seeded species;

- Provide for the retention of incident moisture on disturbed areas during the seed germination and seedling establishment stages;

- Select adapted revegetation species that establish and spread rapidly;

- Discourage weed competition.

Long-term objectives are related to maintenance procedures following construction and include:

- Select species that will provide a self-perpetuating community that is compatible with post-construction land use;

- Maintain revegetated areas to promote the continued vigor of seeded species and permit establishment of desirable species from adjacent native plant communities.

The three major factors that influence the options available in a revegetation program include pipeline construction and maintenance procedures, environmental factors (climate, soils, topography), and the revegetation performance standards imposed by affected landowners. Figure 1 illustrates an ordering of these major constraint factors as a part of the process of evaluating site conditions to produce a revegetation prescription for the site. The relative priority of these constraints is not fixed. In a state with a detailed set of pipeline revegetation performance standards, state and local regulatory constraints and requirements would take precedence in defining revegetation options. In a state with few performance standards, the climatic and ecologic factors would be given greatest weight in selecting revegetation practices. Differences in constraint precedence are illustrated in a later section.

An initial sorting mechanism for environmental constraints is to define routine revegetation situations in which the probability of success is very high, assuming that proper seeding methods and adapted species are used. The remaining situations are then considered atypical & special problems in which one or more environmental factors limit the likelihood of sucess.

The following section describes the specific factors that must be considered within each of the major constraint classes, and sources of information that can be used to evaluate the importance of these factors.

Pipeline Construction and Maintenance Methods

The principal sources of disturbance on any pipeline project include the pipeline trench, the construction right-of-way adjacent to the trench, and access roads and lay down areas in the vicinity of the right-of-way. Important factors that define the zone of disturbance and revegetation practices include:

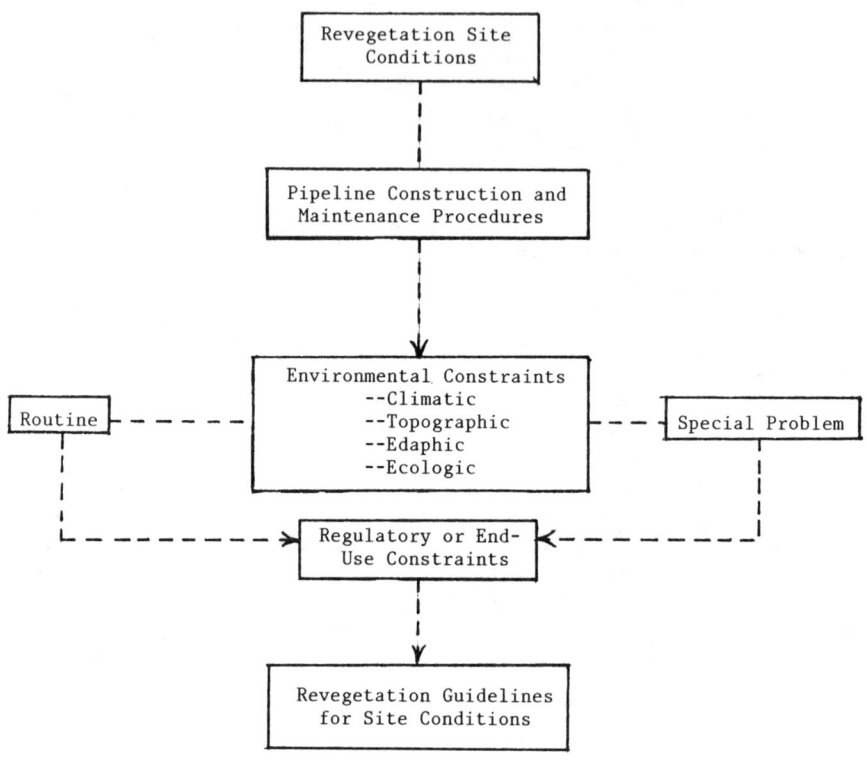

Figure 1. Constraint analysis pathway for preparing pipeline revegetation guidelines.

- Right-of-way width. The maximum right-of-way width primarily defines the potential area of disturbance.

- Excavating machinery. Soil separation capabilities of excavators define the extent to which topsoil can be separated and then reapplied over subsoil when the completed pipeline is backfilled.

- Construction schedule. The completion of different segments of the pipeline will occur at different times of the year, influencing what types of soil protection measures can be applied.

- Right-of-way maintenance requirements. A low, herbaceous ground cover over the right-of-way is favored to permit easy inspection and repair of the pipeline. This requirement may limit the uses of woody plant materials under these circumstances.

Environmental Factors

Environmental factors (climate, soils, grazing animals) greatly influence the potential for long-term revegetation success, and the revegetation species selection process. Baseline information developed for the project can be analyzed to determine where and to what extent environmental factors pose difficulties for revegetation.

Climate

Annual temperature and precipitation patterns dictate the opportunities related to timing of revegetation, and seeding methods and species selected. A simple approach to analyzing the effect of climate is to develop a precipitation zone map of the pipeline (Figure 2). This map quickly segregates the area where precipitation is usually adequate from those where precipitation is often inadequate to promote growth of revegetation species. A general guideline of 10 inches or less of annual precipitation is recommended to segregate difficult from more routine revegetation requirements.

Sources of climatic information include:

- Climatic atlases of different states produced by the National Oceanic and Atmospheric Administration. Atlases include data from individual reporting stations.

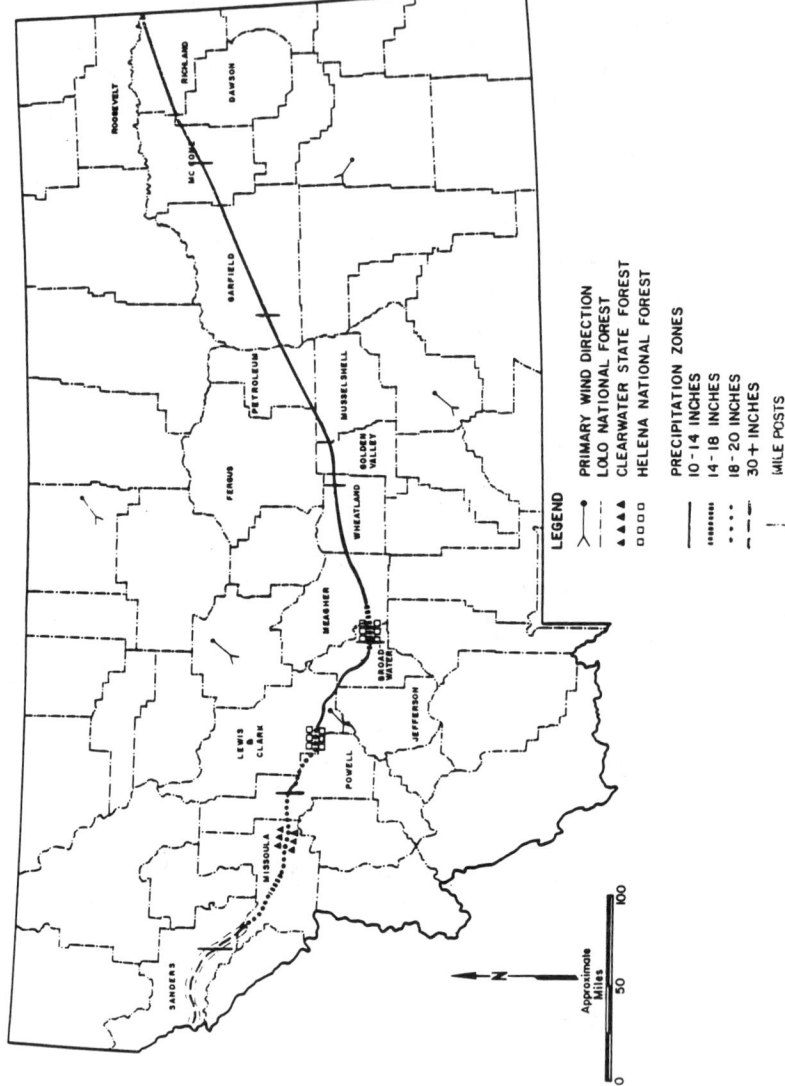

Figure 2. Major Precipitation Zones Along the Northern Tier Pipeline Route in Montana.

- U.S. Soil Conservation Service, or State Extension Service

Other aspects of the climatic factor include:

- Precipitation amounts and frequency through time
- Growing season length (can be obtained from climatic records or local inquiry)
- Temperature variation and extremes
- Snowpack depth and duration
- Wind speed and direction
- Flood frequency

Soils and Topography

The combination of soil texture, soil chemistry, slope angle and slope aspect must be analyzed to identify revegetation problems. Soils data vary in detail and quality from place to place. The Soil Conservation Service usually has published county soil surveys that are valuable for revegetation planning. Soil information must be reviewed to identify specific types of problems and categorize them by name. Table 1 illustrates how soils information was analyzed for soil and topographic related constraints by segment for a portion of the Northern Tier Pipeline route in the state of Washington. This analysis can be very useful in computing revegetation costs, probabilities for success, and timing of revegetation work. Analysis of existing information in the revegetation planning stage identifies gaps in knowledge about soils in different pipeline segments.

Special or Sensitive Habitats

The pipeline right-of-way may inevitably intersect examples of natural communities that are generally recognized as economically, recreationally or aesthetically valuable. These habitats include coastal and inland wetlands, river and stream crossings, and critical wildlife habitats such as wintering or calving areas, and unique natural areas. These types of habitats will usually be identified as sensitive areas during the baseline and impact assessment portions of the project. Revegetation specifications for these types of areas will usually be negotiated with agencies and landowners having jurisdiction.

TABLE I

POTENTIAL SENSITIVE AREAS FOR THE NTPS IN WASHINGTON
(Topsoil Suitability Rated as Poor)

County	Milepost	Soil Association - Series	Rocky Soils (Loose Fragments- Cobbles, Gravel)	Acid & Muck (Poor Drainage, Low pH, High Organics)	High Water Erosion Potential (Steep Slope, Physical Instability)	Shallow Soils
Clallam	9-11.9	Clallam-Agnew	X			
Island	32-41	Whidbey		X		
Island	54-54.7	Puget			X	
Snohomish	56.7-59	Kitsap			X (slump)	
Snohomish	59-63	Alderwood	X	X		
Snohomish	63-68.5	Indionola		X		
Snohomish	94-97, 99-103	Rough Stony Land	X		X	
King	111-112, 115-116	Rough Stony Land	X		X	
King	105-108, 109-111, 113-115, 117-120, 120-123	Alderwood	X	X		
King	128-130	Rough Mountainous	X		X	X

Consumers and Competitors

Consumption of seedlings by herbivores, and weeds that compete with planted species often slow the establishment or cause a seedling failure on a revegetated site. It is important to understand what options are available to protect new seedings from livestock, and balance the composition of seed mixtures to include some less palatable species to wildlife. Weed invasion is a chronic problem on newly seeded rights-of-way, particularly in agricultural areas. The types of weeds present and appropriate control measures usually require discussions with local experts, such as agricultural extension agents.

Regulatory and End-use Constraints

"Regulatory" in this context means the specific rules and regulations enforced by an agency on pipeline construction conducted on land under its jurisdiction. "End-use" means ultimate use (e.g. grazing, wildlife habitat, Christmas tree farm) of the right-of-way after revegetation has been completed.

Government Agencies

Federal and State Government agencies have land management responsibilities for large areas of land in the Western United States. On federal lands, the U.S. Forest Service has jurisdiction over forest lands, the Bureau of Land Management manages rangelands, and the Army Corps of Engineers is responsible for stream crossing permits.

State agencies with land management responsibilities include the Departments of State Lands, and the Fish and Game. In most cases these agencies have developed specifications for revegetation of disturbed sites. These revegetation specifications can be incorporated into revegetation prescriptions for public lands.

Private Landowners

Individual landowners may negotiate to return the right-of-way to its previous use, such as pasture or cropland. Soil management practices are often a key issue because of the need to replace agricultural soils in their preexisting horizon sequence. Individuals may also request vegetative screening to reduce the visibility of the right-of-way. Use of local Soil Conservation Service recommendations is often an acceptable starting point for defining right-of-way species composition and seeding methods.

INCORPORATING CONSTRAINTS INTO A REVEGETATION PLAN

As indicated above, numerous sources of information exist that define the environmental and revegetation constraints, and the solutions to particular revegetation problems. A thorough review of existing information is required, as well as interview with local experts in the field. Our experience indicates that interviews with the District Conservationist of the Soil Conservation Service and County Extension agents are extremely useful in defining local revegetation problems such as problem soils, weed infestations, insect pests, and local agricultural practices. Wildlife, forestry and revegetation considerations can be addressed by interviewing respective state and federal natural resource managers (e.g., Fish and Game, Forest Service, BLM). These individuals provide useful information on local preferences for crops and erosion control measures, and wildlife and livestock needs.

After the literature and regulatory requirements have been reviewed, and local revegetation practices derived from field interviews have been summarized, a framework is developed to express the kinds of routine and special revegetation problems found in a region or state. The terminology used for defining these situations is comparable to that commonly used by agencies that supply revegetation specifications, such as the Soil Conservation Service or Cooperative Extension Service. Depending upon the type of land use, the level of information available from agencies and the literature, and political concerns, the framework for expressing the array of revegetation situations may reflect a classification by political unit (e.g. county), or by ecological or edaphic (soil) unit. Table II and III illustrate both types of organization. Table II shows the classification framework developed for a pipeline segment in the State of Washington. This organization emphasizes revegetation guidelines by county unit. This approach reflects the extensive amount of information available at the county level, and the potential for an extensive amount of local concern about pipeline revegetation in this relatively densely populated and predominantly agricultural region. Table III shows the classification framework for a pipeline segment in the State of Montana. This pipeline crosses sparsely populated, semi-arid rangeland for much of its length. The classification framework for this state is oriented toward climatic and edaphic factors that influence revegetation success.

Species selection for different revegetation sites is a process that is strongly influenced by various environmental and regulatory constraints. Issues that commonly arise are

Table II

Revegetation sites identified for a pipeline route in Washington

Revegetation Situation	Clallam	Island	Snohomish	King	Kittitas	Grant	Lincoln	Adams	Spokane
Routine									
Forestlands below 2,500'	X	X	X	X					
Forestlands above 2,500'				X					
Dryland range (under 15 inches rainfall)					X		X	X	X
Dryland range (over 15 inches rainfall)					X				X
Special Problems									
High Coarse Fragments (Droughty)-moist climate	X	X	X	X					
(Droughty)-dry climate					X				
Steep Slope Erosion Control (moist climate)	X	X	X	X	X				
Steep Slope Erosion Control (dry climate)					X				
Muck Area or Poorly Drained Sites (acid)	X	X	X	X	X		X	X	X
Stream Crossing (moist climate)	X	X	X	X					
Stream Crossing (dry climate)					X	X			
Fine Sands (subject to blowing)					X		X		
Saline Coastal Wetland		X	X						
Freshwater Inland Wetland		X	X						
Shallow Soils Over Basalt					X				X
Alkali						X	X		X
Intermittently Flooded Areas, Potholes						X	X		X
Subirrigated Wet Meadow									X

TABLE III

Revegetation sites identified for one of several precipitation zones for a pipeline route in Montana

Climate	10-14 Inch Precipitation Zone							
Topographic Position	Upland				Stream Crossing			
					Dry Floodplain		Dry Embankment	
Chemistry/ Depth	Nonalkaline	Moderately Alkaline	Strongly Alkaline	Pan Spots	Shallow Soils	Non to Moderately Alkaline	Strongly Alkaline	Non to Moderately Alkaline	Strongly Alkaline
Texture	CLAYEY / LOAMY / SANDY								

106

the conflicts between the use of rapidly establishing exotic species, and slowly establishing native species; the balance between species that are palatable to both domestic livestock and wildlife; the balance between cool and warm season species; and the use of legumes such as alfalfa that add nitrogen and establish rapidly, but may occasionally be toxic to livestock. Tables IV and V illustrate differences in species selection approach between the States of Montana and Washington. In western Washington (Table IV) the emphasis is on the rapid establishment of exotic grasses and legumes to control erosion, provide year-round forage for livestock, and to fix nutrients in well-drained soils where nitrogen is rapidly leached. In Montana (Table V) the emphasis is on reestablishing native range grasses (e.g., western wheatgrass, slender wheatgrass) with a supporting fraction of exotic grasses (e.g., crested and pubescent wheatgrass) that establish rapidly and control erosion. Mixtures that contain "something for everybody" may be politically expedient, but may be failures in actual practice because of differences in species competitiveness.

Figure 3 provides a flow diagram for stages in a pipeline revegetation project that includes monitoring and maintenance components. Gathering and organizing information about revegetation constraints has been discussed previously. A particularly important relationship in this flow chart is the feedback loop from Unsuccessful Revegetation to Identify Constraints. This feedback implies that constraint factors will be redefined as the revegetation program is implemented. As a consequence, identification of important constraints should improve through time.

Another important process is the interface with reviewing agencies, which are responsible for the adequacy of revegetation plans. These reviews identify additional constraints that must be incorporated into seeding methods and revegetation species selection criteria.

SUMMARY

Identification of pipeline construction, environmental, and regulatory constraints during revegetation planning for a pipeline project is helpful in developing a framework for classifying the types of revegetation situations to be considered. The relative importance of these contraints must be evaluated when deciding the level of detail needed to adequately describe revegetation procedures, and when selecting revegetation species that will meet several (and sometimes incompatible) end-use requirements for pipeline rights-of-way. Constraints encountered in the pipeline revegetation process must be continuously reevaluated to determine whether revegetation methods used are meeting revegetation goals.

Table IV

Guideline seed mixtures for revegetating the pipeline right-of-way in Clallam County, Washington

Guideline Mixtures

Soils with High Coarse Fragment Content (Subject to Droughty Conditions)

Tall fescue (Fawn, Alta)	10.0
White clover (New Zealand)	3.0
	13.0 lbs/acre

OR:

Tall fescue (Fawn, Alta)	10.0
Subterranean clover (Mt. Barker, Tallarook)	4.0
	14.0 lbs/acre

(the use of subterranean clover is for extremely droughty conditions)

Slope Erosion Control

Intermediate wheatgrass (Greenar, Oahe)	8.0
Hard fescue (Durar)	2.0
Perennial ryegrass (various)	0.5
White clover (New Zealand)	3.0
	13.5 lbs/acre

Forest Lands

Hard fescue (Durar)	6.0
Tall fescue (Fawn, Alta)	2.0
Perennial ryegrass (various)	0.5
Birdsfoot trefoil (Cascade)	2.0
	10.5 lbs/acre

Muck Areas or Poorly Drained Sites

Reed canarygrass	8.0 lbs/acre
or	
Meadow foxtail	6.0 lbs/acre

Stream Crossing Mixture

Kentucky bluegrass (Troy)	2.0
Red fescue (various)	8.0
Perennial ryegrass (various)	0.5
White clover (New Zealand)	3.0
	13.5 lbs/acre

TABLE V

Species lists and seed mixtures for the 10-14 inch precipitation zone in Montana

A. UPLAND MIXTURES

Grazing (Non-saline/Alkaline)

Sandy			Loamy			Clayey		
Prairie sandreed	3.20	(80%)	Western wheatgrass	2.10	(35%)	Same as for "Loamy"		
Indian ricegrass	1.20	(20%)	Thickspike wheatgrass	1.80	(35%)			
Total	4.40		Green needlegrass	1.00	(20%)			
			Slender wheatgrass	0.60	(10%)			
			Total	5.50				

Alternates: Possibly crested, pubescent, thickspike, bluebunch, Siberian, western and slender wheatgrasses

Alternates: Crested, pubescent, streambank, Siberian and bluebunch wheatgrasses, needle-and-thread, alkali sacaton

Alternates: Streambank and bluebunch wheatgrasses

Moderately Saline/Alkaline Soils

Sandy			Loamy			Clayey		
Prairie sandreed	1.30	(33%)	Western wheatgrass	1.80	(30%)	Western wheatgrass	3.00	(50%)
Indian ricegrass	2.00	(33%)	Thickspike wheatgrass	1.50	(30%)	Thickspike wheatgrass	2.00	(40%)
Thickspike wheatgrass	1.70	(34%)	Streambank wheatgrass	1.50	(30%)	Slender wheatgrass	0.60	(10%)
Total	5.00		Total	5.40		Total	5.60	

Alternates: Few species are adapted to this type of planting. Possible alternates are crested, pubescent, thickspike, bluebunch, Siberian, western and slender wheatgrasses.

Alternates: Crested and Siberian wheatgrasses, green needlegrass

Alternative: Green needlegrass

1/All seed amounts in tables are in terms of pounds of Pure Live Seed (PLS) per acre drilled.

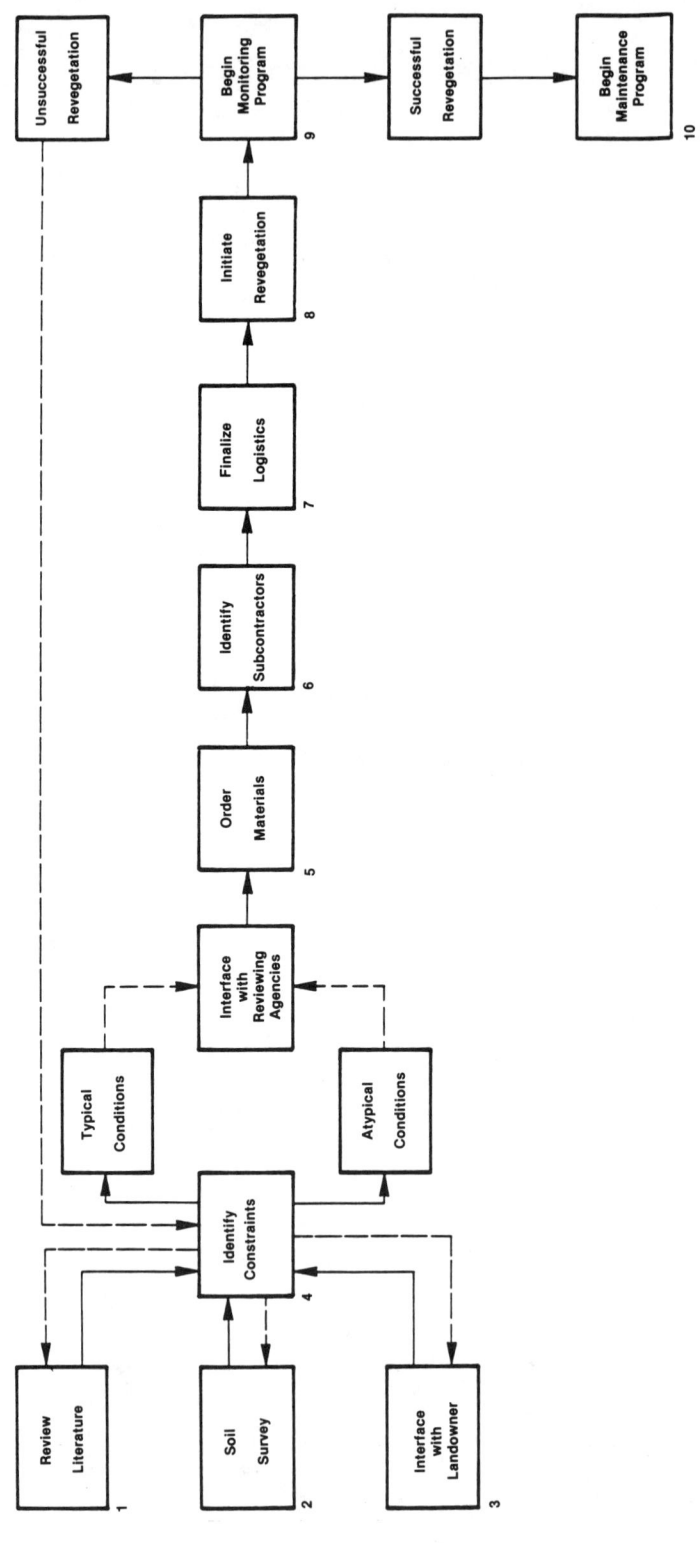

Figure 3. Flow Chart Illustrating Step by Step Procedures for Revegetation of the Northern Tier Pipeline System Right-of-Way in Washington.

FACTORS ASSOCIATED
WITH THE RECLAMATION OF
URANIUM DISPOSAL AREAS

Thomas K. Eaman
  Dames & Moore
  Golden, Colorado

INTRODUCTION

The accumulation of uranium mill wastes or tailings has grown enormously since large scale extraction of uranium commenced during World War II. By 1975 the volume of uranium mill tailings was estimated at 110 million cubic yards [1]. According to a Dames & Moore mining engineer, the height of 100 million cubic yards of tailings, if piled on 160 acres in a conical shape would be 1162 feet or well over twice the height of our Anaconda Tower here in Denver. This illustrates the physical magnitude of uranium mine tailings reclamation problems.

As is true with every kind of reclamation of lands disturbed by mining, the foremost factor is that of reaching an effective stabilization of the area. Uranium tailings disposal areas must first be effectively stabilized against wind and water erosion or the objectives of containment will not be met. The final reclaimed surface must be within the minimal limits of soil losses. Even slight losses of soil will bring about a failure of the cover over the long time periods necessary for protection against radioactive emissions to the atmosphere and surface waters.

Every feasible means must therefore be used to reach the highest degree of stability. Stability actually encompasses many factors which variously aid and detract from achieving reclamation goals depending on characteristics of each uranium disposal site.

This paper focuses on factors associated with the reclamation of uranium disposal areas in the Western United States. It is based on our experience in preparing plans for this type of reclamation in Wyoming, Utah, Colorado and New Mexico.

RECLAMATION OBJECTIVES

Reclamation as a process has generally become accepted as including a vegetative cover capable of controlling erosion and of self-regeneration. Obtaining such a vegetative cover requires an association of plants adapted to the specific location of the site. Thus the use of plant species that are native to the area or agronomic relatives of adapted species are often the most successful. A second criterion for success is to provide a soil suitable for their sustained growth.

SOIL RESOURCES

The first reclamation step, then, is to assess or inventory the soil or soil materials that are available on or reasonably near the site. A detailed soil survey provides an essential data base for reclaiming uranium tailings. The soil survey should provide the quality and quantity of soil that can be used to develop a favorable, long-lasting plant growth medium for achieving vegetation stability.

Guidelines developed by the Wyoming Department of Environmental Quality (WDEQ) contain excellent criteria for the selection of soil materials suitable for use as a plant growth medium. The material presented in Table I is from the Wyoming guidelines [2].

It would be convenient to use the term "topsoil" but in the strict sense of the term, topsoil is too limiting for the use intended in this form of reclamation. The traditional system of soil classification would limit the material to the A horizon; the soil in which most root growth is concentrated, where the percentage of organic matter is the highest and where the soil nutrients for plant growth are most abundant. Because of the limitations related to the term topsoil and because of the limited quantities of true topsoil, in many locations, it is preferable to use soil material which acts as a plant growth medium. Soil materials meeting the criteria in Table I are sometimes found and are available for reclamation use at depths considerably below the depths of the A horizon (topsoil) occurrence.

The reclamation plan must, of course, include the collection, transport, delivery and spreading the medium over the site to be reclaimed. Some of the current reclamation operations of surface coal mining companies follow a procedure of collecting, transporting and delivering of suitable surface soils to the area to be reclaimed in one action. This practice has many advantages. Costs of handling and stockpiling are reduced and the value of the soil is protected. Stockpiled soils rapidly decline biologically because of the loss of soil fauna and flora and the loss of plant nutrients, viability of seeds and plant propagules such as rhizomes and bulbs. Unlike this advantageous approach, the reclamation of uranium tailing disposal sites offers little possibility of using this method, since it is necessary to stockpile in advance the material to be used for

TABLE I

SUITABILITY RATINGS FOR SOILS AS SOURCES OF TOPSOILING MATERIAL [2]

Soil Characteristic	Degree of Soil Suitability			
	Good	Fair	Poor	Unsuitable[5]
pH[1]	6.0 - 8.4	5.5 - 6.0 8.4 - 8.8	5.0 - 5.5 8.8 - 9.0	Under 5.0 Over 9.0
Conductivity (EC) mmhos/cm @ 25°	Under 4	4 - 8	8 - 16	Over 16
Saturation Percentage (SP)	25 - 80		Over 80 Under 25	
Texture class	sl, l, sil, scl	cl, sicl, sc, ls	c, sic, s	
Sodium Adsorption Ratio (SAR)	Under 6	6 - 10	10 - 15	Over 15
Calcium Carbonate[2]	none - slight (0 - 15%)	moderate (15 - 30%)	high (Over 30%)	
Moist Consistence	friable	loose, firm	very firm	
Dry Consistence	loose, soft	slight hard, hard	very hard	
Selenium[3]	2 ppm or less			greater than 2 ppm
Boron[4]	5 ppm or less			greater than 5 ppm

[1]Saturated Soil Paste, U.S.D.A. Handbook 60, pg. 102.
[2]Effervescence with acid of $CaCO_3$ equivalent, U.S.D.A. Handbook 60, pg. 105.
[3]Hot water soluble, A.S.A. Mono #9, Part 2, pgs. 1122-23.
[4]Hot water soluble, A.S.A. Mono #9, Part 2, pgs. 1062-63.
[5]Establishment of vegetation on soils having any of these extreme conditions would be severely restricted.

final surfacing (for example in the event tailings are to be buried beneath the surface).

Locations of disposal sites are often in arid to semi-arid regions where sufficient quantities of soil material are non-existent. To borrow material for final surfaces to be reclaimed has its objections. The borrow area is disturbed and a new area must in turn be reclaimed. An off-site soil collecting technique that has applicability in certain soil survey designated areas is an earth moving scraper equipped with small shovels or plows attached at intervals along the cutting edge of the scraper. Soil is collected from furrows and conveyed into the earth mover for transport of the soil to the area to be reclaimed. The borrow area is only partially disturbed by this method and if the furrows are on the contour there is the added benefit of water conservation. In addition to providing a proper seedbed, the possibility also occurs for plants and propagules to being transported to the reclaimed area from the fresh borrowed material.

DEPTH OF PLANT GROWTH MEDIUM

Debates continue among reclamation planners over the depth of plant growth medium that is needed. Guidance can be taken from the depths of topsoils that have developed naturally on soils in the area of the disposal site. As a general rule, at least as much soil depth should be provided on the reclaimed area as occurs naturally on the site. This statement must be qualified to include the same kind of underlying material. The Soil Conservation Service has developed minimum soil rooting depths for different land capability classes. Table II lists the recommended soil depths for classes I through VI for Colorado.

TABLE II
MINIMUM SOIL ROOTING DEPTHS IN INCHES FOR LAND CAPABILITY CLASSES I TO VI [3]

Land Capability Class	Minimum Soil Rooting Depth (inches)
I	60
II	40
III	20
IV	20
V	40
VI	10

The basic criteria to strive for in the depth of plant growth medium should be at least that listed in Table II for LCC VI.

PLANT SPECIES SELECTION

The question of whether to use introduced plant species in place of natives has become controversial in many instances involving reclamation. The ready availability of a number of outstanding introduced species together with their proven abilities to form a quick cover has given them deserved appeal.

Crested wheatgrass is a prime example of an introduced species that has demonstrated its wide geographic adaptability and effectiveness in establishing a protective cover against wind and water erosion. Opponents to introduced species contend that such species exclude the entry of native species into the stand, are vulnerable to total elimination by insects or disease and add nothing to the diversity of the plant community. Advocates of crested wheatgrass or other examples of adapted introduced species cite advantages that include commercial availability, ease of planting and rapidity of establishment. How long these kinds of plants will continue to form a protective plant cover is secondary to establishing a quick cover for stabilization and for adding organic matter to the soil. Perhaps a significant value of such introduced species is in the pre-conditioning of soils for the eventual growth of a stand of native species.

Plant species diversity has grown to be a common index of the desired native vegetation for many reclamation projects. Diversity is usually characteristic of the natural plant community and therefore an all-grass plant cover is objectionable because of minimal diversity. This is a goal of revegetation in reclamation but the attainment of diversity is often difficult. Shrubs and trees have not been as readily established by direct seeding as have grasses and legumes due to their environmental requirements. Seeds of shrubs and trees are often difficult to obtain and their scarcity can make for higher costs than the seed of many grasses. Planting seedlings, especially through the use of tubelings, has been used successfully in establishing trees and shrubs at a lesser cost. Biodegradable material used as containers for woody seedlings has also greatly facilitated the planting and establishment of trees and shrubs.

DEEP ROOT PENETRATION

Some interesting contradictions arise regarding obtaining plant species diversity by using woody plants. In addition to increased diversity, one of the reasons for specifying trees and shrubs on a reclaimed disposal site is that their roots usually extend downward below the depths of grasses and can thereby obtain water and nutrients from soil horizons out of reach of shallower rooting species. Conversely this rooting habit causes concern when root channels permit the release of radioactive products into the surface environment. Controversy on this issue will no doubt remain in uranium tailings reclamation although establishment and maintenance of a grass stand has a number of disadvantages.

SELECTED LAND USES

Unlike the reclamation of lands disturbed by the mining of coal, gravel, or minerals where the reclamation plan is directed toward land development for a specified land use, disposal of uranium tailings has no permissible land use. In fact, regulations [4] demand that the reclaimed area be tightly fenced to prevent use or trespass. This requirement rules out the establishment of suitable land for forage, timber, crops and by and large even for wildlife. An advantage to such restrictions in land use is the reduction in disturbance. However the threat of vegetation loss to fire is a serious concern and safeguards such as fire breaks and fire control arrangements must be built into the reclamation plan. Fire suppression can be difficult because most disposal sites are in remote areas, far away from fire control facilities.

SOIL NUTRIENTS

Fertilizers are often supplied initially to the seed beds of areas to be reclaimed. This practice is warranted because the plant growth medium may be deficient in essential plant nutrients - especially nitrogen, phosphorus and potassium. Research has revealed that native plant species respond less dramatically than introduced plants to fertilization [5]. Nonetheless the beginning seed bed can be improved by the addition of fertilizer as specified by soil laboratory analyses. Soil microorganisms also require nutrients and in turn are important for chemical recycling processes and therefore a vital part in growth of a renewing plant community.

Nitrogen is generally the element most deficient for plant growth in reclaimed disposal sites. Legumes, (clover, alfalfa) due to their fixation of atmospheric nitrogen, are therefore often included in the reclamation seed mix. Native legumes for use in revegetation of many disposal sites in arid and semi-arid sites are lacking or scarce or have not been studied enough to fully understand the part they play in nitrogen fixation. Regions where uranium tailings are to be disposed usually have few native legumes available for reclamation seed mixtures. Cicer milkvetch, a native to Eurasia, shows promise of being highly important for revegetating.

The amount of nitrogen supplied by organic matter recycling in a plant community is difficult to determine. Lands low in precipitation have organic matter percentages that are so low it can be concluded that available nitrogen for plant growth is minimal. As already mentioned, the nitrogen from fixation in the nodules of legumes is often minimal. It is reasoned that other sources must account for replenishment of nitrogen in the soil for plant growth. One source is that of the fixation of atmospheric nitrogen by lightning. Studies have measured as much as 53 kg/ha of nitrogen produced by electrical storms per year [6]. Lightning may be quite important for supplying nitrogen to maintain plant cover after reclamation tasks are completed and fertilizer applications have been discontinued.

STONINESS OF FINAL COVER

It may be the influence of our farm and agronomic experience and background that has made us so averse to the presence of stones on or in the surface of soils to be revegetated for reclamation projects. The natural surface of many rangelands is characteristically stony having stones ranging in size from cobbles to boulders. Where annual precipitation is low (less than an arbitrary 16 inches) stones on and in the soil act to increase the quantity of water available for plant growth. The soil volume is reduced thereby making a given quantity of water more readily available for plants. Stones also represent miniature watersheds, concentrating water around them. Stones interrupt the plain, featureless landscape of a graded reclamation area and they serve as perches and refuges for birds and small animals. The subject of stones is brought up here because they are often abundant at disposal sites and their removal and burial can become a major cost in reclamation. However, radiation hazard of such materials also needs to be considered. Thus, planned placement of stones on reclamation surfaces can enhance the area through improved water relations and habitat diversity.

ESTIMATING RAINFALL EROSION

Testing for the degree of erosion control resulting from reclamation can be performed during the planning stage and after reclamation practices and measures have been installed. The Universal Number Soil Loss Equation (USLE) [7] has been of valuable assistance in this testing. Having a goal of limiting soil loss to well below the 5 tons per acre per year (national level designated by the Soil Conservation Service) is used to evaluate soil stabilization from reclamation. The elements of the USLE are the rainfall (R), the soil erodibility (K), the length and steepness of the slope (LS), the cover (C) and the configuration of the surface (P). The equation is $A = R.K.LS.C.P$ with A being the soil loss from rainfall erosion in tons per acre per year. It can be seen that the two elements of the equation that reclamation cannot control are the rainfall and the inherent erodibility of the soil. The LS, C and P factors are primarily controllable and can favor stabilization and reduce excessive erosion. Slopes can be controlled both in regard to their lengths and steepness. Slope gradient goals should be no more than 4 horizontal to 1 vertical or 25%. Length of slopes can be controlled by benching or terracing to a limit of 50 feet or less. Cover is a major objective of erosion control and it should be of a type that forms as dense a perennial ground cover as the climate of the area can support. The cover should be diverse including vertical structure or a canopy, again within the bounds of the natural vegetation of the area. A canopy above the ground cover is effective in breaking the impact of rain on the ground. The most useful measure of effective cover is that expressed by the plant production (weight/unit area, eg. pounds per acre). The weight of plant growth combined with the uniformity of cover, or density, is essential for gauging the value of vegetation cover as erosion control. The surface soil configuration factor (P) has importance in stability when it is applied in the form of roughening by pitting and contour furrowing. Smooth, well graded surfaces formed by preparing final reclamation surfaces are highly erodible. It has been demonstrated that a surface configuration of alternating depressions and ridges acts to control surface runoff and soil loss. The entry of the elements of the USLE that are within the range of practical achievement is convincing evidence of the value of their part in stabilizing the reclaimed area.

Only a few of the many factors associated with reclamation of uranium tailings have been presented here. However, our experience at Dames & Moore indicates that these are the primary factors in successful reclamation of uranium mill tailings.

LITERATURE CITED

1. Energy Research and Development Admin. and Nuclear Regulatory Comm. 1975. Controlling the radiation hazard from uranium mill tailings. Report to Congress: Comptroller General of the United States.

2. Wyoming Department of Environmental Quality. 1980. Guideline No.3. Cheyenne.

3. U.S.D.A. Soil Conservation Service 1979. Determining capability class of soils. Agronomy Note No. 56 (Revised) Colorado. Denver.

4. U.S. Nuclear Regulatory Commission. 1980. Final Generic Environmental Impact Statement on Uranium Tailings, NUREG-0706, Vols. I, II, III. Gov't Printing Off. Washington, D.C.

5. Aldon, E.F., H.W. Springfield, and G. Garcia. 1975. Can soil amendments aid revegetation of New Mexico coal mine spoils? U.S.D.A. Forest Service Research Note

RM-306. Rocky Mountain Forest and Range Exper. Sta. Fort Collins, Colo.

6. Pesek, J., Stanford, G., and N.L. Case. 1971. Nitrogen production and use. In: Fertilizer Technology and Use. Edited by: Olson, R.A., Army, T.J., Hanway, J.J. and V.J. Kilmer. Soil Sci. Soc. America. Madison, Wisc.

7. U.S. Environmental Protection Agency and Soil Conservation Service. 1977. Preliminary guidance for estimating erosion on areas disturbed by surface mining activities in the interior Western United States. EPA-908/4-77-005 July.

Land Reclamation Plan and Procedures
Used on the Green Mountain (Wyoming)
Uranium Exploration Project

A. Robert Easter
   Project Coordinator
   Anaconda Copper Company

ABSTRACT

   The Anaconda Copper Company, Pioneer Nuclear, Inc., and Texas Eastern, Inc., are jointly involved in a uranium exploration project in Fremont County Wyoming. Over three-hundred exploration drilling locations, totaling 450 acres, have been built in the eight square mile project area. In 1979 a comprehensive reclamation plan was developed, one aspect of which was the reevaluation of all locations built prior to 1979 for possible reclamation work. In formulating the plan many factors were considered: topography, climate, soil, vegetation, and federal and state regulations. Recontour work, which is initiated as soon as possible after drilling is completed, is followed by the revegetation phase, in which disturbed areas are seeded with eight species of native grasses (or planted with trees), mulched, and fertilized. To date 35,000 lodgepole pine seedlings have been planted; 40,000 will be planted in 1982. Various types of earthmoving and agricultural equipment were used to reclaim over 200 acres in 1981 alone. Equipment and labor account for 85-90 percent of the total costs of reclamation. First, second, and third year growth has been monitored and is quite successful. Grazing by cattle, wild horses, and sheep poses the only serious threat to revegetation efforts. Three half-acre vegetation research areas have been established to permit long-term studies.

INTRODUCTION

Since 1977 the Anaconda Copper Company has had a fifty percent undivided interest and Pioneer Nuclear, Inc., and Texas Eastern, Inc., each have had a twenty-five percent undivided interest in a uranium exploration project in south central Wyoming. Anaconda acts as operator for the approximately eight square mile project area in Fremont County fifteen miles southeast of Jeffrey City.

Green Mountain proper and much of the surrounding area is federal land, a part of the Lander Resource Area, managed by the Bureau of Land Management (BLM). As such, it is a public recreation area, an important factor to be considered in both the construction and reclamation of mineral drilling locations. Highly visible drilling locations can, however, be used to advantage, especially in terms of the quality of reclamation.

Since 1969 over three-hundred exploration drilling locations have been built on all types of terrain, from level ground to $30°$ slopes. During 1979 and 1980 approximately one-hundred fifty acres were recontoured and revegetated; in 1981 slightly more than two-hundred acres were similarly completed.

RECLAMATION PLAN

A simple objective was established in 1979: to evaluate, and, where necessary, recontour and revegetate all areas disturbed by exploration drilling and access road construction from 1969 onward, a total area of about 250 acres. An additional 200 acres were disturbed in 1980.

To achieve this objective, and to consistently apply the criteria set by Anaconda for successful reclamation to all areas disturbed from 1969 through 1981, required many hours of map and reconnaissance work.

Approximately 100 acres of old (i. e., pre-1979) locations were evaluated. Most of the approximately 200 locations were determined to need either no reclamation work (if natural revegetation was successful) or a minimal amount of work (which usually consisted of seeding and fertilizing areas which lacked complete vegetation cover). A few locations, however, required minor recontour work to correct settling, for example.

In the absence of deliberate reseeding efforts, and given these topographic, climatic, and soil conditions, five or six years are necessary for native species to produce enough ground cover to adequately control soil erosion.

Techniques for the safe and efficient use of heavy equipment to construct and recontour drilling locations and access roads were gradually refined; the sequence of events and equipment used for vegetation removal through and including vegetation replacement (seeding or tree planting) became systematic. Achievement of the desired results became less time consuming and more cost effective.

Specific factors which had to be considered in order to form a comprehensive plan were the topography, climate, soil, and vegetation of the area as well as federal and state regulations pertaining to surface disturbance, exploration drilling, drill hole abandonment, reclamation procedures, and land use.

Topography

Green Mountain is a mass of unconsolidated granite, with a maximum elevation of 9081 feet, which lies roughly on an east-west axis about seven miles south of and parallel to the Rattlesnake Range of the Granite Mountains. It was formed over many millions of years from sediments which accumulated from the gradual erosion of the Granite Mountains.

Situated in the Wyoming Basin of the Rocky Mountain System it forms a part of the northern "rim" of the Great Divide Basin, commonly known as the Red Desert, the largest unfenced land area in the United States.

To the north the elevation gradually decreases about 2700 feet over a distance of about five miles, during which foothills become rolling plains; to the south, the elevation decreases more rapidly, about 2000 feet in a distance of three miles, to form the desert floor.

Climate

The area surrounding the Great Divide Basin is a mid-latitude, continental cool, semiarid region which receives 12-16 inches average annual precipitation. The moisture regime is such that most precipitation occurs as snow between mid-October and mid-May, with greatest accumulation between February and April. Frost depth during that period may reach 40 inches. Winter storms are responsible for a mean annual total snowfall of 60-100 inches, an amount common on Green Mountain due to the orographic effects exerted by the mountain itself.

The prevailing winds are from the southwest and most thunderstorms originate over the Great Basin and move in a northeasterly direction. Late spring and early fall rains are not uncommon but are usually of quite limited duration.

Weather conditions, which normally deteriorate rapidly after October 1, combined with poor roads in general and the remoteness of most drilling locations, require that seeding, mulching, and fertilizing be completed by mid- to late September. The planting of lodgepole pine seedlings is withheld until spring in order to take advantage of snowmelt. Roads are usually passable by about June 1.

Soil

The soil of Green Mountain is rather consistent in texture, a sandy loam with pH ranging from 6.0-7.5, usually dry, and characteristically low in organic matter.

Due to low annual precipitation organic decomposition takes place very slowly. Consequently, in nearly all areas above 7500 feet the humus layer rarely exceeds 1.5 inches in thickness although the layer of dead, fallen timber may form a layer of litter as much as 18 inches thick in some places.

The topsoil is rarely more than six inches thick, exceptions being found in broad drainage basins at lower elevations and at the bottom of steep slopes. In these areas topsoil may reach a thickness of 12-15 inches.

Due to the sparseness of vegetation cover and the steepness of slopes, topsoil is quite susceptible to erosion by wind and water. To control erosion in such areas biodegradable matting is applied to the surface to supplement the growth of vegetation in the stabilization process. Another erosion control method commonly used is that of placing intact straw bales on slopes (with their long axis perpendicular to the slope) to function as water barriers.

The dark brown color of the topsoil, in sharp contrast to the beige color of the subsoil, is an advantage which allows clean separation in nearly all instances. Topsoil is the most valuable commodity in this environment; hence, concerted efforts are made to save and reuse all available topsoil material.

Vegetation

The vegetation of Green Mountain is varied and complex but can be easily categorized and studied on the basis of change in elevation (Table I). Note that there is a transition zone between sagebrush steppe and coniferous forest which extends from approximately 7000-8000 feet in elevation and exhibits vegetation characteristics common to both biomes.

In addition, park areas exist at elevations above 9000 feet which are dominated by grasses, sedges, forbs, and shrubs. The absence of trees in these areas is not fully understood although it is suspected that excessive soil moisture is probably the limiting factor.

Federal and State Regulations

Two agencies are responsible for natural resource development on Green Mountain. The BLM is primarily responsible for surface leases, mineral rights, oil and gas rights, grazing rights, forest management, timber sales, reforestation, and land use. The state agency, the Wyoming Department of Environmental Quality (DEQ), Land Quality Division, is the agency to which all applications for permit to disturb the land surface by road construction, exploration drilling, mining, etc., are addressed.

Nearly all regulations under which Anaconda

Table I. Vegetation Change With Elevation

Elevation Range (Feet)	Biome Type	Characteristic Vegetation
6000-7500	Sagebrush Steppe	Sagebrush Low growing shrubs Bunch grasses Various forbs
6000-8000	Riparian	Grasses, rushes, sedges Deciduous trees Coniferous trees Woody shrubs
7000-8000	Transition Zone	Species present are common to both Sagebrush Steppe and Coniferous Forest biomes
7500-9150	Coniferous Forest	Lodgepole pine, Limber pine Deciduous trees Fir, Spruce (of minor importance on Green Mountain) Understory: yews, junipers, various forbs
9000-9150	Park	Sagebrush Bunch grasses, sedges Various forbs

operates are those of the DEQ; they are quite specific as to exploration drill site location and construction, drill hole records and reporting, drill hole abandonment procedures (which are a part of reclamation), and revegetation practices.

PHASES AND PROCEDURES

Most exploration drilling locations are surveyed on a grid system, the location of each hole being a specific distance and direction from all other holes. From a geological standpoint this method of definition drilling aids in the plotting of areas of mineralization. In most cases the locations and access roads are constructed so as to disturb the smallest area possible.

As dictated by the grid pattern of definition drilling many locations built in 1979 and 1980 were built on steep slopes. With deeper holes being drilled by bigger rigs with more equipment, locations were deliberately built larger than necessary so that certain safeguards could be implemented (e. g., higher-than-normal berms were built around the location to contain any escaped fluids; longer and deeper mud pits were dug to prevent any overflows).

After a point has been surveyed for construction of a drilling location, heavy earthmoving equipment is used to remove vegetation, strip and stockpile topsoil, excavate mud pits in the subsoil, and level and compact a "pad" for the drilling rig and its accessory equipment.

When drilling and geophysical hole logging have been completed, the location is prepared for recontouring.

Post-Drilling Clean Up

Mud pits are pumped dry or nearly so after particulate matter has been allowed to settle for several days. If the water is reasonably clear it is pumped out directly onto the undisturbed land surface surrounding the location. If any oil, diesel fuel, gasoline, etc., or any other substance which would be toxic to topsoil is present in the pits, the water level is very carefully lowered by pumping to a point at which only the oil, diesel fuel, etc., and a small amount of water remain in the pits. At no time, regardless of the condition or amount of water present in the pits, are water pumps left unattended.

All trash and any subsoil contaminated by oil, etc., is either buried in the mud pits or removed to a sanitary landfill. This accomplished, a tracked backhoe is used to backfill the pits and a tractor used to compact the area. To account for later subsidence the entire area occupied by the pits is mounded above the surrounding contour. The remainder of the location, however, is worked with equipment to match the surrounding contour.

The bore hole itself is filled with a bentonitic material and the surface casing capped with a cement plug at a point approximately two feet below the original surface contour.

The location, which has now been recontoured, is allowed to dry out for several days before the topsoil, stockpiled earlier, is distributed by caterpillar tractor. Timber, downed as the result of site construction, is removed by backhoe from standing timber and either piled as "slash," for later burning by the BLM, or spread over the location to match the accumulation of litter on the surrounding area.

After all recontour work has been completed on a location the various phases of revegetation are begun, usually in early September.

Recontour work and revegetation work are done concurrently in late summer. In this manner the time during which topsoil is exposed to potential erosion is minimized.

Seeding

A special purpose grass seed mixture, based upon BLM recommendations and consisting of five species of cool season grasses (Table II), has been used with success since 1979 on approximately 200 acres of disturbed land. This year three additional species, all of which are native, have been added to the mixture. The 200 acres which were reclaimed this year have been seeded with 2,500 pounds of this new improved mixture.

Seeding is either done mechanically, by Rangeland Seed Drill (see Figure 1), at a rate of 10-12 pounds of bulk seed per acre, or by hand, broadcast seeded at a rate of 15-20 pounds of bulk seed per acre. The latter method, although less

Table II. Special Purpose Grass Seed Mixture

Common Name (Variety)/ Scientific Name

Western Wheatgrass (Montana Native)/
    Agropyron smithii

Smooth Bromegrass (Lincoln)/
    Bromus inermis

Streambank Wheatgrass (Sodar)/
    Agropyron riparium

Kentucky Bluegrass/
    Poa pratensis

Green Needlegrass (Lodorm)/
    Stipa viridula

Thickspike Wheatgrass/
    Agropyron dasystachyum *

Slender Wheatgrass/
    Agropyron trachycaulum *

Timothy/
    Phleum pratense *

* Added to the 1981 mixture.

---

efficient and economic than the former, is used on all areas which are either extremely steep or inaccessible to agricultural equipment.

Mulching

Barley straw (Hordeum, spp.), grown in the Riverton, Wyoming, area is the only material used by Anaconda as mulch on Green Mountain. All straw is accepted only after being inspected and certified as being free of noxious weeds by the Fremont County Weed and Pest Control District agent.

Approximately 210 tons of mulch were applied to reclaimed areas this year at the rate of one ton per acre. Nearly all areas except extremely steep slopes were power mulched by one of two Finn Mulchers (see Figure 2). Areas inaccessible to this equipment were mulched by hand at the same rate.

Applied mulch was then crimped by either of

two methods, both mechanical: a Finn Krimper was used on relatively smooth areas whereas a caterpillar tractor was used on rough, rocky areas to "walk-in" the mulch.

Experimentation using pine bark and wood chips is being undertaken on a limited basis. Various logging companies which operate on Green Mountain use bark peeling machines to produce planed fence posts. The by-products of this process are obtained at no cost and can be used as an organic mulch but will likely tend to decrease soil pH slightly if used on areas dominated by grasses; if used on timbered locations, however, no change in pH is expected. This material has been applied to both types of locations and its effects will be monitored.

The cost of equipment necessary to load, transport, and unload this material, the availability of which is uncertain, make its use uneconomical at present.

Figure 1. The Rangeland Seed Drill in operation. A fertilizer spreader is being towed behind the drill. Note the straw mulch in the foreground.

Figure 2. A recontoured drilling location being power mulched. Most of the location lies outside the field of view to the right. (Helen P. Stewart)

Fertilizing

A special purpose prilled fertilizer was developed and applied either mechanically, by spreader (see Figure 1), or by hand at a rate of 100 pounds per acre. The 40-0-20 mixture (40% Nitrogen, 0% Phosphoric acid, 20% Potash, and 40% Inert materials) was based upon soil sample analyses using various parameters. No phosphoric acid was deemed necessary since the soil is somewhat acidic naturally.

Tree Planting

Each year the BLM collects cones from mature healthy lodgepole pines on Green Mountain and sends them to a government greenhouse and nursery in Montana for the extraction and germination of seeds and the raising of seedlings. In this manner, the year-old seedlings planted each year are germinated from seeds produced by parent trees which are adapted to the soil and climatic peculiarities of

Green Mountain.

Although the BLM is responsible for selection of a planting contractor, Anaconda assumes all costs of cone collection, germination of seeds, raising seedlings, transportation, and actual planting.

Seventeen thousand seedlings were planted by hand in June, 1980; this year an additional 18,000 were planted. Approximately 40,000 will be planted in June, 1982. At that time roughly 150 acres of land will have been replanted with trees. In the meantime, however, much of that same area has been seeded, mulched, and fertilized in order to stabilize the topsoil as quickly as possible.

Five-hundred seedlings per acre is the average planting rate for locations in clear cut areas; the rate varies considerably with an increase in slope, and therefore erosion potential, and with the proximity of the area to be planted to an immediate seed source (i. e., a drilling location surrounded by mature standing timber).

Biodegradable Erosion Control Matting

Deep cuts and large fills on a newly constructed permanent road and certain drilling locations built on extremely steep slopes were subject to immediate erosion by either wind or water, or both. Biodegradable matting (Hold-Gro) was applied to these areas directly over seed, mulch, and fertilizer, and attached to the ground surface with wire staples. The function of the matting is to hold seed, mulch, and fertilizer in place until seed germination can take place, then gradually decompose as the expanding vegetative cover stabilizes the topsoil.

Hold-Gro applied to such slopes in October, 1979, has almost completely decomposed; vegetative cover is now sufficient to prevent erosion.

Some experimentation on cuts and fills has been undertaken with this material by applying it in various lengths and at different angles to the horizontal. Approximately 20,000 square feet have been applied this year. More may be applied in the future, depending upon the evidence of erosion.

EQUIPMENT AND USAGE

Various types of heavy earthmoving and agricultural equipment were used during <u>all</u> phases of reclamation work, from the building of access roads to the obliteration of those same roads, and from the construction of drilling locations through the recontouring and revegetation of those same locations.

Table III lists by category the types of equipment used and presents a brief description of their respective functions. Nearly all of the equipment listed, as well as support vehicles (pickups, fuel and lube truck, dump trucks, and welding truck), was operative during the eight-week period of concentrated reclamation work (August 1-September 30, 1981).

COSTS

The overall costs of reclamation on Green Mountain for the years 1979, 1980, and 1981 are shown in Table IV. Equipment and labor account for 85-90 percent of the total costs for each of the three years while the cost of materials ranges 7-8 percent, and other expenses only about 4-7 percent.

Since heavy equipment and labor are the most expensive aspects of reclamation, both should be given careful consideration in planning and in deployment on site. To achieve maximum productivity and efficiency requires full cooperation between the company representative and the contractor; a "system" for the work can be mutually developed.

The "people" aspect of a project has direct influence on efficiency and, therefore, costs. Praise and a personal compliment to an operator or laborer for a job well done will go far to encourage initiative and the desire to do excellent work.

RESULTS

First (1980) and second (1979) year growth is

Table III. Earthmoving And Agricultural Equipment

Equipment Type	Function
Earthmoving	
D8H Caterpillar Tractor D8 Caterpillar Tractor Fiat-Allis 21-C	Vegetation removal Location construction Access road construction Ripping Recontouring
Liebherr 921HD Tracked Backhoe (1 Yd^3) John Deere 690B Tracked Backhoe (.75 Yd^3)	Recontouring steep slopes Excavate mud pits Backfill mud pits Pull slash and topsoil out of standing timber "Pull in" roads
140G Caterpillar Road Grader	Scarify roads Scarify locations Ditching, blading, and crowning permanent roads
Bucyrus-Erie 30-B Dragline (1.75 Yd^3)	Recontouring steep slopes
930 Caterpillar Wheel Loader	Minor recontour work Loading Back ripping Spreading topsoil and slash

Table III. Earthmoving And Agricultural Equipment (continued)

Equipment Type	Function
Earthmoving	
Ford 555 Rubber Tired Front End Loader/ Backhoe (1 Yd^3/.5 Yd^3)	Minor recontour work Surface casing excavation "Pull in" roads Loading
Agricultural	
Finn Power Mulcher, B250 Finn Power Mulcher, BMS.24	Mulching
Rangeland Seed Drill, 12'	Drill seeding
McCormick International Spreader, 10'	Fertilizer spreading
Finn Krimper, KR25.8	Crimping mulch
David Brown Diesel Tractor, 1210 McCormick Farmall Tractor, 450 D4 Caterpillar Tractor	Towing mulchers, seed drill, spreader, crimper, and mulch trailers

Table IV.   Total Reclamation Costs

1979	1980	1981
$143,000	$244,000	$220,000

Bar chart showing % of Total Costs:
- 1979: Equipment and labor 89, Materials 7, Other 4
- 1980: Equipment and labor 85, Materials 8, Other 7
- 1981: Equipment and labor 87, Materials 7, Other 6

 Equipment and labor

 Materials (seed, mulch, fertilizer, etc.)

 Other (salaries, field expenses, maintenance, tools, parts, etc.)

being monitored on many drilling locations. Where practical, third year (1978) growth is also being monitored.

Revegetation efforts by Anaconda have proven to be quite successful. So successful in fact, that a commonly encountered situation is that of cattle and wild horses grazing the palatable grasses on reclaimed locations in preference to adjacent, undisturbed native vegetation. Areas which have been subjected to such grazing over the past two summers have withstood the pressure amazingly well. Since most of Green Mountain is open range, however, this situation will likely persist.

Grazing by sheep is by far the most serious threat to emergent vegetation. In August, 1980, twenty-one drilling locations were completely denuded of new growth by sheep. This damage occurred primarily because of poorly maintained fencing, although three sheep companies do have grazing rights throughout most of the project area.

Sheep grazing in 1981 has been confined to areas unaffected by reclamation work. The locations mentioned above were reseeded, mulched, and fertilized in October, 1980; vegetative cover now looks quite encouraging.

In order for emergent vegetation on reclaimed areas to withstand sheep grazing pressure, a minimum of two years of undisturbed growth is necessary.

Three half-acre research areas have been established (and fenced to prevent grazing) which will allow long term (5-6 years) studies of undisturbed vegetation. Although no elaborate, detailed experimentation is planned, close observation and careful taxonomic study should provide any evidence needed to modify the seed mixture, indicate which species will most quickly increase biomass, and should provide information which will have potential application to other high elevation projects.

The magnitude and duration of reclamation work on Green Mountain has afforded the opportunity to work with various independent contractors and, with their interest and cooperation, achieve results far above expectations. The importance of striving for excellence in reclamation work is fully realized by Anaconda; management has fully supported all aspects of the project.

The Bureau of Land Management and the Wyoming

Department of Environmental Quality are both satisfied with the efforts put forth by Anaconda to achieve successful revegetation. A good working relationship with both agencies has been maintained since development of the reclamation plan in 1979.

The company will continue to demand work of the highest quality from all involved; that attitude will not be compromised. The quality of reclamation work--no matter the topography, soil, or climate--is the final picture presented to the general public. What they see is what they remember.

Future reclamation work will consist of tree planting in 1982, recontouring any areas which might have settled, and monitoring locations on a biannual basis.

When, and if, the uranium market improves to a point at which a reasonable rate of return on investment can be expected by Anaconda, then exploration work and development toward an operating mine will resume. When that occurs, reclamation of areas disturbed will be continued, using the equipment and methods described here.

TECHNICAL INVESTIGATIONS IN
SUPPORT OF THE FEDERAL COAL
PROGRAM

<u>Bureau of Land Management</u>
  Division of Special Studies
  U.S. Department of the Interior
  Denver, Colorado

The Bureau of Land Management (BLM) initiated in 1975 a program to assess the rehabilitation potential of Federal coal lease lands using basic hydrology, soils, overburden, and vegetation data, analyses and interpretations. Originally called EMRIA (Energy Minerals Rehabilitation Inventory and Analysis), the program is now known as the Technical Investigations in Support of the Coal Program.

The Technical Investigations in Support of the Coal Program is an integral part of the Secretary of the Interior's Coal Management Program. Specifically, it provides information needed to delineate and rank potential coal lease tracts in all the major coal regions. Principal users are BLM land managers and staffs, but state agencies, local planning groups, coal companies, and Environmental Impact Statement (EIS) groups are also using these data.

The Technical Investigations staff at the BLM's Denver Service Center, Division of Special Studies, is active in eleven different states (Utah, Colorado, North Dakota, Wyoming, Montana, New Mexico, Alabama, Arkansas, West Virginia, Oklahoma, and Alaska). The Technical Investigations staff at DSC contracts with other Federal and State agencies, universities, and consulting firms for collecting this information. Hydrologic data are collected and analyzed for use by the BLM land manager in assessing the probable impacts of surface mining on surface and groundwater hydrology, which is used in ranking potential lease tracts. Unlike the functions EMRIA performed, reclaimability is no longer analyzed in these activities. It is now BLM policy that the vast majority of reclaimability data ought to be collected <u>after</u> coal leases are issued, and that data collection is therefore the responsibility of the lessee and the permitting authority. The BLM will continue to collect hydrologic information necessary to determine the environmental impacts of coal mining.

Study reports also provide summaries of the data gathered and analyses made and summaries of existing information utilized in the study and an explanation of how the conclusions were reached.

The following is the sequence for a typical study:
1) the study area is identified by the land manager
2) availability of pertinent information is determined
3) study parameters are defined
4) the pre-study planning meeting is conducted
5) schedules and work plans are prepared
6) data collection and analyses are conducted
7) final report is prepared.

The Technical Information staff works closely with the U.S. Geological Survey, Water Resources Division hydrologists and other technical personnel, approving procedures, reviewing reports, etc. Reports are aimed at providing information for land management technical specialists.

The Technical Investigations staff at DSC initiated a Technical Information Program in fiscal 81. The program's mission is to disseminate useful, current information to coal program managers in the Bureau and to serve the public by providing current information on the Bureau's role in the Federal Coal Management Program.

The Technical Information Program is currently developing an automated information system to monitor the progress of BLM's Coal Program Technical Investigations. This system will also change to emphasize data and literature bases pertinent to the assessment of hydrologic impact. It will aid coal program managers in developing lease stipulations to avoid or minimize adverse impacts of mining on natural resources if coal development should occur. Information contained in the system will include geography, project objectives, methods and progress, published results, and names and addresses of personal contacts. These people will be able to provide project information before it is available in published reports. Information contained in the system will be available to other State and Federal agencies, educational and research institutions, and the general public upon request from the Division of Special Studies, Denver Service Center, BLM.

THE ENDANGERED SPECIES ACT:
PLANNING AND THE SECTION 7
CONSULTATION PROCESS

G. A. Reyes-French
    Environmental Research &
    Technology, Inc.
    Fort Collins, Colorado

T. G. Shoemaker
    Environmental Research &
    Technology, Inc.
    Fort Collins, Colorado

ABSTRACT

    The Endangered Species Act of 1973, as amended, has proven to be one of the most far-reaching and controversial of U. S. environmental laws. The Act affects nearly all projects through the Section 7 Interagency Cooperation requirements. This paper reviews the requirements of Section 7 of the Act and advocates integration of the Section 7 consultation process into general project planning for major developments. The paper introduces the concept of a threatened and endangered species management system which is cooperatively developed during project planning to avoid adverse effects on listed species. It is important to stress that no general approach is sufficient for threatened and endangered species planning; unique solutions are needed for unique problems. Examples of problem solving approaches are provided from experiences with the proposed Northern Tier Pipeline System.

The Endangered Species Act of 1973, as amended, has proven to be one of the most controversial of U.S. environmental laws. Much of the controversy has resulted from a few major construction projects in which irresolvable conflicts with threatened or endangered (T&E) species were encountered late in the planning phases or during construction. Unfortunately, the controversy still lingers from cases such as Tellico Dam and the conservation needs of T&E species continue to be viewed as impediments to major construction projects.

Our experience with numerous development projects, primarily in the western United States, has demonstrated that the protective requirements of the Act need not impede major new projects. The Act's requirements certainly do impact industry. However, we believe that both the conservation needs of T&E species and the needs of industry can generally be satisfied if T&E species concerns are considered early in the development planning process and if cooperative problem solving is used to resolve issues which do arise.

This paper and a companion paper (Shoemaker and Reyes-French in prep.) expand on the early planning/problem solving philosophy expressed above. This paper (1) briefly reviews the requirements and processes associated with Section 7 of the Act, (2) suggests an approach for integrating studies of T&E species into a generalized pipeline planning process, and (3) illustrates the types of solutions which may be needed to resolve potential conflicts. ERT's experiences on the Northern Tier Pipeline are used for examples.

In order to plan for threatened and endangered species concerns and the Section 7 process, it is important to understand the major steps and requirements of the Act.

OVERVIEW OF SECTION 7 CONSULTATION REQUIREMENTS

Two sections of the Endangered Species Act of 1978 (as amended) have important influences on the activities of the petroleum industry. Section 9 prohibits "taking" of threatened and endangered animals. Section 7 affects the petroleum industry through the process referred to as the Section 7 consultation process. Section 7 consultation is essentially a means of assuring that federal agencies carefully consider threatened and/or endangered (T&E) species' concerns in carrying out their actions or activities. The petroleum industry is involved because the Act defines agency actions as including activities authorized, funded, or actually carried out by the agency. This includes any permits or leases for development on federal lands.

Section 7 consultation refers to the process in which federal agencies conform to Section 7(a)(2), which specifically requires federal agencies to ensure that their actions do not jeopardize the continued existence of a T&E species or destroy or adversely modify designated critical habitat. Agencies are to meet this requirement in consultation with the assistance of the Secretary of Interior or Commerce, as represented by the U.S. Fish and Wildlife Service (FWS) and the National Oceanic and Atmospheric Administration (NOAA). The Act imposes several procedural steps, time limitations, and restrictions on consultation.

The 1978 amendments included a second process, exemption, which may be entered following consultations which fail to resolve conflicts between the protective requirements of Section 7 and a needed agency action. Exemption is essentially a last resort for projects which are of regional or national significance and cannot be implemented without jeopardizing the continued existence of a T&E species. The exemption process has not been widely used since its creation. The steps involved have been previously reviewed (Shoemaker 1979) and are contained in the Code of Federal Regulations in Chapter 50, subchapter C, "Endangered Species Exemption Process."

The steps in the consultation process are not as well defined as those of the exemption process. This is partly due to the fact that the regulations for consultation (found at 50 CFR 402) predate the 1978 amendments, but also relates to the nature of the process. In our view, consultation is a process for cooperative early planning. It is a negotiation and problem solving process which must respond to diverse situations which involve biological, engineering,

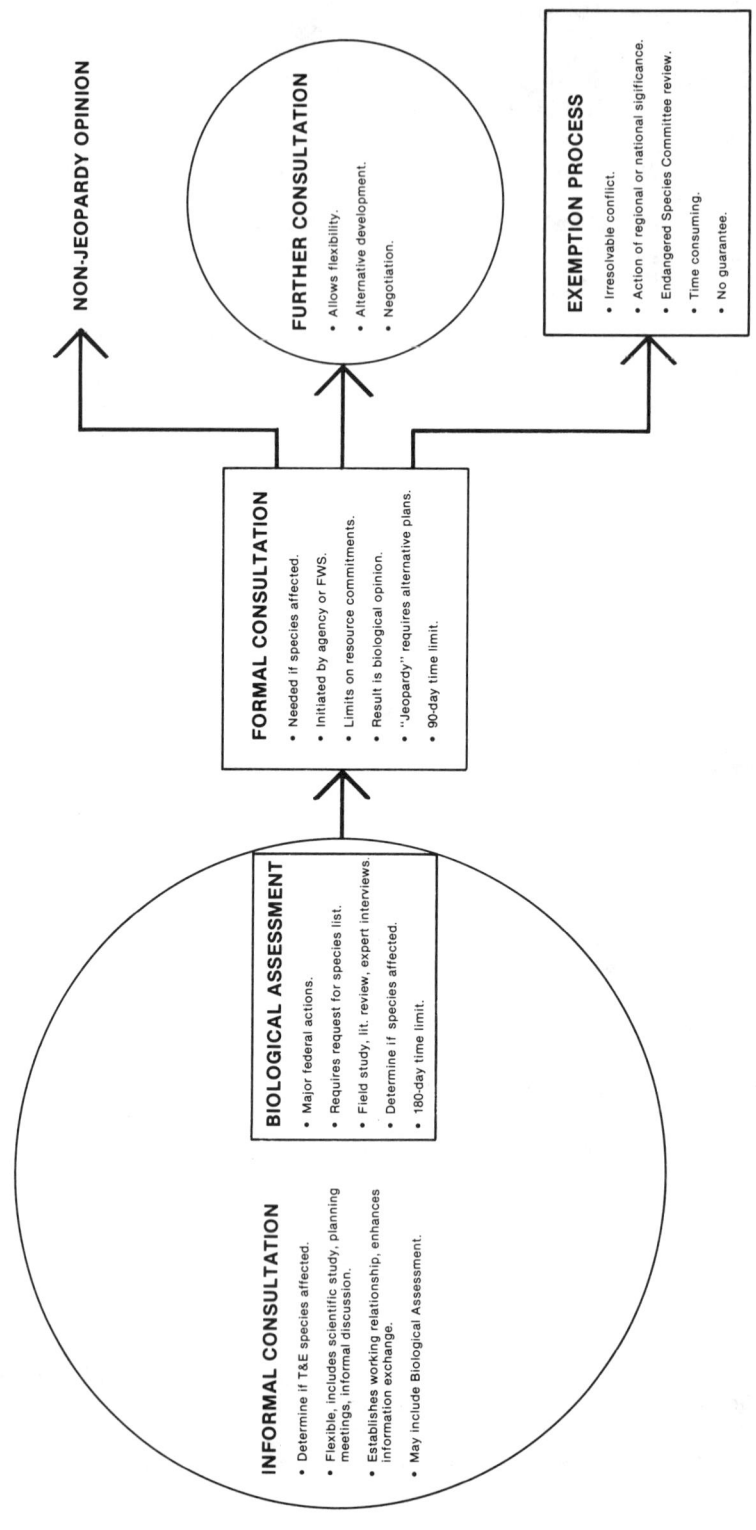

Figure 1. Overview of the Section 7 Process.

political, and economic considerations. The problems and concerns vary, so the process must be adaptive. At the same time, the process must be sufficiently defined that participants can count on receiving timely input to planning and decision-making efforts. Based on past experience in consultations, and on extensive discussions with FWS representatives, there are three major phases involved in consultation (Figure 1). Together the three phases provide (1) the restrictions needed to assure input to decision-making and planning, and (2) the flexibility needed to adapt the process to diverse circumstances.

Informal consultation precedes formal consultation and consists of any written exchanges, meetings or discussions regarding potential effects of an agency action on T&E species. Informal consultation may be initiated at any time by the federal agency or nonfederal applicant. Informal consultation may include a variety of activities which are used to: (1) establish working relationships, (2) enhance information exchange to expedite formal consultation, (3) promote early development of options and project modifications, (4) identify potential problem areas, and (5) utilize FWS expertise in planning processes.

Informal consultation may also include several steps required by the Act, if the project is a "major federal action significantly affecting the quality of the human environment." In these instances, the federal agency must (1) formally request (from FWS) a list of species potentially occurring in the project area, and (2) conduct a biological assessment of the potential effects of the project on those species. The biological assessment requires field study, literature review, interviews with species' experts, and a variety of other information-gathering activities used to determine if T&E species are present in the project area and if they may be affected. The biological assessment must be completed within 180 days following receipt of the list of species (unless an extension is agreed to). In addition, the Act prohibits the agency or nonfederal applicant from beginning a construction project or entering into a contract for construction until the biological assessment is completed.

Formal consultation is the most defined phase of the process. It is required when an agency action may in any way affect a T&E species or critical habitat. During consultation, the agency (and nonfederal applicant) is prohibited from making any irreversible or irretrievable commitments of resources which would foreclose alternatives or options to the planned action. Formal consultation follows a review of the potential effects of the agency action and may be initiated by written request of the federal agency to FWS or NOAA. NOAA has jurisdiction over marine species including the endangered whales. Formal consultation provides further review of the nature and magnitude of the effects of the agency action on T&E species. Often consultation will require meetings of FWS, agency, and industry representatives to consider alternatives to prevent or minimize adverse effects on T&E species.

Formal consultation must be completed within 90 days unless an extension is mutually agreed to. Following completion, FWS or NOAA issues a Biological Opinion on the projected effects of the action. The opinion reviews the nature of the effects, discusses actions which may be taken to reduce adverse effects, and reaches one of three conclusions: (1) the action will promote the conservation of the species, (2) the action is not likely to jeopardize the continued existence of the species; (3) the action is likely to jeopardize the continued existence of a species. If a jeopardy opinion is issued, the federal agency may not proceed with its action without modifying it to preclude jeopardizing the species or obtaining an exemption from the requirements of section 7(a). In any jeopardy opinion, FWS must include a discussion of reasonable alternatives which are available to meet the objective of the agency action and preclude jeopardy to the species.

Further consultation may follow formal consultation if a jeopardy opinion is issued. This simply is a time when federal agencies may consider additional options or alternatives and may continue to meet with the FWS.

PROJECT PLANNING

Most large projects involve careful step-by-step chronological planning to accomplish the final goal. The planning approaches and information needs are similar although the objectives may vary from development of an oil or gas field, to construction of a power generating station, or a new pipeline. Implementation of a major development is a complex job and many companies make the task more manageable by identifying distinct planning processes or "tracks" such as engineering, environmental, or permitting. Usually these "tracks" are managed and monitored somewhat independently and integrated at critical project phases or milestones.

We suggest that the Section 7 consultation process can be viewed as a distinct project "track" and should be integrated into the overall planning scheme for major developments (Figure 2). The T&E species track should begin very early in planning and continue throughout the life of the project. The product of the program should be development and implementation of a T&E species management system which meets two goals:

1. It should be compatible with other project systems.

2. It should be consistent with the protective requirements of the Act.

Frequent and effective interface with project engineers is essential if the first goal is to be achieved. Representatives of industry, the FWS, and the involved federal agency must participate in a cooperative planning and review process (encompassing informal and formal consultation) to achieve the second.

We developed Figure 2 as an illustration of one example approach to integrating planning for T&E species management into overall project planning. It is important to stress that the process must be adaptive. The phasing we have suggested will need to be modified for each project and the information needs and study efforts will be unique. There are, however, six assumptions incorporated into Figure 2 which we feel are generally applicable.

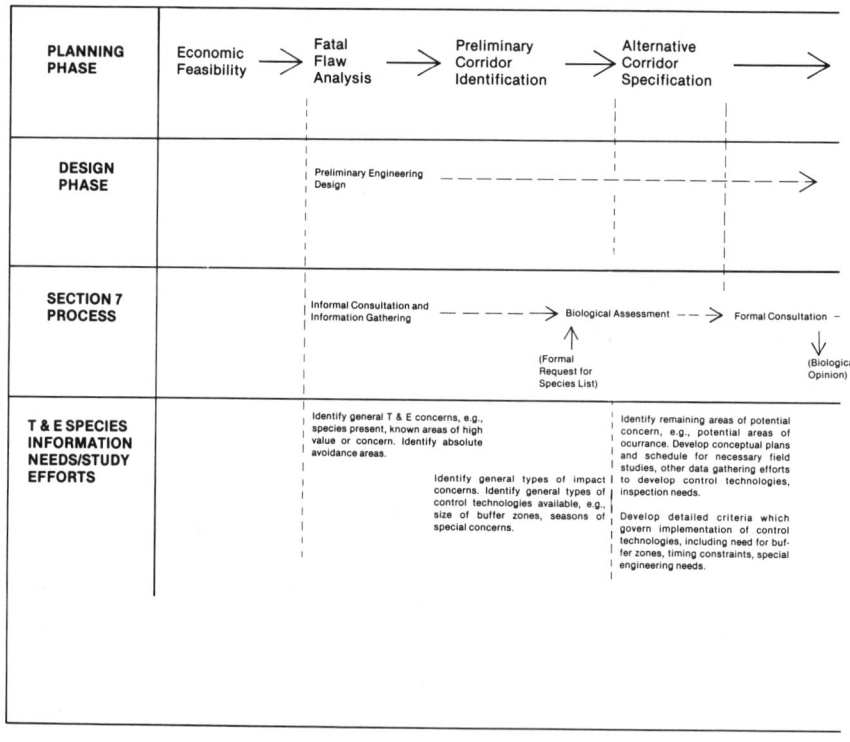

Figure 2. Incorporating T & E species concerns into the pipeline planning process.

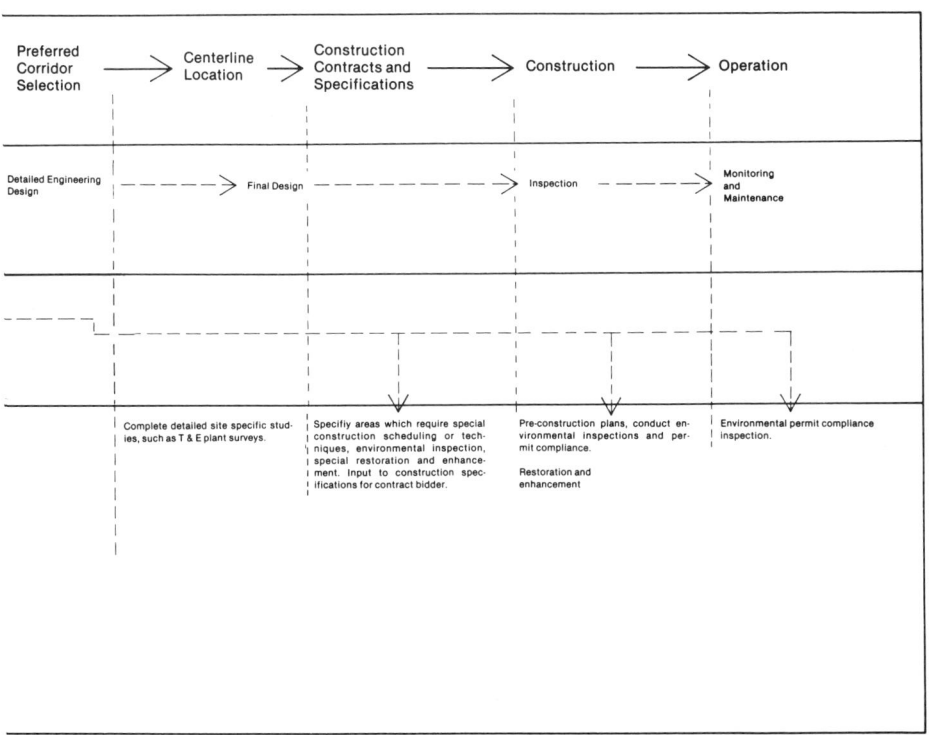

Figure 2. continued

Assumption one is that T&E studies are not required for input to Economic Feasibility Analysis. T&E species concerns should not determine economic feasibility and will not if project planning is effective.

Assumption two is that T&E studies should begin with Fatal Flaw Analysis. Certain critical T&E species areas should be considered as absolute avoidance areas. They are fatal flaws in the same sense that not considering a National Park Class I air quality area could be a "fatal flaw" for a power plant. Locating a facility at the only recorded nest site for the bald eagle in a western state is also a "fatal flaw" not in the sense that it will necessarily stop the project, but it reflects poor planning and will result in unnecessary conflicts. Identification of fatal flaws is essential at this stage to avoid irresolvable conflicts or to identify early the case in which an exemption may be needed. Fatal Flaw Analysis usually will not involve field work because information on T&E species critical areas is readily available from Regional or Area FWS offices or state agencies.

The third assumption is that potential concerns related to T&E species may be preliminarily identified based on existing published information, discussions with FWS and state agencies, and interviews with T&E species experts. Once a general geographic location is identified, it is possible to identify species which may be encountered, their habitats, and the general types of concerns which may arise due to the proposed development. This is valuable input to the process of selecting alternative corridors.

Assumption four is that a Biological Assessment can and should be conducted for more than one alternative. This assumption imposes constraints on the level of resolution possible in the analysis and on the manner in which potential problems are addressed. For example, intensive field study to document species presence would be prohibitive if conducted on each corridor or alternative. The analysis, then, must be based on habitat-level information and directed at determining potential areas and seasons of occurrence. The information must be specific enough to realistically define the nature of T&E species concerns associated with each alternative and to develop stipulations, or conservation measures, which would be applied if a given alternative is selected. The objective is to develop a management strategy which assures compliance with the Act, but maintains the flexibility needed in selection of a preferred corridor or centerline. Development of alternative management strategies at this stage offers additional benefits in that it allows the relative costs (time and money) of T&E management measures to become a factor in evaluating the merits of several alternatives.

The fifth generalization is that the Biological Assessment Report is an important means of documenting the entire informal consultation process. If our approach is followed, informal planning and discussions will have been ongoing for some time. The Biological Assessment Report can be used to document the prior planning efforts completed to develop alternatives for more detailed analysis.

There are no formal regulations regarding content of biological assessments, so the following suggestions may be useful: The document should include the results of:

- 1) Efforts completed to determine if T&E species or suitable habitats are present; 2) interviews with recognized experts on the subject; 3) a literature review of pertinent data; 4) a review and analysis of the effects of the action on the species; and 5) an analysis of needed alternative actions or measures.

- The biological assessment document should include a brief project description outlining the scope of the development. This discussion should include enough detailed information on the project to reveal the general phases and schedule of the project and potential impacts or conflicts with T&E species.

- The assessment should include a discussion of the life history and habitat requirements of the species as they relate to their presence in the study area and the potential impacts on the species.

- Photographs of onsite conditions may add significantly to the discussion.

The final assumption regarding the planning process is that it may be possible and desirable to complete formal consultation prior to selection of a preferred corridor or centerline. This will vary considerably with the project and location. Consultation at this stage may be possible if: (1) issues and concerns have been sufficiently defined in a corridor level analysis, and (2) the project developer, federal agency, and FWS can agree to stipulations which can be applied when site specific conditions warrant, to preclude harm to T&E species. Completion of consultation at this stage may be desirable in the sense that FWS' official views can be obtained on more than one alternative. This offers flexibility in project planning and provides valuable input to industry and agency decision making. It also contributes to the thorough analysis of alternatives required in the environmental impact statement process.

EXAMPLES OF ENDANGERED SPECIES MANAGEMENT TECHNIQUES

We have stressed in the previous section that effective early planning can prevent conflicts between developments and T&E species. The concept of a T&E species management system is important, but it is difficult to provide a general approach that will work to solve the unique problems encountered each time a new proposal is developed. We believe there are many possible solutions that may be appropriate, depending on individual circumstances. Useful examples of some approaches are provided by the Northern Tier Pipeline System (NTPS).

The Northern Tier Pipeline Company proposes to build a common carrier crude oil pipeline across Washington, Idaho, Montana, North Dakota, and Minnesota for the purposes of moving Alaskan North Slope and foreign crude oil to inland states to make up for anticipated shortages and increased demand. The BLM issued their final impact statement on Crude Oil Transportation Systems in November of 1979. In accordance with the Endangered Species Act of 1973, Environmental Research and Technology, Inc. (ERT), prepared the Northern Tier Pipeline biological assessment for the U.S. Bureau of Land Management (BLM). The biological assessment covered the entire 1,491-mile route from Port Angeles, Washington, to Clearbrook, Minnesota. The assessment dealt with 10 threatened or endangered species and/or groups, both the marine and terrestrial environments, NOAA, two regions of Fish and Wildlife Service, as well as the concerns of several state agencies. This study provides an excellent case history and opportunity to examine the range of conflicts a pipeline or other similar corridor development may encounter with threatened and endangered species and perhaps suggest some strategies for project planning.

The primary concerns associated with major pipeline construction and operation and the potentially affected species occurring along the NTPS are depicted in Figure 3. This impact matrix also illustrates the various approaches identified and recommended in the biological assessment for avoiding or minimizing potential impacts.

One of the best means for avoiding conflicts with endangered species is to utilize a "corridor concept" for

Endangered Species potentially affected by NTPS	Biological Concerns and Potential Impacts									Mitigation Recommendations for NTPS
	CONSTRUCTION					OPERATION				
	loss of habitat from clearing	disturbance from noise and other human activity	stream crossing degradation	reduced productivity or suitability of habitat	barrier to movement	contamination of habitat from oil leaks	disturbance from r-o-w inspection	disturbance from r-o-w maintenance	increased access on r-o-w	
Bald Eagle	1	1,2	2,3			3				No construction Jan.-July. If nesting is successful, reroute ½ mile from new nests. Protect fisheries.
Peregrine Falcon	1	1,2				3				Oil Spill Contingency Plan (OSCP) Construct July-Feb. Reroute ½ mi. from eyrie.
Whooping Crane	1			3		3				Avoid wetlands; ditch plugs to prevent draining affected wetlands. OSCP
Grizzly Bear *	1	3	3	4	3	3			3	Avoid critical habitats, control trash, no guns, reveg. riparian zones, "skip" section in trench, OSCP, locked gates.
Black-footed Ferret	1					3				Avoid prairie dog towns, OSCP.
Plants	1					3		3		Avoid plant habitats and populations, OSCP, maintenance confined to right-of-way.
Whales **		o			o	3				No recommendations.
Sea Turtles **		o				o				No recommendations.
Brown Pelican **						o				No recommendations.
Gray Wolf **				o						No recommendations.

* Present on one alternative corridor.
** Present along route but not affected significantly.
o Minor potential impact.
1 Avoidance (centerline alignment)
2 Timing constraints (avoid critical periods)
3 Environmental Control Technology (inspector, OSCP, sediment control)
4 Restore or Enhance

Figure 3. Biological concerns and proposed mitigation for the potential impacts to threatened and endangered species affected by the Northern Tier Pipeline System.

routing and site selection. The "corridor concept" advocates following established corridors such as pipelines, railroads, roads, and transmission line rights-of-way. This usually limits the severity of impacts associated with pipeline construction and operation to wildlife, including T&E species. Wildlife living in or near disturbed areas (early successional habitats) are more adapted to dealing with disturbance than are those species or populations that have evolved in wilderness or more pristine, undisturbed, mature habitats. Employing the corridor concept also concentrates the disturbance to one general area instead of fragmenting habitats unnecessarily. The NTPS follows existing corridors from about 40 percent of the preferred route. However, pipeline economics, site conditions and project objectives may not always allow the corridor concept to be used. This is often true for gas collecting pipelines and for distribution pipelines supplying oil and gas or other products to new markets.

Another method for limiting potential impacts is to avoid areas where threatened and endangered species occur. It is very difficult to avoid crossing the potential ranges of over 100 threatened and/or endangered species found in the western United States. However, slight realignments of the centerline can be used to miss populations of threatened and endangered plants, avoid critical habitats, or avoid important biological features, such as feeding areas. For example, the NTPS centerline will be located to avoid prairie dog towns (habitat for black-footed ferret) in North Dakota. Pipelines and other linear developments have a distinct advantage in dealing with potential environmental impacts because of their relatively small area of disturbance, and their flexibility in routing and timing.

Timing constraints or restrictions are another method of limiting impacts on threatened and endangered species, especially those impacts associated with human activity and presence. For instance, bald eagles are especially sensitive to human disturbance during nesting. Disturbance early in pair bond formation, egg laying, incubation, or rearing can result in reduced breeding success, nest failure and loss of eaglets. Parental care of eaglets begins to decrease after 8-9 weeks when the young begin wing exercises. Around 13 weeks the young are flying and ready to leave the nest. In Washington state young eaglets fledge in July (Grubb 1976). Therefore, nesting eagles are less sensitive to disturbances after July. ERT recommended that Northern Tier not construct in Washington from January to July within 660 feet of known nest sites. If, however, the eagles do not "successfully" nest (that is, eagle incubating eggs or hatched live young) by May 15, construction within the buffer zone could be initiated. All recommendations for timing constraints must reflect biological criteria, sound reasoning, and if possible, documentational support. For instance, the 660 foot buffer zone is consistent with measures recommended in the U.S. Fish and Wildlife Service (1977) Washington-Oregon Bald Eagle Guidelines.

Environmental control technology includes a variety of mitigating measures that deal with specific impacts. Along with the classic engineering control technologies such as sediment basins, water bars, stream "diapers," entrainment screens, etc., other types of environmental impacts can be dealt with using other types of "control technology." For instance, the concept of an "environmental inspector" can be a means for dealing with changes in site specific conditions. The environmental inspector may be a qualified biologist representing government and/or industry. For the NTPS, the environmental inspector will survey the route prior to construction to insure no new threatened and/or endangered species are present. The inspector will also insure that permit stipulations are complied with and the company is not vulnerable to expensive delays because of permit infractions. Other specific control technologies include; limiting disturbance by controlling the size and activities of the construction work force (such as controlling trash disposal, use of guns, and limiting access in grizzly bear habitat), and using ditch plugs to prevent draining of whooping crane wetlands.

By employing these types of strategies or technologies, most T&E problems can be solved. Restoration and enhancement should be considered in incidents where the impact cannot be avoided. For example, if an area containing snags used as perches for bald eagles is affected by construction, this area could be restored by replacing these snags. Or if riparian vegetation in grizzly bear habitat is being cleared, revegetation plans could include similar species and/or species preferred by grizzly bear to enhance food and cover. In some cases such as the unavoidable destruction of endangered plant habitat, the species could be removed prior to construction and donated to a local or regional herbarium. In extreme cases where irreconcilable differences have arisen, funding for study and future habitat maintenance has been negotiated. Both of these suggestions would require extensive interaction with FWS and should not be viewed as "standard" approaches to solving T&E species conflicts.

CONCLUSIONS

Many options are available to the environmental professional, especially the planner involved in route or site selection studies. Pipelines and other linear developments have a distinct advantage over other developments. They have the flexibility inherent in routing and construction timing. However, for this flexibility to be realized, environmental concerns, especially effects on threatened and endangered species, must be considered early in project planning.

The Endangered Species Act of 1973, as amended, contains sufficient flexibility to allow for consideration of project planning goals and objectives while considering environmental concerns for T&E species. Industry will be serving its own best interest by taking an active role in identifying potential future conflicts. The informal consultation process and the biological assessment encourage early input from FWS, allowing industry to participate in development of future stipulations on federal permits. We have tried to present a brief overview of the different environmental control technologies available for mitigating impacts and an approach to incorporating these measures and appropriate study into project planning. However, as in all cases dealing with biological systems, there is no one "right way." We hope by providing one approach we will encourage development of new and better approaches and techniques to inventory resources, evaluate impacts and alternatives, and recommend conservation measures which are beneficial to endangered species and responsible to the economic and practical constraints of industry. There is no "cookbook" approach; for each project it is important to seek unique solutions to unique problems.

LITERATURE CITED

Grubb, T. G., Jr. 1976. A survey and analysis of bald eagle nesting in western Washington. M.S. Thesis, University of Washington, Seattle. 83 pp.

Shoemaker, T. G. 1979. The new Endangered Species Act. Environmental Science and Technology 13(9):1058-1061.

Shoemaker, T. G. and G. A. Reyes French. In preparation. The Endangered Species Act and rights-of-way management: requirements, results, and implications of the Section 7 consultation process. Accepted for publication at the Third Symposium on Environmental Concerns in Rights-of-Way Management. February 15-18, 1982, San Diego, California.

U.S. Fish and Wildlife Service. 1977. Bald eagle management guidelines, Oregon-Washington. Published handout. 8 pp.

WETLAND CLASSIFICATION -
TRUTH AND CONSEQUENCES

Virginia Carter
U. S. Geological Survey

A new wetland classification system, "Classification of Wetlands and Deepwater Habitats of the United States", by L. M. Cowardin, Virginia Carter, F. C. Golet, and E. T. LaRoe (1979), was recently developed for use by the U. S. Fish and Wildlife Service in their current National Wetlands Inventory. The new system, has three primary objectives: (1) to group ecologically similar habitats so that value judgements can be made, (2) to furnish habitat units for inventory and mapping, and (3) to provide uniform terminology throughout the United States. The system is hierarchical in structure to allow users to select the level of detail appropriate for their needs.

Five systems are defined: marine, estuarine, riverine, lacustrine, and palustrine. Subsystems, such as subtidal and intertidal in the marine and estuarine systems, are defined at the second level of the hierarchy following traditional ecological concepts. The third level has 13 classes based on life form of plants for vegetated areas or broad substrate types for unvegetated areas. Third-level classes are separated into subclasses at a fourth level based on plant characteristics such as needle-leaved evergreen, or on detailed substrate type, such as sand (Figures 1-6). The fifth and lowest level is based on dominance type as defined by the dominant or codominant plant or animal species present. The system is open ended at the dominance type level so that new types may be added as needed or further subdivision may occur.

Figure 1. Broad-leaved Deciduous Forested Wetland, Merchants Mill Pond, North Carolina.

Figure 2. Gravel Unconsolidated Shore, Point Lobos, California.

Figure 3. Persistent Emergent Wetland, Nanticoke, Maryland (Chesapeake Bay).

Figure 4. Moss-Lichen Wetland, Campobello Island, Maine and New Brunswick, Canada.

Figure 5. Aquatic Bed, Rhode Island.

Figure 6. Broad-leaved Evergreen Scrub-Shrub Wetland, North Carolina.

In addition to the classification hierarchy, there are modifying terms for water regime, water chemistry, and soils, special modifiers for man-made disturbance, and regional locators. The wetland definition recognizes the complex interaction between hydrology, soils and vegetation included in the concept "wetland". It expands the traditional definition or concept of wetland as a vegetated or open-water area to include (1) areas with hydrophytes and hydric soils such as marshes, swamps, and bogs; (2) areas without hydrophytes, but with hydric soils such as mud or salt flats; (3) areas with hydrophytes, but non-hydric soils such as impoundment or excavation margins; (4) areas without soils, but with hydrophytes such as seaweed covered rocky shores; and (5) wetlands without soils and without hydrophytes such as gravel or sand beaches.

In developing the new classification, the authors recognized that most of the information needed to characterize wetlands accurately is not available, especially pertaining to wetland hydrology. Thus the classification system itself represents "truth" at one brief instant in time according to the state-of-the-art in wetland research. It provides a somewhat simplistic framework in which to place the enormous diversity and variety of wetlands found both in the United States and elsewhere in the world. For energy and mineral development, as for other activities which potentially affect or alter the landscape, wetland class is really not as important (unless there is a scarcity of a particular type) as an understanding of the interaction of hydrology, vegetation, and soils in the wetland(s). These interactions are closely related to wetland value and wetland resiliency (ability to return to normal conditions after stress). For preparation of environmental impact statements and assessments, and for purposes of planning and mitigation, better and more detailed information is still required. Whereas the wetland classification has provided us with a common framework and language, the real "consequences" of the development of this system and of its increasing acceptance by government agencies, private institutions, and individuals is the increase in productive wetland research into all aspects, including function and value. This research, especially into wetland hydrology and the relationship between hydrology and other wetland components, should improve planning and management capabilities related to this important but poorly understood resource.

DETAILED MAPPING FOR
DEVELOPMENTAL PLANNING

Sue A. Degler
  Sohio Alaska Petroleum
Company

Detailed mapping of topography, landforms, soils and vegetation of the arctic coastal plain has long been sought by industry as well as governmental agencies. Some work of this type was published as early as 1975; however, detailed information appeared in June of 1980 in CRREL's Geobotanical Atlas of the Prudhoe Bay Region, Alaska [1]. Since that time, Sohio Alaska Petroleum Company has prepared detailed mapping of larger North Slope areas and has found such maps to be a valuable planning tool.

The maps, at a scale of 1" = 500', use detailed topographic maps (5' contour intervals) as a base. Information on landforms, soils and vegetation community types obtained through photointerpretation with on-site verification is then added to create a detailed image of the study area. From base information, derived maps may be prepared reflecting sensitivity of any given area to, among other things, off-road travel, pollution, or construction.

For purposes of discussion here, a site was selected that contained a drained lake complex, a small stream, and, like all of the coastal plain, numerous small ponds and lakes. In general, the area is one of moist to saturated soils; vegetation in the area includes aquatic grasses, sedges, dwarf willows, and mosses. A scar from vehicle movement across the tundra meanders from northeast to southwest across the area of interest. Drainage in the area is generally to the north. The proposed project is placement of a gravel drill pad, similar to those used for oil and gas production on the North Slope.

Two sites within the area of interest were analyzed in detail:

Site One--basically located in the center of the area of interest, about one half of Site One is in a drained lake complex. The site is oriented east and west and covers an area 750' X 2,000'

Site Two--oriented with the topographic features in a north northwesterly-south southwesterly direction, Site Two is located on better drained soils and is nearer the stream. Both sites are the same size.

By placing overlays of the two sites over detailed maps, the following comparisons may be drawn with regard to viability of each site:

Table I. Comparison of Siting Considerations

Consideration	Site One	Site Two
Orientation with topography	Perpendicular to topography	Aligned with natural features
Maintenance of natural drainage patterns	Potential drainage problems	Good drainage
Potential conflicts with breeding bird populations	4-6 pairs	0-4 pairs
Wetness of area	70% lakes and wet tundra	60% moist tundra
Oil spill recovery potential	very good to excellent	moderate to very good
Use of area already damaged by man's activities	2,000' corridor	750' corridor
Proximity to lakes and streams	Covers some small ponds; well away from stream	Covers less standing water, but closer to stream

Objective measurements of specific types of terrain may be made using this approach. It allows for consideration of sensitive terrain or valuable habitats during the site selection process. Whether or not areas of concern may be avoided totally, decision makers may at least quantify their judgements.

The mapping can present detailed information concerning potential construction sites to regulatory agencies. With such facts in hand, developers may meet to analyze real impacts of any project. Though philosophies regarding relative values of any particular resource may vary substantially, detailed mapping allows for a common basis for discussion.

Present alternatives to detailed mapping are site selection followed by on-site investigation (which is time consuming and can result in selection of a site only because there is no time to locate another); use of Landsat (which at the present time provides insufficient resolution to do more than select general corridors which must be carefully analyzed for specific construction locations); and the use of low altitude aerial photography where it exists (requires a good knowledge of photointerpretation and familiarity with the area of interest).

LITERATURE CITED

1. Walker, D.A., K.R. Everett, P.J. Webber and J. Brown. 1980. Geobotanical Atlas of the Prudhoe Bay Region, Alaska. CRREL Report 80-14. UNESCO MAB Project 6. International Biological Program, U.S. Tundra Biome. U.S. Army Corps of Engineers, Cold Regions Research and Engineering Laboratory, Hanover, New Hampshire, U.S.A., 69 p.

RAPID ESTIMATION OF THE
DIVERSITY OF POLLUTED COMMUNITIES

E. C. Pielou
Biology Department,
University of Lethbridge,
Lethbridge, Alberta T1K 3M4

ABSTRACT

The diversity of an aquatic community is often treated as a rough measure of the quality of the water it inhabits. But because a diversity index is inevitably an imprecise measure of water quality, it is not worth while using an index that is difficult and time-consuming to obtain. The trouble is compounded by the need to base a diversity estimate for a whole community on a large number of sample collections. For many purposes there is no need to compute a traditional diversity index, based on counts of all the species in each sample collection. Instead, it suffices to record the relative abundances of only the three most abundant species in each. The three relative abundances constitute a triplet of proportions that sum to unity and the triplet can therefore be represented by a single point in a ternary graph. Thus one can represent the diversity of a single sample collection by a single "diversity point" in a triangle, and the diversity of n sample collections from a community by a swarm of n diversity points in a triangle. The location and spread of the swarm gives a graphic picture of the overall magnitude, and the internal variability, of the diversity of an extensive community.

INTRODUCTION

Many ecologists who work in the field of environmental protection, and especially those concerned with water quality, have tried using ecological diversity indices of various kinds as inverse measures of so-called environmental stress. The literature is large; some recent examples, and a host of references can be found in Grassle, Patil, Smith and Taillie

(1979) (referred to below as GPST). Among the communities whose diversities have been measured are freshwater macroinvertebrates (Resh, in GPST), marine zooplankton (Dennis and Rossi, in GPST), marine phytoplankton (Patten, 1962), diatoms (Moncrieff and Perrin in GPST; Patrick, 1961, 1963) foraminifer assemblages (Lagoe, 1976) and the marine benthos of "a crowded harbor with oil terminals" (Flôres, Barreto and Da Costa, in GPST).

The use of diversity measurements for biomonitoring has given rather disappointing results, however. The correlation between ecological diversity and water quality is seldom high. Scatter diagrams relating diversity and, say, pollutant concentration, are usually very scattered indeed. This is so whether diversity is measured simply by S, the total number of species in the "monitor" community, or by well-known Shannon-Wiener index $H' = -\Sigma\, p_i \log p_i$; here p_i for $i = 1, 2, \ldots, S$, is the relative abundance of the ith most abundant species.

For useful results to be obtained, therefore, it is necessary to determine the diversities of a very large number of sample collections at each point along the pollution gradient being monitored; or of a very large number of sample collections from the supposedly pure, and supposedly polluted, bodies of water being compared. Diversity is affected by so many factors besides pollution level that the signal-to-noise ratio of diversity measurements is unavoidably low.

Usually it is extremely laborious to obtain estimates with acceptably small standard errors, of H' or S, for many-species communities of small (or microscopic), taxonomically difficult organisms. Identifying and counting the individual organisms in a sample collection may require hours of work by a skilled taxonomist. Even when that has been done, the problem of estimating S or H' is only half solved; the statistical difficulties of obtaining an unbiased estimate, and of estimating its standard error, are formidable (Routledge, 1980), a fact overlooked by those who blithely substitute observed p_i values in the famous Shannon-Wiener formula and imagine that the resulting number means something.

Indeed, the effort required to correctly estimate S or H' (or any of the other diversity indices that require all the individuals in a sample collection to be identified and counted) is usually prohibitively great having regard to the usefulness of the estimate obtained. As already remarked, the correlation between diversity and water quality is not great. The paltry amount of information yielded by a few diversity values is a poor reward for their cost. Diversity measurements are simply not worth having unless they can be quickly and easily obtained.

DIVERSITY POINTS IN DIVERSITY TRIANGLES

A measure of the diversity of a community that, though, crude, is adequate for most purposes, may be obtained quickly and easily using the relative abundances of only the three commonest species. Denote these relative abundances by q_1, q_2 and q_3; the q's are so scaled that $\Sigma_1^3 q_i = 1$, and the species are so labelled that $q_1 \geqslant q_2 \geqslant q_3$. One would expect to find the three q values more nearly equal in a community of high diversity with many species than in one of low diversity with few (see Figure 1). This expectation stems from the accumulated experience of many ecologists; it is not a logical necessity. Thus the lefthand panel of Figure 1 shows typical species-abundance histograms for a high diversity community (A) and a low diversity community (B). The species-abundance histograms for the three most abundant species only, for each community, are shown in the center panel (in each of the four histograms, the heights of the bars sum to unity).

Clearly, one can replace any 3-bar histogram by a single point in a ternary graph, as has been done in the righthand panel of Figure 1. In the upper triangle is the point for community A, and in the lower triangle the point for community B. Hereafter, these points will be called "diversity points". One expects to find the diversity point of a highly diverse community to be closer to the center of the triangle (marked X), at which $q_1 = q_2 = q_3$, than the diversity point of a less diverse community.

Note next that, by definition, $q_1 \geqslant q_2 \geqslant q_3$. Hence (see Figure 2) diversity points are necessarily constrained to lie in the small right-triangle XYZ which constitutes the one-sixth part of the triangle xyZ in which these inequalities hold; triangle xyZ is the triangle for an "ordinary" ternary graph, in which the three proportions defining any point are not ranked in order of magnitude. The small triangle XYZ will be called a "diversity triangle". The values of (q_1, q_2, q_3) for the vertices of the two triangles are:

$$(q_1, q_2, q_3) = \begin{cases} (0, 0, 1) & \text{at } x \\ (0, 1, 0) & \text{at } y \\ (1, 0, 0) & \text{at } Z \\ (½, ½, 0) & \text{at } Y \\ (1/3, 1/3, 1/3) & \text{at } X \end{cases}$$

Obviously, the diversity points of any number of sample

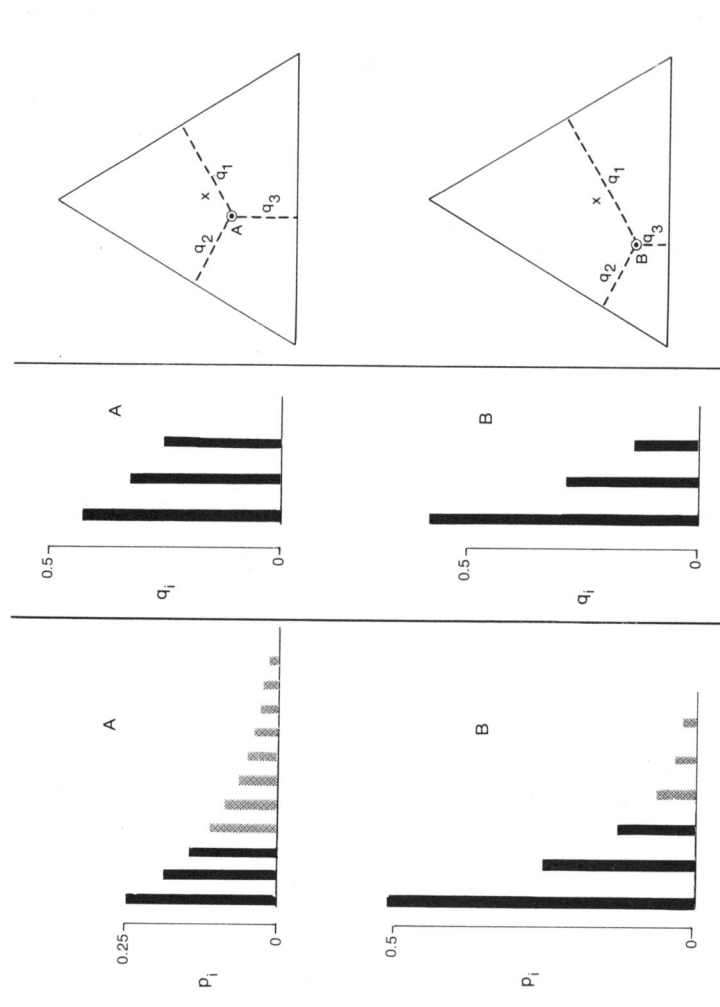

Figure 1. *Left*: The relative abundance distributions of two communities, A and B; A, with 11 species, has high diversity; B, with 6 species, has low diversity. *Center*: The relative abundance distribution of the three most abundant species in communities A and B. *Right*: Ternary Graphs showing the diversity points of communities A and B.

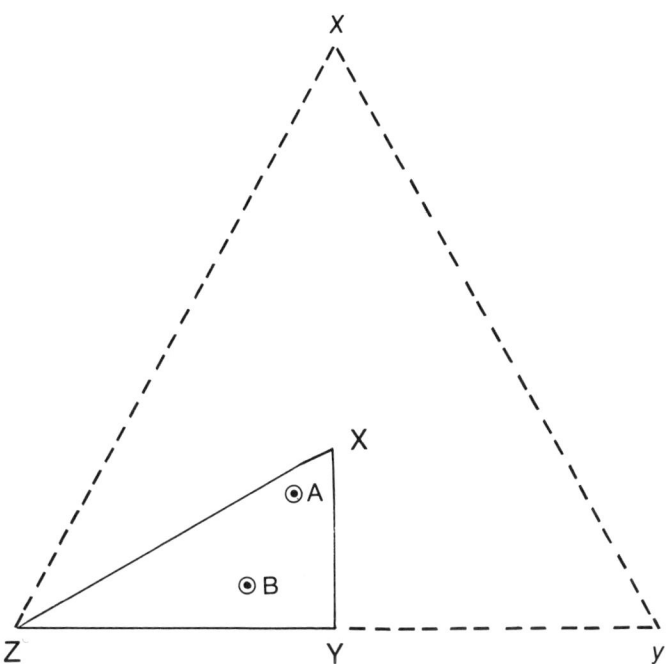

Figure 2. To show the "diversity triangle" XYZ, which is one-sixth of the equilateral triangle xyZ. The points A and B are the diversity points of the communities shown in Figure 1.

collections can be plotted in a single diversity triangle. Figure 2 shows an ordinary ternary graph, the triangle xyZ, and within it, the diversity triangle XYZ; the points A and B are the diversity points for communities A and B of Figure 1.

AN EXAMPLE

Because of the inevitably high sampling variation, one should never attempt to infer the overall diversity of a large natural community from a single sample-collection taken at one point. To characterize a community's "diversity" and hence, by inference, the "quality" of an extensive body of water, one needs numerous sample-collections from scattered points within it. Each such sample-collection can be represented by its own diversity point; the diversity of the whole water-body is then represented by a swarm of diversity points, one for each sample collection, plotted in a diversity triangle. A clear picture of the precision of one's data is conveyed when they are shown as a swarm of points; nothing is gained by attempting to combine the data so as to make them yield some single diversity index.

Figure 3 shows two examples. Both diversity triangles show the diversity points of assemblages of benthic foraminifera in sediment samples from the sea-floor (Pielou, 1979). The swarm of points in the upper diversity triangle represents 134 sediment samples collected in shallow (< 25 m) water at high latitudes (within the polar circles); the swarm in the lower triangle represents 354 sediment samples from deeper water (> 100 m) at lower latitudes (between 45°N and 45°S).

The contrast between the two swarms is obvious on inspection. In the triangle representing warm, deep water, the points tend to be higher, that is, nearer to the upper vertex, at which $q_1 = q_2 = q_3$. The evidence suggests that foram assemblages from warm, deep water are more diverse than those from cold, polar, shallow water.

The null hypothesis that the two swarms do not differ can be easily tested, as follows. Divide each triangle into halves of equal area with a horizontal line, as shown in the figure. Count the number of points above and below the dividing line in each triangle. Then construct the following 2 x 2 table. The upper cell entries are the observed numbers of points in the categories defined by the row and column headings; the lower, bracketed entries are the expected numbers, assuming the null hypothesis.

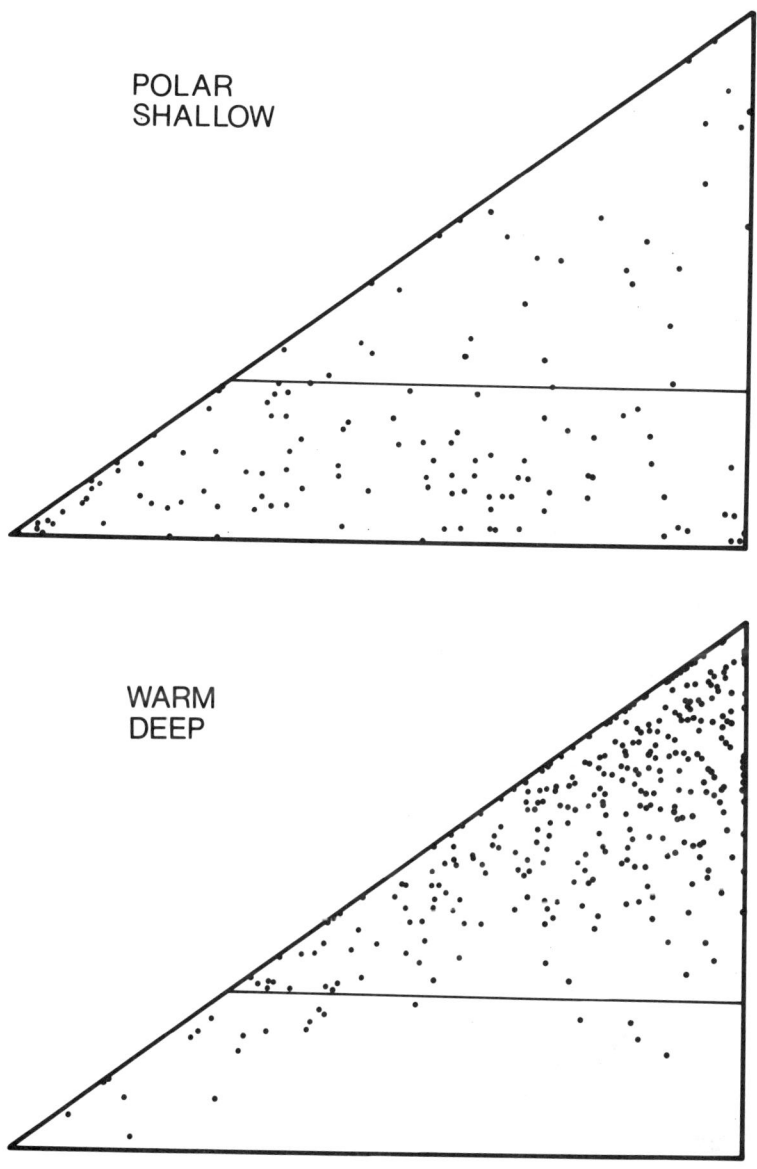

Figure 3. Diversity triangles showing the diversity points of assemblages of benthic foraminifera in sediment samples from shallow seas at polar latitudes (upper triangle) and deep seas at low latitudes (lower triangle). The horizontal line across each triangle divides it into upper and lower halves of equal area. (From Pielou, 1979).

		Sediment Source		Totals
		Polar shallow	Warm deep	
Position in Triangle	Upper	32 (100.2)	333 (264.8)	365
	Lower	102 (33.8)	21 (89.2)	123
	Totals	134	354	488

The test criterion for a χ^2 test, adjusted for continuity, is $X^2 = 250.3$. The evidence that foram diversity is greater in the warm, deep water is thus convincing: $P(\chi^2_{(1)} > 250) < 0.000001$. It accords, also, with the result yielded by using the number of species, S, rather than the diversity point, to measure each sample-collection's diversity (Pielou, 1979).

The wide scatter of the diversity points in the two triangles in Figure 3 shows how imprecise a diversity index based on only a few sample-collections is apt to be. For comparisons to be reliable, they must be based on numerous sample-collections; this would often be a herculean task, if one were to try to identify all the individuals, of every species, that the collections contained. Concentrating on the three most abundant species only, and ignoring the rest, has three big advantages: (1) it makes the work much quicker; (2) because rare species are excluded, taxonomic problems are evaded, and tallying a collection need not be done by an expert in the group concerned; (3) the three proportions used in defining a diversity point, since they are the largest, are therefore those with smallest standard errors.

LITERATURE CITED

("GPST" denotes the book referenced as Grassle, Patil, Smith and Taillie, 1979).

Dennis, B. and O. Rossi. (1979). Community composition and diversity analysis in a marine zooplankton survey. GPST:309-316.

Flôres, R.G., M.K. Barreto and H.R. Da Costa. (1979). On the relationship between ecological and physico-chemical water quality parameters: a case study at Guanabara Bay, Rio De Janeiro. GPST:195-206.

Grassle, J.F., G.P. Patil, W. Smith and C. Taillie (eds.). (1979). Ecological Diversity in Theory and Practice. International Cooperative Publishing House, Fairland, Maryland.

Lagoe, M.B. (1976). Species diversity of deep-sea benthic foraminifera from the Central Arctic Ocean. Geol. Soc. Am. Bull. 87:1678-1683.

Moncrieff, R. and S. Perrin. (1979). The effects of increased water temperature on the diversity of periphyton. GPST:207-217.

Patrick, R. (1961). A study of the number and kinds of species found in rivers of the eastern USA. Proc. Acad. Natural Sciences Philadelphia 113:215-258.

Patrick, R. (1963). The structure of diatom communities under varying ecological conditions. Ann N.Y. Acad. Sci. 108:359-365.

Patten, B.C. (1962). Species diversity in net phytoplankton of Raritan Bay. Jour. Marine Res. 20:57-75.

Pielou, E.C. (1979). A quick method of determining the diversity of foraminiferal assemblages. J. Paleontol. 53: 1237-1242.

Resh, V.H. (1979). Species diversity indices, and taxonomy. GPST:241-253.

Routledge, R.D. (1980). Bias in estimating the diversity of large, uncensused communities. Ecology 61:276-281.

APPARATUS FOR SOLUBILIZATION OF
CRUDE OIL IN SEAWATER

R. G. Kanter,
R. C. Miracle,
C. J. Foley,
M. L. Sowby
MBC Applied Environmental
Sciences, Costa Mesa, CA

ABSTRACT: As part of MBC Applied Environmental Sciences (MBC) commercial and sport fish oil toxicity studies for the Bureau of Land Management, MBC developed an apparatus for producing solutions of water soluble fractions (WSF) of crude oil. An important module incorporated into this system is a solubilizer design utilized by NMFS Tiburon Laboratory. This design maximizes oil-water contact and exchange of water soluble components of the crude oil into seawater. The Tiburon design, however, proved unviable for Santa Barbara crude oil which readily formed a "mousse" when mixed with seawater. Modification of the initial design and addition of an apparatus for separation of the "mousse" from seawater containing the WSF are presented. The apparatus allows for a greater diversity of crude oils to be tested and assists researchers toward determination of crude oil component effects on aquatic life.

MBC Applied Environmental Sciences is presently conducting a study for the Bureau of Land Management (BLM) through the Pacific Outer Continental Shelf Office. The study objective is to examine the biological effects of chronic low-level crude oil exposure on eggs, larvae and adults of selected California commercial/sport fish and shellfish. The experimental regime is designed to be realistic in terms of exposure concentration yet controlled with regard to the number of variables tested. In formulating their study plan, the BLM established four requirements:

1. the crude oil selected for experimentation should have a high probability of entering the southern California marine environment;

2. the experimental toxicant concentrations administered to test organisms should be realistic in terms of potential field concentrations;

3. the experimental exposure period should be long-term (up to 120 days);

4. the experimental regime should be flow-through.

These requirements necessitated a research and development program directed at the design and construction of a system that was capable of continually supplying water soluble fractions (WSF) of crude oil for continuous toxicant dosing.

Continuous dosing of single compounds (e.g. benzene) can be accomplished rather easily by the purchase and administration of stock solutions through a commercially available dosing apparatus. Working with single compounds presents fewer problems than those associated with complex hydrocarbon mixtures such as crude oil. In particular, differential solubilities of crude oil components, and the physical characteristics of the parent crude interact to create variables which must be controlled.

Conceptual design of a solubilizer for the BLM experiments required that:

1. the solubilizer must be capable of handling a viscous crude oil;

2. the solubilizer must be capable of providing a continuous supply of the WSF;

3. contamination of the WSF by emulsified oil droplets be controlled;

4. the WSF concentration produced by the solubilizer be sufficiently high to meet required experimental dosage levels;

5. the system be capable of extended operation with limited maintenance.

Initially, a solubilizer designed by the National Marine Fisheries Service laboratory at Tiburon, California [Nunes and Benville 1978], was constructed (Figure 1).

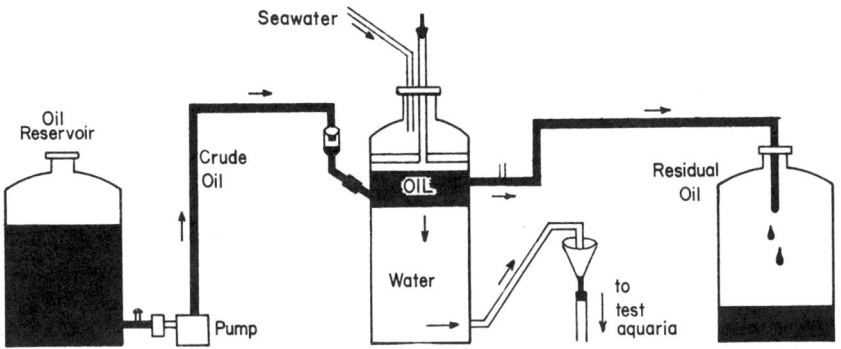

Figure 1. "Tiburon" solubilizer [Nunes and Benville 1978].

The "Tiburon" design generally consists of a glass tank in which a continuous layer of crude oil is floated over a constant water volume. Seawater is introduced through the top of the solubilizer where it disperses in droplet form from a perforated plate above the oil. These droplets then pass through the oil absorbing the WSF enroute. The WSF solution accumulates at the bottom of the vessel and is discharged through a line to a reservoir. The insoluble oil residue is continually removed through a discharge on the side of the solubilizer.

Use of the "Tiburon" solubilizer with viscous Santa Barbara crude (SBC) oil proved unsuccessful due to the formation of a "mousse" (i.e. apparently not a problem for Nunes and Benville with Cook Inlet Crude). The "mousse" limited the quantity and quality of the WSF that could be obtained and several modifications of the "Tiburon" solubilizer failed to solve problems associated with this "mousse" formation.

Because of these difficulties, a two-part solubilizer apparatus was designed by MBC (Figure 2). In this design the first stage provides for the initial mixing of seawater and crude oil, while the second stage separates the WSF from insoluble oil and the "mousse" residue.

The solubilizer description that follows is based on the use of Santa Barbara crude oil at 16°C (60°F). Oil and seawater flow rates and the resultant WSF test solution concentrations are oil and temperature dependent. Experiments with these variables can be conducted prior to using different oils.

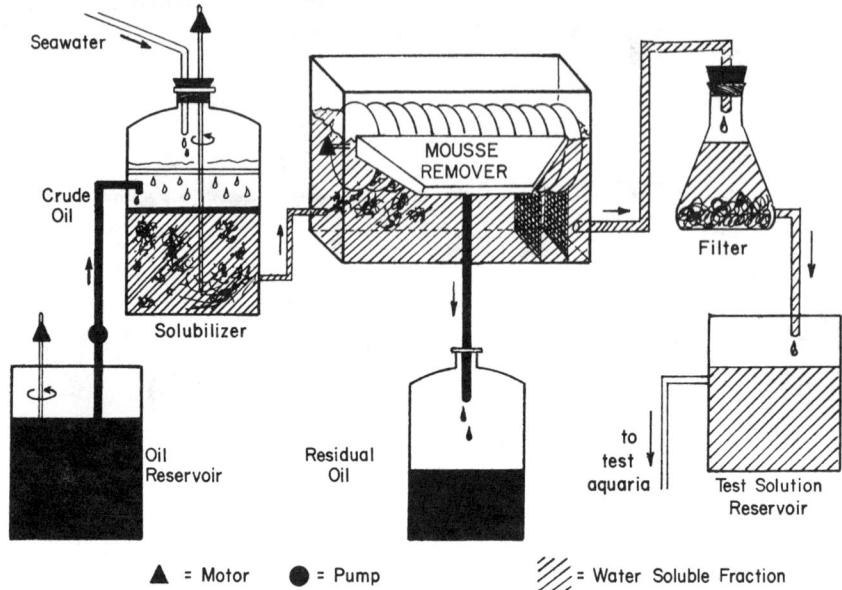

Figure 2. MBC crude oil solubilizer for continuous production of water soluble fraction dosing solutions.

Seawater and oil enter the MBC solubilizer at flow rates of 0.44 ℓ/min and 0.6 mℓ/min, respectively. As in the "Tiburon" design, water is dispersed by a perforated plate positioned above a layer of influent oil. The water droplets percolate through the oil absorbing WSF; and simultaneously a "mousse" is formed. A stirring assembly moving at 4 rpm aids in mixing the oil and water, and promotes accumulation of the "mousse" into globules. the WSF and "mousse" exit through an opening at the bottom of the solubilizer and enter the second stage "mousse" remover.

The "mousse" remover is modeled after commercially available oil spill cleanup equipment. It is composed of a rotating bank of discs which move at a speed of 6 rpm. As the discs pass through the oil-water interface, the "mousse" adheres to the discs. The discs then rotate through a comb which cleans them prior to the next revolution. Residual oil flows by gravity from the combs to a collector barrel. The WSF passes through a glass wool scrubber prior to entering a reservoir.

WSF solutions obtained from the Santa Barbara crude oil contained approximately 3 to 5 ppm total hydrocarbons. This stock solution was subsequently diluted with seawater to selected experimental dosing levels. Monitoring of the test solutions occasionally detected higher molecular weight hydrocarbons; however, since these components are not soluble at the ambient temperatures used, it is suspected that occasional oil micelles pass through the system.

The variable nature of crude oil necessitates regular monitoring of test solutions. Fluctuations in WSF concentrations can be minimized by observing the following precautions:

1. oil for a particular experiment should be obtained just prior to an experiment and in sufficient quantity;

2. oil should be refrigerated in closed containers to minimize loss of components and degradation by bacterial action;

3. the solubilizer system should be as "closed" as possible to reduce volatile loss; and

4. the system should be periodically cleaned to reduce contaminant accumulation on all surfaces.

Crude oils vary significantly in toxicity. The MBC solubilizer allows a greater diversity of crude oils to be tested in the laboratory. With this flexibility, geographic specific oils can be tested on endemic species, and more informed decisions can be made concerning hazards associated with each oil, oil related development and oil transport.

Literature Cited:

Nunes, P., and P. E. Benville, Jr. 1978. Acute toxicity of the water-soluble fraction of Cook Inlet crude oil to the Manila clam. Mar. Poll. Bull. 9:324-331.

PRELIMINARY CHARACTERIZATION AND
DETOXIFICATION OF TAILINGS POND
WATER AT THE SYNCRUDE CANADA LTD.
OIL SANDS PLANT

M. D. MacKinnon,

J. T. Retallack
  Environmental Affairs Dept.,
  Syncrude Canada Ltd.,
  Edmonton, Alberta

ABSTRACT

The chemical and physical properties of the tailings pond at Syncrude's oil sands extraction plant in Northern Alberta have been determined. The tailings pond is a complex system and the tailings area must be considered an environmental hazard. Syncrude is required to reclaim this pond or leave it as a viable water body. As part of this eventual reclamation, methods for the cleanup of the toxic waters in the tailngs pond are required. Various physical and chemical methods have been examined for the detoxification of these waters.

Preliminary results indicate that the tailings pond waters can be relatively easily rendered non-toxic, at least on a bench scale. The quality of the waters resulting from such treatments was significantly improved.

INTRODUCTION

Syncrude Canada Ltd. operates an oil sands plant in northern Alberta at Mildred Lake in the Athabasca oil sands deposit about 400 km north of Edmonton (Figure 1). The Syncrude project is based on open-pit mining. Construction began in 1973 and the plant became operational in mid-1978. Its design capacity is 129,000 barrels of synthetic crude oil per day. Over the 25-year mine life, more than one billion barrels of oil will be produced.

With the Syncrude process, production of synthetic crude from oil sands requires three basic steps - oil sands mining, bitumen extraction, and bitumen upgrading (Figure 2). Before mining, muskeg and overburden are removed. The oil sands are mined by draglines, piled into windrows and transported by conveyor belts to the plant where the bitumen is extracted using a caustic hot water extraction process. The extracted sand is transported as a 50% slurry to the tailings disposal area. The extracted bitumen is upgraded to synthetic crude which is sent by pipeline to Edmonton for refining.

Syncrude follows a "zero discharge" policy so that all process affected waters and liquid wastes are retained on the site within a tailings pond. The pond is designed to contain all aqueous wastes for 25 years of operation. The tailings pond also acts as a settling pond where the high solids content of the tailings is reduced by settling. The low solids water is then recycled to the plant for use in the extraction process.

The tailings pond is located in the northern part of Lease 17 in the former Beaver Creek valley (Figure 1). Local drainage patterns were altered so that most surface waters flowing towards the site were diverted either north along the West Interceptor Ditch or south through Beaver Creek Reservoir and the Poplar Creek Spillway to the Athabasca River. Since the tailings pond is built on relatively flat ground, a system of dykes is required to enclose it. The dykes are constructed of compacted tailings sand delivered as a slurry by pipeline from the plant (Nyren et al, 1978). The present and projected features of the tailings pond are summarized in Table 1.

The caustic hot water process used for the extraction of bitumen from the oil sand requires large quantities of water (Table 1). At full capacity, about 300,000m^3 of

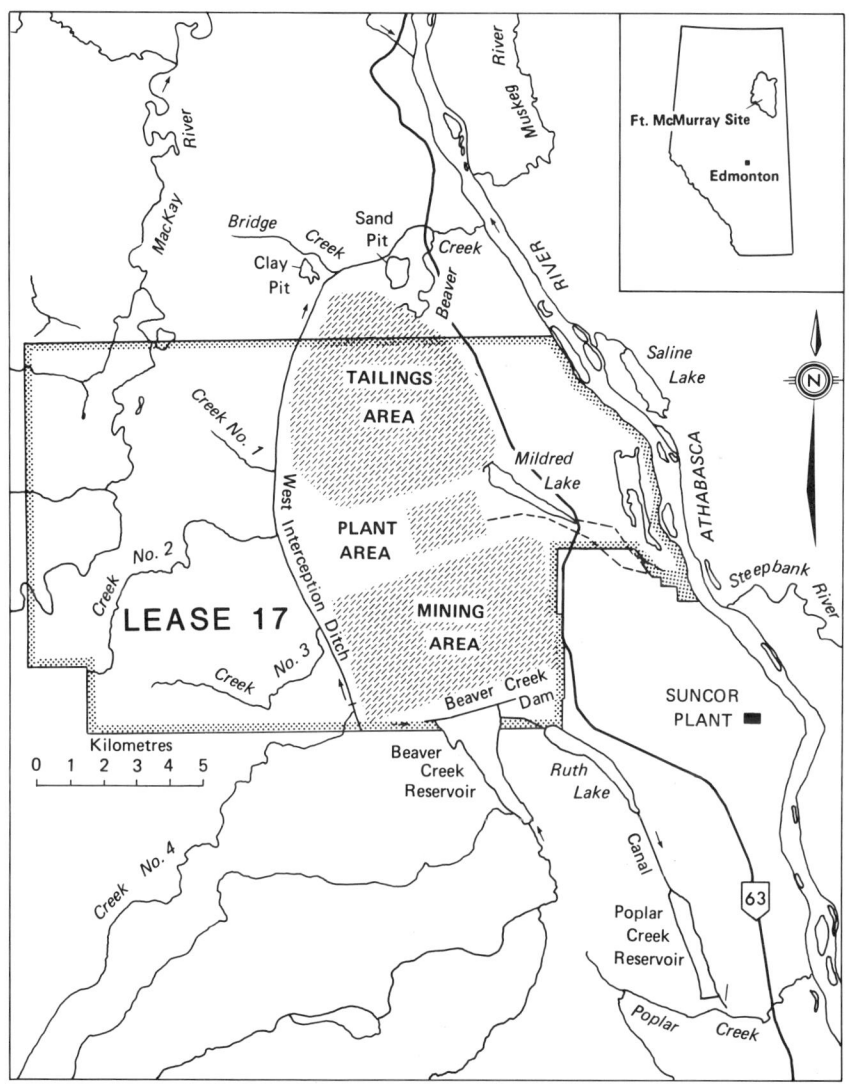

Figure 1: Surface drainage area at the Syncrude oil sands plant on Lease 17, north of Fort McMurray, Alberta.

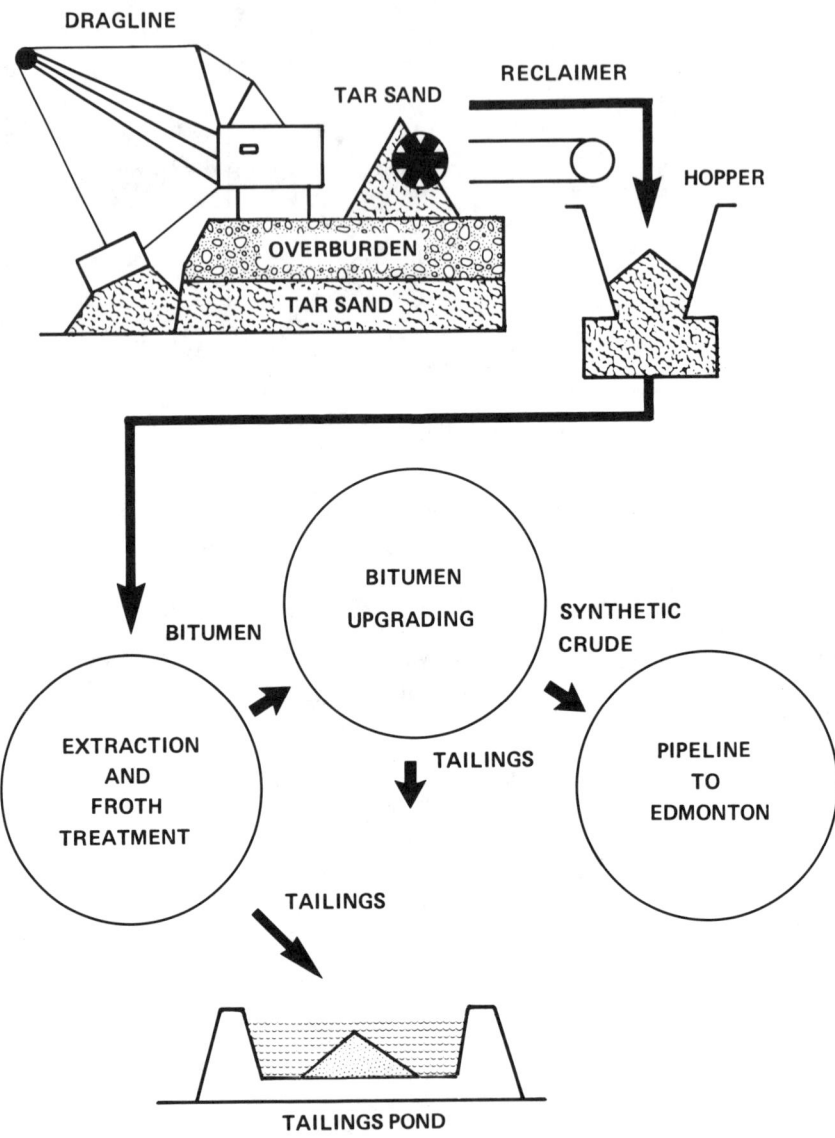

Figure 2: Outline of Syncrude's process for production of synthetic crude from tar sand.

water per day will be required. Process waters are drawn from two sources: raw water from the Athabasca River, and recycle water from the low solids zone of the tailings pond. Under present conditions about 60% of process water is recycle water from the pond. Eventually, this source will account for more than 75% of the water requirements. The tailings pond is expected to increase in volume at a rate of about $3 \times 10^6 m^3$ per month.

Table I. Physical Characteristics of Syncrude's Tailings Pond

	Present (1981)	Projected
Synthetic Crude Production	8.5×10^4*	1.3×10^5*
Tailings Pond		
• Area	15 km^2	28 km^2
• Volume	$150 \times 10^6 m^3$	$10^9 m^3$
• Water Elevation	310 m	344 m
- Mean	10 m	35 m
- Deepest	25 m	60 m
• Dyke Elevation	315 m	347 m
- Height	up to 50 m	up to 80 m
- Length	11 km	20 km
• Rate of Water Input		
- Raw Water from Athabasca River	$10^5 m^3$ day^{-1}	$10^5 m^3$ day^{-1}
- Tailings		
Water	$2.5 \times 10^5 m^3$ day^{-1}	$2.5 \times 10^5 m^3$ day^{-1}
Sand	$1.5 \times 10^5 T$ day^{-1}	$2.0 \times 10^5 T$ day^{-1}
Bitumen	1.5×10^4*	10^4*

* bbls per day

The tailings produced by extraction are delivered to the pond as a slurry of water, solids, and non-extracted bitumen (50:50:1 by weight) (Table I). As well as these materials, other wastes are added to the pond. These include process and laboratory chemicals, byproducts of upgrading, and cleaning solvents. The tailings pond contains a variety of potential environmental pollutants. It must be operated in an environmentally safe manner during the operational phase of the Syncrude project, and must be reclaimed after abandonment. Some present and potential environmental hazards include:

- Physical Hazard - floating bitumen poses a hazard to waterfowl in the area, particularly during migration.

- Chemical Hazard - water quality is poor in relation to natural waters. Thus, loss of contained waters must be prevented.

- Potential Groundwater Recharge - input into local aquifiers could result from the high elevation of the pond in relation to the local water table.

Under the "Conditions to Development and Reclamation Approval No. OS-1-78," Syncrude is required to conduct research into the reclamation of tailings disposal areas so that they can be reclaimed as a viable water body after abandonment or safely discharged into the surrounding surface waters. Current environmental management as well as the development of future reclamation options require that the chemical and physical properties of the tailings pond be understood. The distribution of variables within the pond and the effect of seasonal changes on the properties of the pond waters have been examined. The waters in the tailings pond are highly toxic to aquatic organisms, with 96-hour LC_{50} values of less than 10% for fish (rainbow trout and fathead minnows) and about 20% for the crustacean Daphnia magna. The EC_{50} values obtained for bacteria using the Beckman Microtox analyzer were about 25%.

In this paper, results of the distribution of properties within the pond and a summary of the general structure of the tailings pond are presented. The toxicity of this water has been examined and treatment for its detoxification by physical and chemical methods tested. Preliminary results indicate that on a bench scale, the highly toxic waters of the tailings pond waters can be relatively easily detoxified.

PROPERTIES OF SYNCRUDE'S TAILINGS POND

The properties of tailings pond waters, collected from various depths and locations throughout the pond, were determined during the ice-free periods of 1980 and 1981. On each sampling period about 7-13 stations were sampled. At each station, water samples for later analysis were collected at selected depths (for example: 0, 1, 2, 3, 5, 7, 10, 13, 16m, bottom) using a pump system. Temperature, pH and conductivity were measured with a water quality monitor at 1m intervals. The analytical scheme and methods are described in MacKinnon (1981). The variables analyzed included total and suspended solids, suspended bitumen, major anions, major cations, organic carbon, nutrients, total phenol and trace metals. Several of the tailings pond samples were analyzed in more detail by GC/MS and surveyed for priority pollutants using the 1977 U.S. Environmental Protection Agency protocol (Park and Maynard, 1980).

The results of the 1980 study averaged into depth zones (0-5, 6-9, 10-15, 15m) are summarized in Table II. During 1980, six complete surveys of the pond were conducted and over 300 samples were analyzed. In Figure 3, the depth profile of the mean suspended solids values and bitumen concentrations are plotted. Similar profiles were found on each of the sampling days. The solids and bitumen are added as a slurry to the pond (about 50% solids, 50% water, 1% unextracted bitumen). Low solids values (<1%) were observed in the surface zone (top 6-8m). This low solids, low density zone overlies a high solids zone (>3%) of high density, in the deeper waters below 10m. Between these two zones lies a well defined pycnocline, which is also the area where the thermocline of the pond is established. Some of the bitumen added to the pond forms into mats which are evident on the surface. While the density of bitumen is about 1.02 $g.cc^{-1}$, the bitumen added to the pond is less dense since it is hot (about 60°) and well aerated. These bitumen mats will float until deaerated. The residence time of floating mats depends on temperature and wind conditions.

The distribution of temperatures within the pond provides an indication of the structure of the pond and the dynamics of mixing processes. In Figure 4, the variation of mean temperatures averaged into depth zones of 0-9m and 10-15m over a 15-month period is plotted. Surface zone temperatures were highest in mid-July and fell steadily to about 4°C just before ice formation in December. During this study, the waters in the surface zone showed the widest range in values (3-22.5°) and reacted rapidly to air temperature changes. The water in the deep zone remained relatively warm through the whole period, with a much smaller range in temperature (9-16°) and slower changes. This is an indication that mixing between the zones is poor. As can be seen in Figure 4, the pond was thermally unstable (colder water over warmer) for much of

Table II. Mean Values of Variables Averaged into Depth Zones in Tailing Pond during April to November, 1980

Variable*	0-5 m	6-9 m	10-15 m	15 m
Total Solids %	0.72 ± .13	1.29 ± .68	3.16 ± 1.26	13.49 ± 9.43
Suspended Solids %	0.58 ± .13	1.18 ± .63	2.97 ± 1.50	12.44 ± 8.7
Bitumen	300 ± 120	710 ± 485	2100 ± 2590	13500 ± 18000
pH	8.10 ± .2	8.20 ± .2	8.20 ± .2	8.30 ± .3
Conductivity $S\ cm^{-1}$	1160 ± 225	1200 ± 150	1170 ± 136	1150 ± 100
Organic Carbon mg CL^{-1}	47.6 ± 8.1	51.0 ± 11.4	51.3 ± 10.4	57.5 ± 15
Phenol	0.28 ± .06	0.28 ± .04	0.30 ± .15	
Major Anions				
Chloride	77.6 ± 6.0	76.8 ± 7.1	73.9 ± 6.7	73.3 ± 11.0
Sulphate	195 ± 15.1	180 ± 23.3	142 ± 23	92 ± 53
Bicarbonate	452 ± 368	469 ± 38	496 ± 28	514 ± 50
Major Cations				
Sodium	282 ± 17	286 ± 18	280 ± 24	269 ± 31
Potassium	14.0 ± 6.1	17.4 ± 6.8	20.1 ± 9.3	26.6 ± 15
Calcium	6.30 ± .6	5.60 ± .9	4.60 ± .5	4.50 ± .86
Magnesium	4.50 ± .6	4.80 ± 1.0	4.90 ± 1.2	5.90 ± 2.4
Nutrients				
Nitrite & Nitrate mg NL^{-1}	0.07 ± .06	0.06 ± .07	0.06 ± .07	0.10 ± .12
Ammonia mg NL^{-1}	5.60 ± 1.5	5.50 ± 1.3	4.20 ± 3.0	
Phosphate mg PL^{-1}	0.12 ± .11	0.17 ± .18	0.18 ± .15	0.26 ± .20
Trace Metals				
V	0.028 ± .031	0.042 ± .03	0.038 ± .02	0.066 ± .05
Ti	0.297 ± .32	0.350 ± .26	0.319 ± .18	0.255 ± .17
Fe	3.080 ± 3.13	5.440 ± 4.21	7.490 ± 4.44	11.63 ± 9.10
Mn	0.051 ± 0.021	0.065 ± .05	0.069 ± .04	0.084 ± .05
Al	11.86 ± 13.9	20.75 ± 17.3	26.09 ± 16.4	36.65 ± 27.6
Cd	0.026 ± .03	0.034 ± .04	0.029 ± .02	0.034 ± .02
B	0.883 ± .21	0.870 ± .25	0.920 ± .22	0.970 ± .02
Pb	0.037 ± 0.03	0.046 ± .05	0.044 ± .03	0.054 ± .04
Zn	0.010 ± .02	0.012 ± .013	0.020 ± 0.03	0.025 ± .02
Mo	0.066 ± .04	0.094 ± .09	0.081 ± .04	0.084 ± .04

* Concentrations are in mL^{-1} unless shown otherwise.

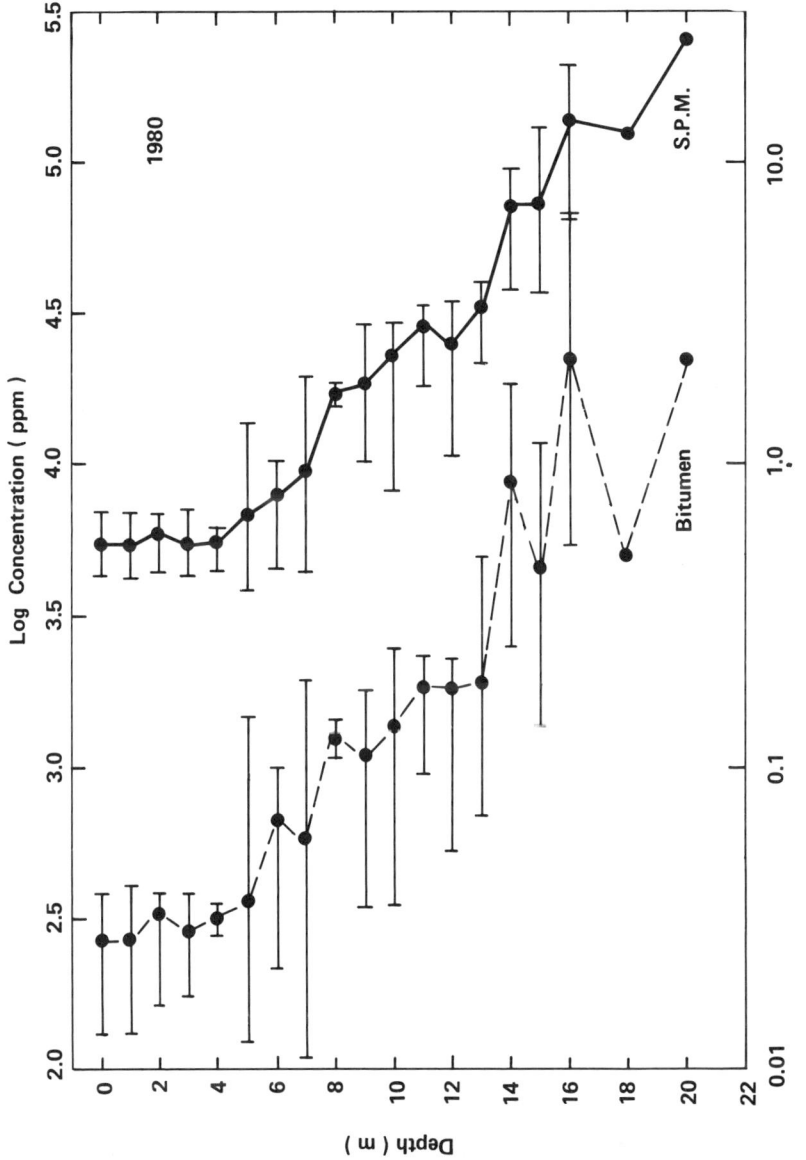

Figure 3: Depth profile of log concentrations of mean values of bitumen and suspended solids in tailings pond during 1980. Bars show ranges in values. Lower scale shows concentrations expressed as percent (w/w).

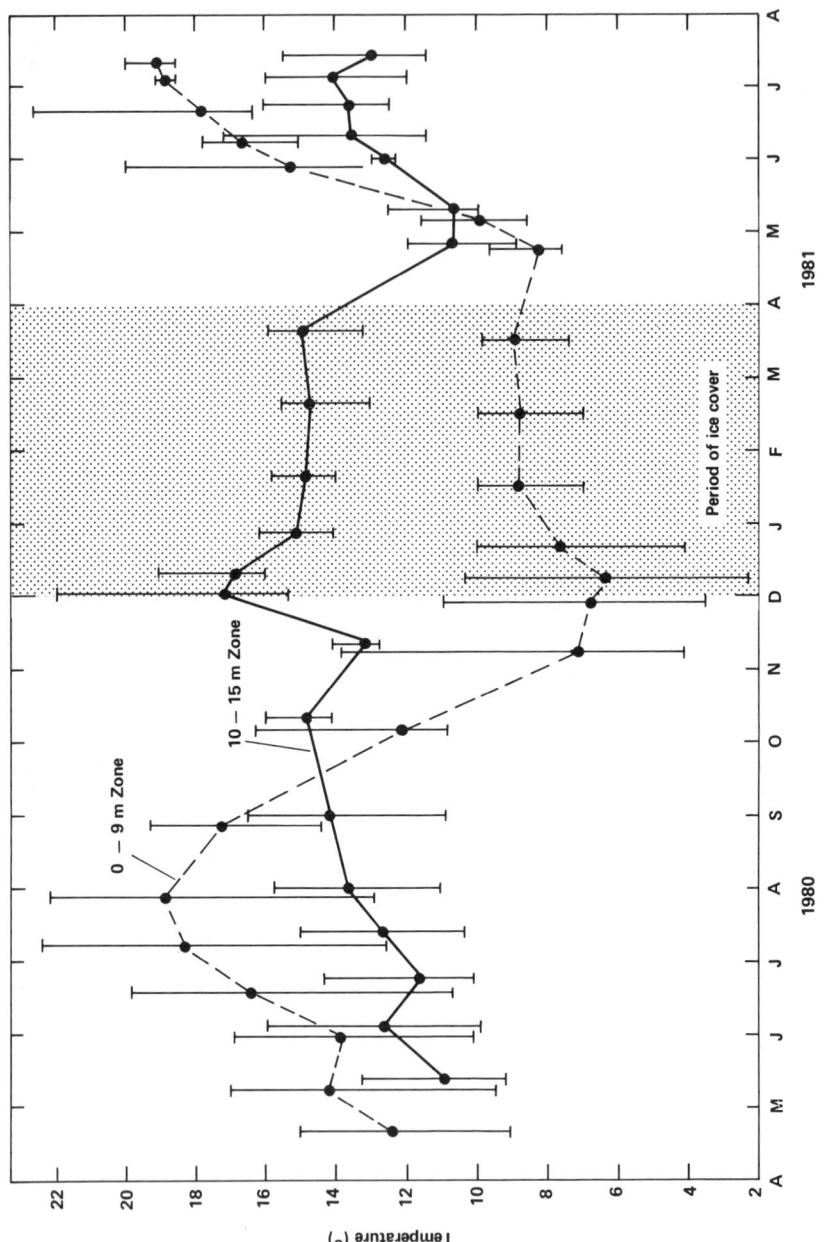

Figure 4: Trend of mean temperatures with time in the surface zone (0–9m) and the deep zone (10–15m) of the tailings pond during 1980–1981. Bars show ranges in values.

the year. However, the high density gradients set up by the high solids content of the deeper waters form a barrier which prevents complete mixing during the periods of spring or fall overturn. The tailings pond is acting as a heat reservoir and maintains higher temperatures during the year in relation to natural water bodies in the area.

Variations in the concentrations of variables such as suspended solids, bitumen, chloride, sodium and sulphate during the period of this study are shown in Figures 5 and 6. Some seasonal relationship in the SPM and bitumen values were evident, particularly during 1980 (Figure 5). However, the correlation between these measured concentrations and the plant extraction rates was poor. In the surface zone, sodium, chloride and sulphate, which originate from both the oil sands and process chemicals, show a steady increase in concentrations over the study period (Figure 6). These variables can be considered conservative, with little chemical, physical or biological interactions.

The composition of the dissolved organic fraction of tailings pond waters was determined by GC and GC/MS methods. Some of the organic compounds characterized in pond waters are shown in Table III. The acidic organic compounds, including phenols, acids and phenolic acids, make up the largest group of dissolved organics (Strosher, 1981). However, this group has not been well characterized. The hydrocarbon fraction in pond waters is a complex mix of normal, cyclic and unsaturated aliphatic compounds. The aromatics, which are made up mainly of alkylated and substituted benzenes with only traces of higher molecular weight polyaromatics, were only about 15-20% of the total hydrocarbon fraction. Tailings pond waters were surveyed for priority pollutants following the protocol of the U.S. E.P.A. (1977) (Park and Maynard, 1980). Other than phthalate esters, simple phenols, and traces of polyaromatic hydrocarbons, no measurable quantities of organic compounds on the priority pollutants list were reported. During this study, no halogenated compounds were identified in tailings waters.

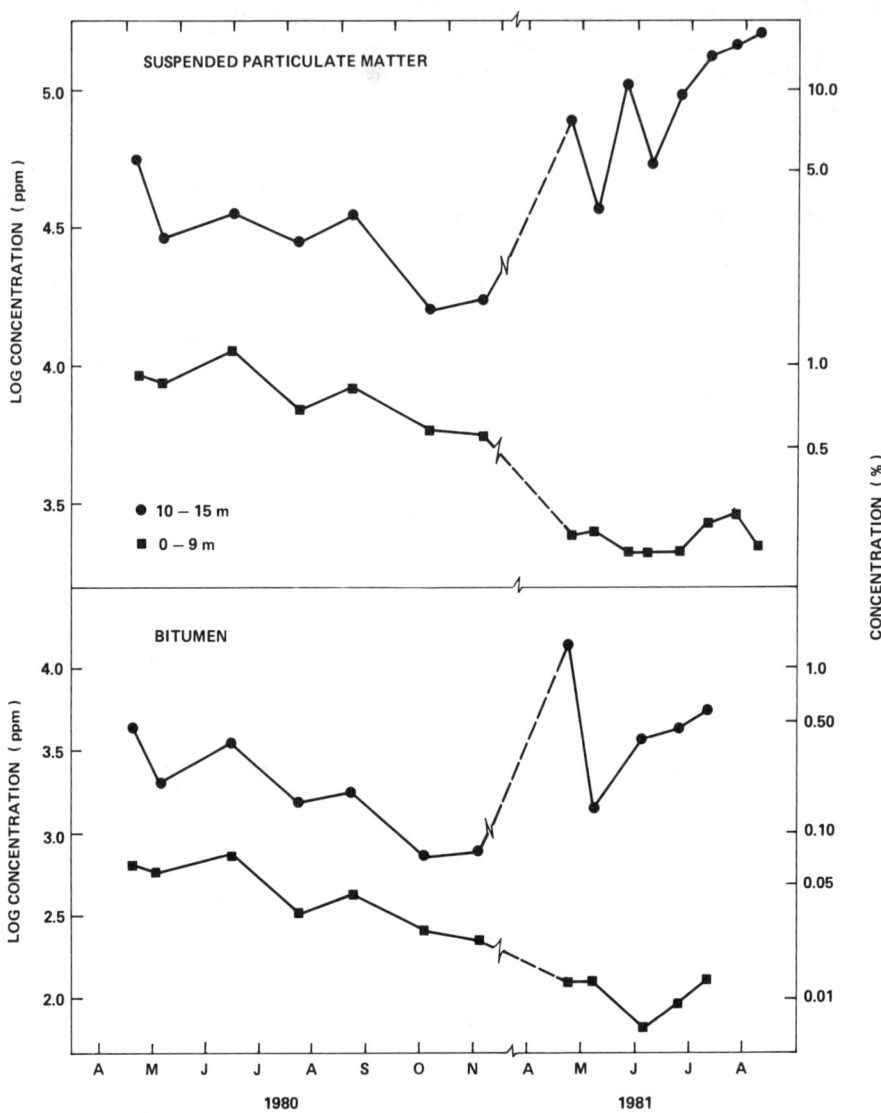

Figure 5: Trend of mean concentrations of suspended solids and bitumen with time in the surface (0–9m) and deep zone (10–15m) of the tailings pond.

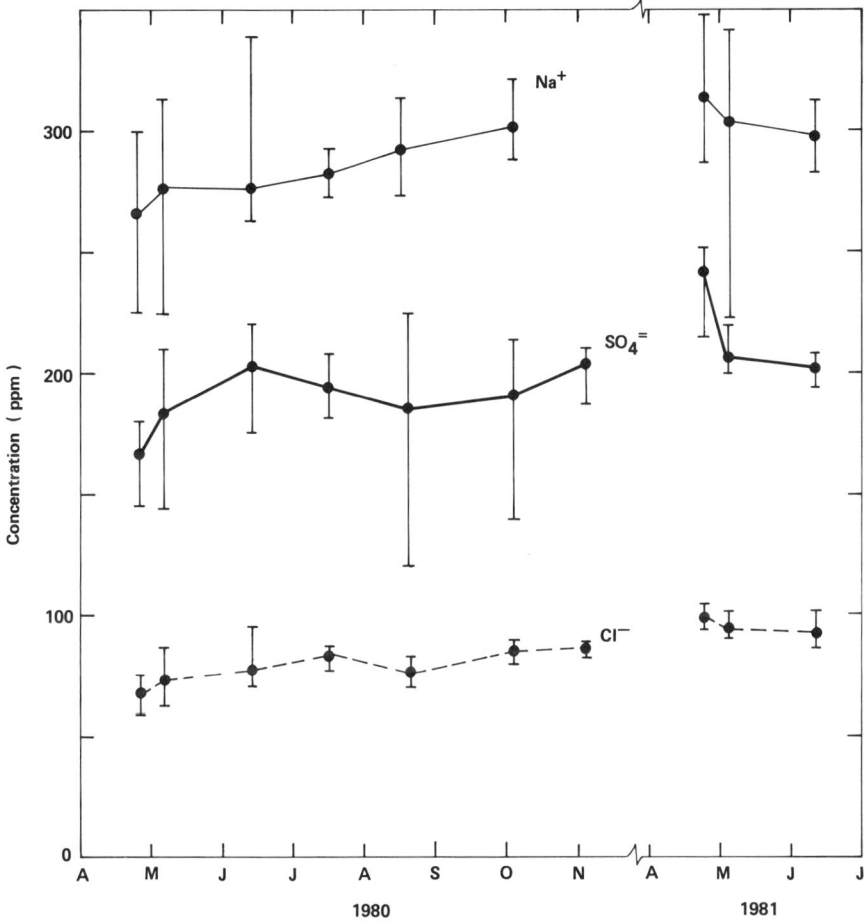

Figure 6: Trend of mean concentrations of sodium, sulphate and chloride with time in the surface zone (0–9m) of the tailings pond during 1980–1981. Bars show ranges in values.

Table III. Organic Compounds and Classes Identified in Tailings Pond Waters

HYDROCARBONS:	• aliphatic	- straight chain alkanes in nC_{16} to nC_{28} range - complex mix of substituted, unsaturated, cyclic and branched chain
	• aromatic	- simple, alkylated, and substituted benzens and toulenes - polyaromatics from naphthalenes to substituted benzanthracene
PHENOLS:	• low molecular weight	- simple and alkylated phenols
	• high molecular weight	- complex mix, including phenolic acids and polyphenols
ORGANIC ACIDS:	• most abundant organic group	- few individual compounds identified
OTHER CLASSES:	• asphaltenes	- make up about 10% of organic carbon
	• organic sulphur compounds	- complex mix - mainly thiophene substituted compounds
	• organic nitrogen compounds	- low concentrations
	• phthalates	- dibutyl and diethylhexyl phthalates

While trace metal concentrations in the tailings pond waters were generally high in relation to surrounding waters (Table II), many of the trace elements have concentrations close to or below the limits of detection with the methods used. The tailings pond waters have a high fines content, and incomplete removal of colloidal forms of trace metals such as Al, Si, Fe and Ti may explain the high concentrations found. The levels of trace elements on the EPA priority pollutants list (U.S. E.P.A. 1977) were low, often below detection levels. At present, the trace metal levels in the tailings pond do not pose a major hazard.

SUMMARY OF PRESENT CONDITIONS OF THE TAILINGS POND

As a result of this study, the present state of the tailings pond can be described as follows. The tailings pond is a complex system undergoing changes caused by internal interactions (physical, chemical) and external factors (seasonal, extraction rates and efficiency). When the result of these effects is simplified, a picture of a highly stratified water body with large gradients over depth is obtained. In Figure 7, a cross section of the tailings pond showing the depth zones and their relationships is schematically outlined. The condition and properties of the pond after three years of production can be summarized as follows:

- Size

  - area: about 15km^2
  - volume: about 150 - 170 x $10^6 m^3$
  - rate of growth: about 3 x $10^6 m^3$ per month
  - depth: up to 25m

- Structure of pond with depth

  - Surface - covered with slicks of oil. Oil films with thickness from several cm thick mats to monolayers are formed. These may influence potential evaporation of water from pond surface.

  - Surface Zone (0-6m) - well-mixed zone of generally low and uniform concentrations of variables. Waters have low solids content with relatively low density of about 1.002 g.cc^{-1}. Waters react quickly to seasonal changes and are usually isothermal.

  - Mid Zone (6-10m) - usually a well-defined zone where rapid changes in variable values are found. Thermocline and pycnocline are noted in this zone. Depth of this zone changes with season.

  - Deep Zone (10-15m) - low sludge zone with high solids content and high densities (mean densities greater than 1.02 g cc^{-1}). Mixing is slow and exchange with surface zone is poor. Subsurface bitumen mats are found at 11-13m. Temperatures remain relatively high through the year with little variability (9-16°).

  - Sludge Zone (>15m) - very high solids (up to 37%) and bitumen values. Water temperatures show little variability during the year.

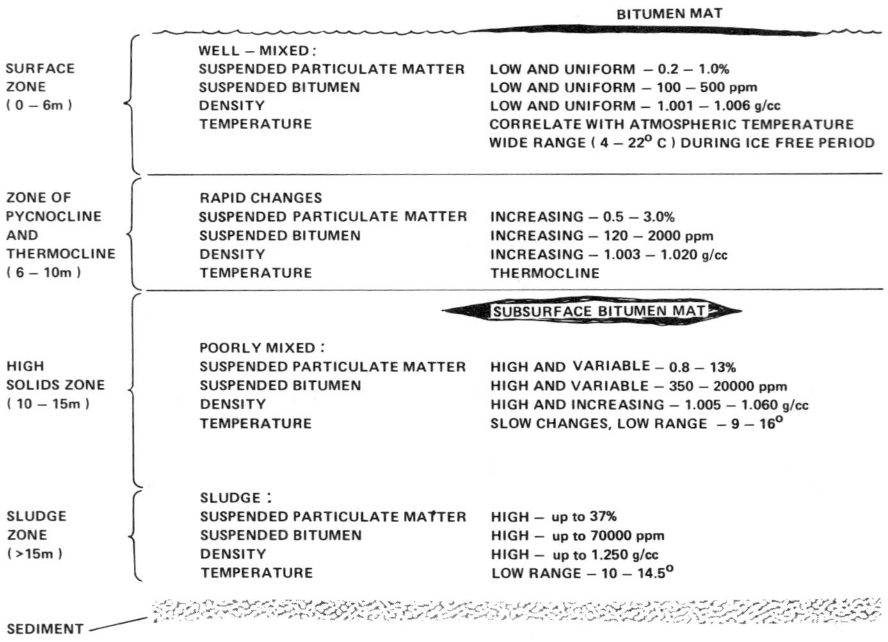

		BITUMEN MAT
SURFACE ZONE (0 – 6m)	WELL – MIXED: SUSPENDED PARTICULATE MATTER SUSPENDED BITUMEN DENSITY TEMPERATURE	LOW AND UNIFORM – 0.2 – 1.0% LOW AND UNIFORM – 100 – 500 ppm LOW AND UNIFORM – 1.001 – 1.006 g/cc CORRELATE WITH ATMOSPHERIC TEMPERATURE WIDE RANGE (4 – 22° C) DURING ICE FREE PERIOD
ZONE OF PYCNOCLINE AND THERMOCLINE (6 – 10m)	RAPID CHANGES SUSPENDED PARTICULATE MATTER SUSPENDED BITUMEN DENSITY TEMPERATURE	INCREASING – 0.5 – 3.0% INCREASING – 120 – 2000 ppm INCREASING – 1.003 – 1.020 g/cc THERMOCLINE
		SUBSURFACE BITUMEN MAT
HIGH SOLIDS ZONE (10 – 15m)	POORLY MIXED: SUSPENDED PARTICULATE MATTER SUSPENDED BITUMEN DENSITY TEMPERATURE	HIGH AND VARIABLE – 0.8 – 13% HIGH AND VARIABLE – 350 – 20000 ppm HIGH AND INCREASING – 1.005 – 1.060 g/cc SLOW CHANGES, LOW RANGE – 9 – 16°
SLUDGE ZONE (>15m)	SLUDGE: SUSPENDED PARTICULATE MATTER SUSPENDED BITUMEN DENSITY TEMPERATURE	HIGH – up to 37% HIGH – up to 70000 ppm HIGH – up to 1.250 g/cc LOW RANGE – 10 – 14.5°
SEDIMENT		

Figure 7: Generalized cross–section of the tailings pond, showing depth zones and their characteristics, at this stage of the pond's development.

- Chemical Properties

    - Particulates - includes suspended solids (fines and clays) and bitumen. Tailings from extraction processes are input to the pond as a slurry. Concentrations increase rapidly with depth to highest values in sludge zones.

    - Dissolved Organic matter - high concentrations were found. Characterization is difficult because of complex nature. Several classes of compounds have been identified. The levels of EPA priority pollutants are low and generally below detection levels.

    - Trace metals - Concentrations were variable and generally higher than natural waters.

    - Sodium, chloride, sulphate - In surface zone, seem to act as conservative species. Concentrations are steadily increasing; chloride about 25 ppm per year, sodium about 50 ppm per year, sulphate about 25 ppm per year. Major source to pond from the extraction process.

    - Nutrients - nitrite, nitrate and phosphate concentrations are low. Ammonia and bicarbonate concentrations are relatively high. Nutrients do not appear to be limiting to biological activity.

- Biological Properties

    - Toxicity - tailings pond waters are acutely toxic with LC 50 of less than 10% for rainbow trout by standard 96-hour bioassay.

    - Biological activity - except for microbial activity, pond waters appear to be abiotic (I. Forrester, pers. comm.)

    - Dissolved oxygen - surface zone waters show low levels of oxygen (about 20-30% saturated) while deeper waters below 10m are anoxic.

Considering the above conditions, the tailings pond must be considered an environmental hazard. However, the quality of waters contained in the tailings pond, at this stage of its development, has not deteriorated to a level where it cannot be treated and reclaimed.

Detoxification methods have been tested and the results of these tests indicate that, with the present condition of tailings pond waters, methods for their cleanup are available.

METHODS TESTED FOR THE DETOXIFICATION OF TAILINGS POND WATERS

Tailings pond water used for testing was collected from the surface zone. Physical and chemical treatment methods were evaluated for their effectiveness in the detoxification of this water. These treatments were chosen because they were methods known to be effective for the removal or reduction of the suspended solids (sand, clay, bitumen) from tailings waters (Baptista, 1972).

The treatments tested on tailings pond waters are listed in Table IV. The chemical composition of the waters before and after each of these treatments was determined using the methods described by MacKinnon (1981). Also, the effects of these treatments on the toxic character of the tailings waters were examined by bioassay methods (Nix and Salahub, 1980). An outline of the bioassay methodologies used in this study is given below.

- Static Fish Bioassays

    A 96-hour LC50 for fish (rainbow trout and fathead minnows) was determined using the guidelines established by Alberta Environment (1978). Test waters were diluted to five concentrations using dechlorinated, carbon-filtered tap water. Twenty litres of each dilution were aerated and cooled to 15°C before fish were added. Five fish were used in each dilution and the fish were observed for mortality over a period of 96 hours. Dissolved oxygen concentrations and pH were monitored daily, and dead fish were removed immediately after detection. Control tests (dilution water only) were conducted in each case.

Table IV. Physical and Chemical Treatments Tested for Detoxification of Tailings Pond Waters

Treatment	Procedure	Supernatant
Physical Methods		
1. Settling	Store undisturbed for about one month.	Non turbid but yellowish color.
2. Centrifugation	Centrifuge at 8000 RPM for ten minutes.	Non turbid but yellowish color.
3. Sand Filtration	Filter slowly through column of florists potting sand.	Turbid.
4. Aeration	Purge vigorously with air for 72 hours.	Turbid. Noticeable decrease in odor.
5. Combined	Combinations of 1-4.	Variable.
Chemical Methods		
A. Acidification	Adjust pH to less than 4 with sulphuric acid. After several hours settling, pH readjusted to 8 with NaOH.	Non turbid, colorless. Settling relatively rapid.
B. Calcium Oxide	Adjust pH to greater than 11 with CaO. After several hours settling, pH readjusted to 8 with H_2SO_4	Non turbid, but yellowish color. Settling is slower than with acidification.
C. Combined	Treatment A or B plus aeration of supernatant.	Non turbid. Volatile organics removed.

- Daphnia Bioassays

    Daphnia bioassays involved five dilutions of test water. These were set up in 200 ml beakers containing 100 ml of test water and ten Daphnia. Dissolved oxygen concentrations and pH were monitored daily and all tests were conducted at room temperatures.

    For all organisms, 96-hour LC50 values were determined by plotting the log of percent dilutions versus the percent mortality. The dilution value at which 50% of the organisms survived was recorded as the LC50.

- Bacterial Bioassay

The Beckman Microtox (Model 2055) Toxicity Analyzer is a bioassay procedure which measures bioluminescence of marine bacteria. The amount of light produced by the organism is an index of the state of their health. Addition of a toxicant results in a measureable reduction in light emission. Comparison of the initial light levels and the "affected" or resultant light levels at different concentrations of toxicant produces an EC50 value (EC50 is the concentration of effluent or toxicant in percent by volume causing a 50% reduction in light emission). In general EC50 values from the Microtox procedure are slightly higher than LC50 values produced in standard fish bioassays. However, they are similar to results obtained in bioassay tests using Daphnia magna.

RESULTS OF DETOXIFICATION EXPERIMENTS

Chemical Effects of Treatments

A summary of concentrations of variables in tailings pond waters before and after various treatments is shown in Table V. In most of these treatments of tailings water, a significant reduction in suspended solids was observed. Physical treatments such as settling and centrifugation resulted in the removal of the solids and bitumen fractions. The chemical composition of the remaining waters was not significantly altered. In the chemical treatments (based on low and high pH conditions), the suspended solids and bitumen fractions were almost completely removed. Also, the concentrations of nonconservative variables were reduced substantially. Some of these variables included dissolved organic carbon (30-35% decrease) and trace metals such as Ti, Fe, Mn, Al, and Pb (up to 100% removal). It appears that the treatment of the tailings waters by chemical means altered the suspended solids so that they became active surfaces for adsorption of hydrophobic species within the sample. After the physical and chemical treatments used in this study, no significant alterations (unless added by a reagent in the test) in concentrations of conservative variables such as sodium, chloride or boron in the resulting waters were observed. There were differences in the color of the waters resulting from the various physical and chemical treatments. In all treatments

Table V. Chemical and Physical Properties of Tailings Pond Water Before and After Various Treatments

Variables*	Untreated Tailings Water (Oct. 80)	Treated Tailings Water**						
		Physical Treatments				Chemical Treatments		
		1	2	3	4	A	B	
Total Solids (%)	0.63	0.13	0.12	0.49	0.65	0.16	0.12	
Suspended Solids (%)	0.52	0.01	.01	0.39	0.54	.01	.01	
Bitumen	210	5	5	130	190	5	5	
Dissolved Organic Carbon (mg CL^{-1})	54	49	45	50	48	33	32	
Major								
Chloride	83.0	8.10	81.5	84.0	91.5	85.0	86.0	
Sulphate	183	180	197	195	200	594	600	
Bicarbonate	535	480	600	630	610	55	65	
Sodium	280	270	296	287	301	325	350	
Potassium	23.2	22.2	27.1	19.6	42.1	20.5	23.9	
Calcium	6.1	9.3	7.4	13.3	6.7	63.3	67.6	
Magnesium	4.4	4.2	4.9	5.7	4.1	0.03	0.03	
Nutrients								
Nitrite & Nitrate (mg NL^{-1})	0.024	0.25	0.018	0.020	0.020	0.021	0.28	
Ammonium (mg NL^{-1})	7.8	3.9	7.2	7.0	7.6	7.8	7.7	
Phosphate (mg PL^{-1})	0.05	0.01	0.09	0.03	0.06	0.05	0.01	
Trace Metals								
Vanadium	0.003	0.000	0.020	0.000	0.007	0.017	0.017	
Titanium	0.200	0.071	0.369	0.215	0.233	0.000	0.010	
Iron	1.20	0.183	3.93	0.643	1.15	0.001	0.013	
Manganese	0.034	0.029	0.053	0.038	0.025	0.001	0.000	
Aluminum	4.23	0.37	16.6	2.32	4.30	0.081	0.069	
Lead	0.015	0.001	0.058	0.008	0.15	0.002	0.000	
Cadmium	0.010	0.001	0.034	0.005	0.010	0.002	0.001	
Boron	1.185	1.200	1.135	1.084	1.084	0.197	1.251	
Zirconium	0.015	0.000	0.036	0.007	0.009	0.003	0.000	
Copper	0.001	0.001	0.001	0.001	0.001	0.001	0.001	

* All concentrations in mgL^{-1} unless shown otherwise.
** Treatment numbers as described in Table 4.

except acidification, the resulting supernatant water displayed a distinctive yellowish or brownish tinge. With the acid treatment, the supernatant waters were not only particle free, but were also colorless. Thus, the colored material appeared to be adsorbed under acid conditions.

Effects of Treatments on Toxicity

Untreated tailings pond water is acutely toxic. Rainbow trout (Salmo gairdneri) displayed 96-hour LC50 values between 4.0 and 6.0 percent tailings water by volume. Fathead minnows (Pimephales promelas) were slightly more tolerant and showed 96-hour LC50 values of 6.0 to 8.5 percent (Figure 8). The crustacean Daphnia magna was a more tolerant test organism, exhibiting an LC50 value of between 16 and 27 percent. The results of the Beckman Microtox analyzer showed EC50 values for the bacteria of about 25%.

After the physical treatments (except sand filtration or aeration), most of the suspended solids (more than 95%) were removed. However, these treatment procedures were slow. The resulting supernatants were non-turbid but had a yellowish color. Only small decreases were noted in the toxicity of the treated water to fish (Figure 8). With combined treatments in which aeration was also used, a significant reduction in toxicity was observed. This may be an indication that the purgeable organics are a potential toxicant in these waters. Daphnia magna, a crustacean commonly found in northeastern Alberta, is considered to be more tolerant to toxins than fish. However, high sediment loads which in themselves are not normally toxic to fish, are acutely lethal to Daphnia. Therefore, those treatments that reduced suspended sediments resulted in a significant reduction in toxicity to Daphnia (Figure 8).

With all the chemical treatments, almost complete removal of suspended solids occurred in relatively short times. The time required for clarification was much shorter than with the physical treatments. The effect of the various chemical treatments on the water composition were similar and no significant differences in the levels of dissolved constituents were measured (Table V). However, differences in the toxicity of the resulting waters were noted. Treatments with calcium oxide alone resulted in only small reductions in toxicity to rainbow trout and fathead minnows (Figure 8). The combined treatment of calcium oxide and aeration resulted in only

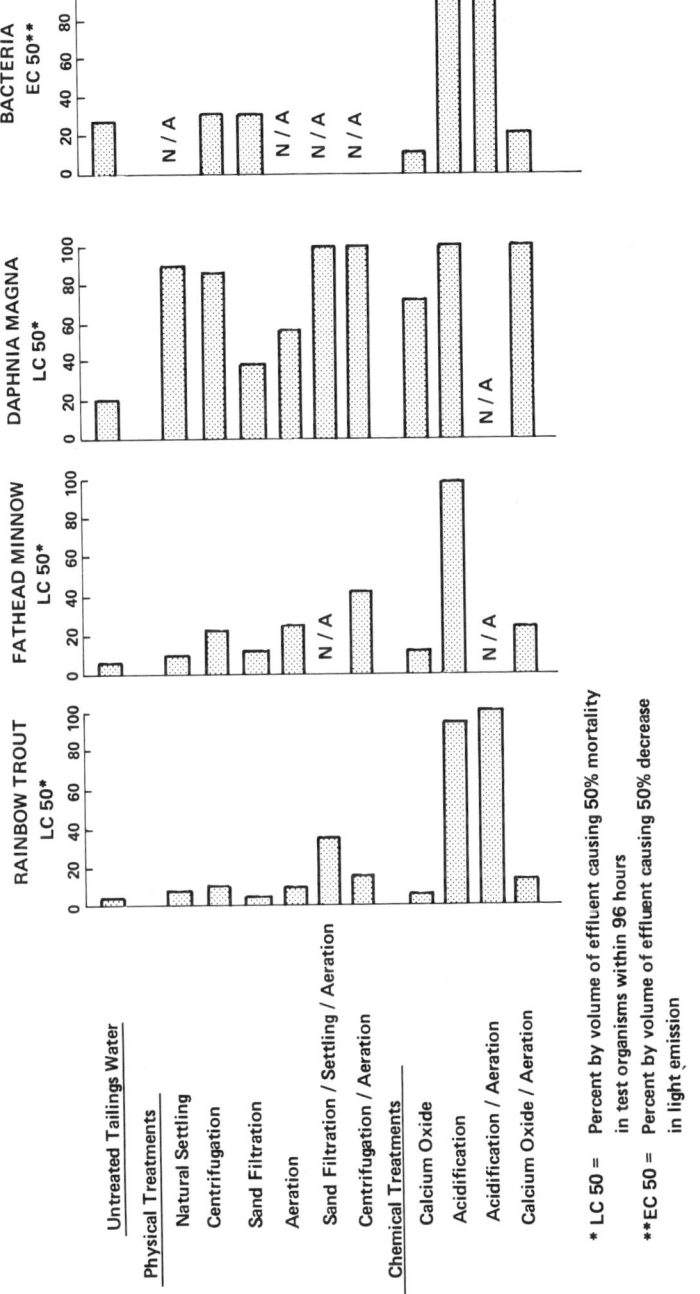

Figure 8: Effects of various chemical and physical treatments on toxicity of tailings pond waters.

minor improvement. In contrast, treatment by acidification resulted in the complete removal of toxic effects on both fishes. The _Daphnia_ and bacterial bioassays gave similar results to those with the fish species (Figure 8).

POSSIBLE CHEMICAL CHANGES AFFECTING TOXICITY

The physical treatments removed the solids fraction by a settling process based on density differences. In these treatments the "water-wet" character of the solids in the tailings pond water was not affected. No significant change in toxicity was noted using physical treatments. However, in the chemical treatments, this "water-wet" character of the solids was broken down and accelerated the settling of the solids. This should also have produced an increase in the surface activity of the suspended clays. The increased surface activity resulted in the adsorption of hydrophobic materials at both high and low pH. Similar levels of reduction of dissolved organic carbon (about 30%) was observed during both the acid and alkaline treatments. However, there should be qualitative differences in the actual compounds being adsorbed. Base-neutral compounds should be adsorbed at high pH (<10) while acid-neutral compounds are expected to be adsorbed at low pH (>4). Such processes would result in the selective removal of acidic or basic surface-active organics during the flocculation process. This may explain the differences in toxicity obtained by the fish bioassay with the acid and base treatments. The treatments with acid produced nontoxic waters while the treatment with CaO produced water that was still acutely toxic. This may be an indication that the toxic components that were removed are acidic organic compounds such as phenols or organic acids. Under low pH conditions, the acidic components will exist in a molecular form and would be more likely to be adsorbed by surface active solids. Another possible explanation for the significant reduction in toxicity after acidification is that the toxicant is extremely labile under low pH conditions.

CONCLUSIONS

The tailings pond at the Syncrude oil sands plant is a complex system undergoing many changes. It is growing rapidly and will ultimately contain about a billion m^3. The tailings pond water is acutely toxic to all organisms tested. The initial attempts at detoxifying tailings pond waters have been encouraging. Under laboratory conditions, tailings water can be rendered essentially non-toxic. Physical treatments resulted in only minor improvements in the water quality. Chemical methods which removed solids by flocculation produced mixed results. High pH treatments with calcium oxide resulted in little change in the toxicity while low pH treatments with sulphuric acid resulted in non-toxic waters. Flocculation under low pH conditions should lead to the adsorption of acidic and neutral organic compounds. Such organic compounds could be the possible major toxic component of the tailings pond water. While the tailings pond water must be considered an environmental hazard, it appears that at this stage of its development, reclamation options are available for treatment so that a non-toxic water of acceptable quality could be produced.

LITERATURE CITED

1. Alberta Environment; Pollution control Division, 1978. Waste water effluent guidelines for 96 hour multiple concentration static bioassay using rainbow trout. 4 pp.

2. Baptista, M.V., 1972. Water clarification and sludge disposal: a review. Syncrude Research Report. 17 pp.

3. Environmental Protection Agency, 1977. Sampling and analysis procedures for screening of industrial effluents for priority pollutants U.S. EPA. monitoring and support Lab., Cincinnati, Ohio.

4. MacKinnon, M.D., 1981. A study of the chemical and physical properties of Syncrude's tailings pond, Mildred Lake 1980. Syncrude Environmental Research Monograph 1981-1. 81 pp.

5. Nix, P. and R. Salahub, 1981. The toxicity of raw and treated tailings pond water. Prepared for Syncrude Canada Ltd. 28 pp.

6. Nyren, R.H., K.A. Haakonson and H.K. Mittal, 1978. Disposal of tar sand tailings at Syncrude Canada Ltd. pp. 54-74 in Proc. 2nd International Tailings Symposium. May, 1978. Denver, Colorado. 600 pp.

7. Park, J.M. and A.W. Maynard, 1980. Survey of contaminants in tar sands wastewaters. Report to the Environmental Protection Service, Water Pollution Control Directorate, Gov't. of Canada. 11 pp.

8. Strosher, M.T., 1981. Organic constituents of oil sands wastewater: An evaluation for possible organic tracer compounds to groundwater system. Draft Report for Syncrude Canada Ltd. 65 pp.

UTILIZATION OF GEOTHERMAL EFFLUENTS
TO CREATE WATERFOWL WETLANDS

V. W. Kaczynski
M. A. Wert
D. J. LaBar
 CH2M HILL
 Portland, Oregon

ABSTRACT

A generic research study was performed to determine the feasibility of using spent geothermal fluids to create waterfowl wetlands. Aspects studied included water quality, biology, ecology, toxicology, ground-water hydrology, geology and soils, wastewater treatment, economic, socioeconomic, and legal constraints.

Results indicate that some geothermal effluents can be used directly with no treatment to create waterfowl wetlands. Many geothermal effluents can be used to create wetlands with relatively minimal pretreatment; this category is economically more attractive than injection. The wetlands themselves will effectively further cleanse the effluents for possible cascading resource use (such as irrigation water or surface water enhancement). Finally, some effluents require extensive pretreatment before wetland use. Economics in this latter category favor injection.

INTRODUCTION

The U.S. Fish and Wildlife Service has adopted the goal of developing methods to enhance or mitigate fish and wildlife resources in the development of new energy technologies in rapid growth areas, primarily the western states. This position recognizes that western energy sources will be developed and used, and that an innovative approach to recycling of energy-related wastes must be found to achieve a positive ecosystem balance.

Many geothermal resources in arid western states have effluent with water quality as good as surrounding surface waters. In these arid states, even water of marginal quality has optimum value for fish and wildlife populations because of limited existing supplies.

Wetlands are dwindling in acreage across the United States, primarily because of dredge and fill activities. Wetlands are extremely important because they offer a wide variety of habitats for wildlife, especially waterfowl. Development of wetlands in arid states could have a considerable positive impact on wildlife populations by offering resting stops, forage, and cover during migratory periods, possibly attracting permanent populations.

Under sponsorship of the U.S. Fish & Wildlife Service, with funding through the U.S. Environmental Protection Agency, a study was performed to determine the potential to develop wetlands habitat to enhance fish and wildlife populations using spent geothermal process water. The project study area included the states of Idaho, Montana, Nevada, New Mexico, Northern California, Oregon, and Utah.

APPROACH

Evaluation of the feasibility of using geothermal effluents to create waterfowl wetlands involved a multidisciplinary phased approach, in which a number of tasks were completed in a logical sequence. An initial physical and chemical characterization of geothermal sources in each state produced a total of 203 individual geothermal sources in 91 geographic areas (Figure 1).

ECOLOGICAL CRITERIA

Ecological criteria were established that defined maximum concentrations of individual constituents in geothermal effluent that are acceptable for the development of a healthy wetlands ecosystem. These ecological criteria were applied to each geothermal source, resulting in the identification of 21 directly usable geographic areas, 47 potentially usable geographic areas, and 3 geographic areas requiring sophisticated pretreatment before use in a wetlands.

In developing such criteria, it was determined that two levels were necessary because of the wide variety of toxicity concentrations reported in the literature.

Directly applicable toxicity information is extremely limited, especially on waterfowl. Because of this, it was determined that conservative criteria should be used to define constituent levels known to be safe (Level 1). Level 1 was established at

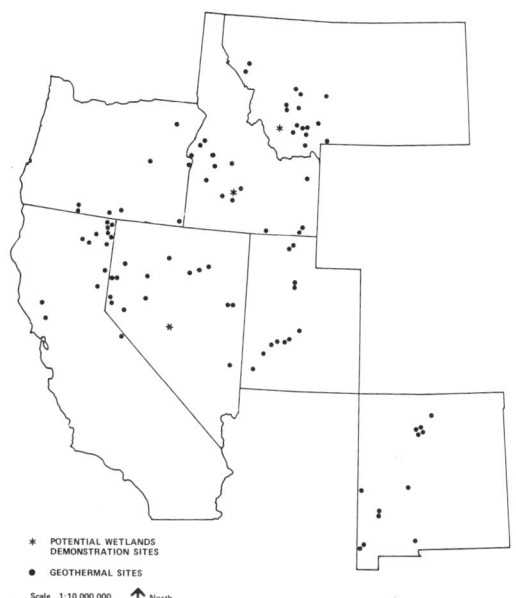

FIGURE 1
LOCATION OF INITIAL GEOTHERMAL SITES
IN SEVEN STATE PROJECT AREA/
LOCATION OF POTENTIAL GEOTHERMAL
WETLANDS DEMONSTRATION SITES

extremely low concentrations to be harmless to relatively sensitive organisms. Criteria are based on toxicity information contained in the traditional scientific literature (mostly invertebrate LD50 data, which are highly variable), irrigation standards, and water quality standards set for livestock. It is anticipated that these conservative criteria can and will be adjusted as more relevant information, especially on waterfowl toxicity, is obtained.

Level 2 criteria, levels that appear possible for waterfowl wetlands, were set at the highest concentration of each constituent that would still allow growth of desired marsh species (these include alkali bulrush, three-square bulrush, hardstem bulrush, Olney's bulrush, sago pondweed, spike-rush, and widgeongrass). Also used as a basis in establishing Level 2 criteria were water quality information at existing waterfowl refuges in the seven project states and information from natural geothermal springs (an average of parameter data from

23 springs in the seven project states). Natural wetlands processes appear to be able to treat Level 2 parameter concentrations sufficiently to cause no significant environmental degradation when the effluents leave the wetlands.

Concentrations of various constituents were established on the basis described above and are listed in Table 1.

Wetlands were found to exhibit potential for purification of wastewater and to be capable of tolerating high concentrations of heavy metals. The flora and fauna that will be able to use spent geothermal fluids will be determined mainly by the temperature and chemical composition of the waters. Waters with high total dissolved solids (TDS) and large quantities of toxic elements (such as mercury, cadmium, arsenic, and hydrogen sulfide) will probably not be able to support a productive waterfowl ecosystem.

ECONOMICS

Wetlands can be economically competitive with other methods of disposal of geothermal effluent, provided minimal or no pretreatment of the effluent is required. Non-site-specific, order-of-magnitude cost estimates were developed for wetlands development and operation and maintenance (Figures 2 and 3). Costs were developed based on other marsh development projects and conceptual layouts of systems for assumed site conditions. A 20 percent contingency and an 18 percent overhead, engineering, and administration allowance were added to the construction cost estimates.

Factors affecting costs include the quantity and quality of effluent, application rates, land area requirement, geographic location, land use, effluent requirements, and system design. Capital costs are presented both for enhancing existing wetlands and for creating new wetlands. Costs were also differentiated by whether or not discharge was allowed from the wetlands.

Also plotted are the costs for three systems that are probably near the high and low ends of the cost range for wetlands enhancement systems. The Incline Village General Improvement District (IVGID) project (CH2M HILL 1980) is expected to cost about $5.6 million for disposal of about 1,500 gpm of sewage effluent on 700 acres total area. For this project, the assessed land cost was $2,000 per acre; the system could allow no discharge of the effluent; springs on the existing wetlands had to be diverted off of the site; and extensive diking and storage for floodwaters had to be included. This project both enhances existing and creates new wetlands. These factors all

TABLE 1
PARAMETER SPECIFIC CRITERIA FOR HYDROTHERMAL EFFLUENT TO BE USED IN WATERFOWL WETLANDS

PARAMETER	LEVEL 1 (DIRECT USE) (ppm)	LEVEL 2 (POTENTIAL USE) (ppm)	AVERAGE OF 22 NATURAL GEOTHERMAL SPRINGS
TDS	1,000	15,000	6,500
pH[c]		6.0 9.0	
Ag	0.003	0.5	0.04
Al	0.01	13.0	2.5
As	0.10	2.0	0.04
B	4.0	25	4.2
Ba	1.0	13.0	1.5
Be	1.0	12.0	0.04
Bi	ISD[a]	ISD	—[b]
Br	2.0	20	—
Ca	200	500	542
Cd	0.01	0.1	0.003
Cl	500	2,500	2,400
Cr	0.05	0.5	0.07
Cu	1.0	1.0	0.1
F	5.0	100	—
Fe	1.0	1.0	0.03
H_2S	0.07	1.0	—
Hg	0.007	0.12	0.006
I	2.85	28.5	—
K	5.0	200	86
Li	0.26	2.6	1.3
Mg	25	300	13.0
Mn	0.05	1.5	0.03
Mo	0.5	1.0	0.1
Na	500	1,400	1,350
NH_4	0.13	2.0	—
Ni	0.4	4.0	0.09
NO_3	90	420	—
Pb	0.15	1.5	0.04
PO_4	20	60	0.2
Ra	ISD	ISD	—
Rb	IDS	ISD	—
Sb	1.0	10.0	0.04
Se	0.05	3.5	0.12
Sn	2.5	25	1.5
SO_4	150	1,000*	167
V	0.1	1.0	0.23
Zn	0.2	5.0*	0.04

NOTE:
*POSSIBLY HIGHER

[a]INSUFFICIENT DATA
[b]NO DATA AVAILABLE
[c]NOT IN ppm

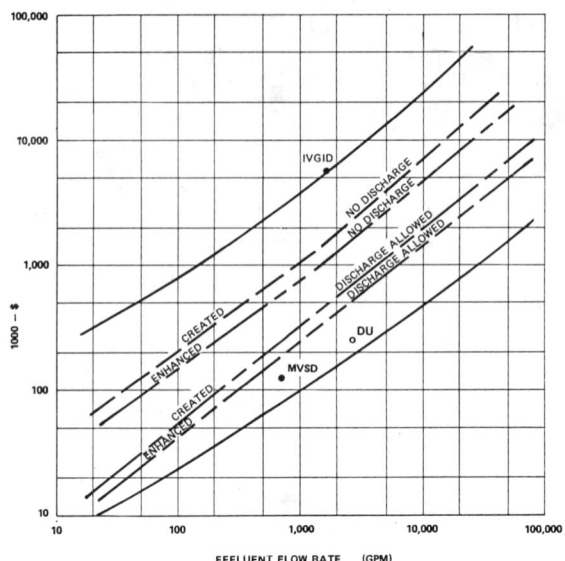

POSSIBLE RANGE IN COSTS
IVGID Incline Village General Improvement District
MVSD   Mountain View Sanitary District
DU   Ducks Unlimited, Pitt Marsh (Barnes 1980)

FIGURE 2
WETLANDS CONSTRUCTION COST

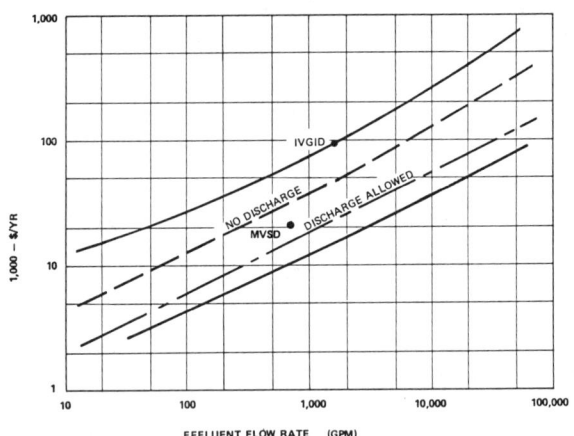

POSSIBLE RANGE IN COSTS
IVGID Incline Village General Improvement District
MVSD   Mountain View Sanitary District

FIGURE 3
WETLANDS OPERATION AND MAINTENANCE COST

tend to increase the cost of a wetlands enhancement system, placing this facility at the high end of the range for wetlands systems.

At the low end of the range is the Mountain View Sanitation District system at Martinez, California (Demgen and Nute 1979). Here, a much smaller volume (800 gpm) of effluent is discharged onto a previously owned parcel of land, and few water control structures are necessary. Discharge is allowed from this system. Also at this end of the price range is a 300-acre ($1,700/acre) wildlife enhancement project near Vancouver, B.C. This project utilizes an existing marsh (Barnes 1980).

As is apparent from Figure 2, the costs are about four times higher for a zero-discharge system compared with a system where discharge is allowed. This is predominantly because of the larger land area requirements (about five times). However, some economy of scale is apparent on the larger systems. Created wetlands will normally cost more than existing wetlands, mainly because of the need for more construction of dikes and water control facilities. This assumes that where existing wetlands are used, existing topography will allow the use of less diking. This also assumes that created wetlands will have to have their boundary areas controlled. If wetlands bounds are not a concern, diking costs will be substantially less but land costs could be substantially more. The cost estimates on Figure 2 must be viewed only as order-of-magnitude estimates. This information does indicate, however, that systems where discharge is allowed will be much less costly.

Operation and maintenance costs will also be greater for zero-discharge systems (see Figure 3). Again, this is because five times more land is required for a no-discharge system. Consequently, labor and maintenance costs will be considerably higher. The IVGID project cost estimate is shown as the probable maximum because of the complications of requiring diversion of existing onsite waters, flood control, and water storage. On very simple systems, where relatively small amounts of water are discharged into large marshes, operation and maintenance costs may be greatly reduced in comparison with the IVGID system.

INSTITUTIONAL CONSIDERATIONS

Federal and state regulations were examined to determine whether there are major stumbling blocks to wetlands development with geothermal effluent. No specific Federal laws or regulations exist for disposal of geothermal effluents. Governing criteria are state regulations relating to the quality of surface and ground waters. State water quality standards have not established allowable concentration levels for most constituents in geothermal effluents. As a result, direct surface disposal of geothermal wetlands drainage to streams and rivers may be possible in some cases.

Public acceptance of wetlands enhanced by geothermal effluents was examined and appears good. It was concluded that few socioeconomic constraints exist beyond those that apply to the initial development of the geothermal resource.

Two potential problems associated with use of geothermal effluent to develop wetlands are (1) resource conservation and (2) subsidence potential as they are affected by fluid withdrawal and nonrecycling to the geothermal reservoir. These potential problems are generic to all geographic areas designated for direct use. No site examined was found to be without these potential problems. The potential for surface-water migration to ground-water aquifers was evaluated for all sites of low to moderate subsidence potential. Ten sites appeared to be moderately favorable to favorable for use in developing a wetlands.

Availability of area for wetlands development, potential for waterfowl use, flyway patterns, and surface ownership were also examined. Three areas (Camas, Idaho; Beaverhead, Montana; and Nye, Nevada) were selected out of the 91 initial areas as having the characteristics most desirable for wetlands development.

CONCLUSIONS

The following general conclusions were reached as a result of the study:

-- Utilization of spent geothermal effluents to create or enhance waterfowl wetlands appears to be technically and economically feasible.

-- A general lack of directly applicable information (especially on waterfowl toxicity) exists in the traditional literature for defining effluent criteria for waterfowl wetlands. Therefore, criteria for directly usable areas had to be established very conservatively. As more information is developed, it may be possible to lower these levels to less stringent concentrations.

Potentially usable criteria are less stringent and are based in part on water quality parameter levels observed in western waterfowl refuges. These criteria are conservative enough that natural wetlands processes would probably reduce any pollutants to levels that would not significantly affect subsequent ground- or surface-water use.

-- A number of geothermal sites appear directly usable for wetlands creation/enhancement; many additional sites may be usable, especially with some pretreatment; some sites are not usable without sophisticated pretreatment.

-- Highly technical pretreatment is not economically feasible; sites requiring such treatment for wetlands use are currently (and in the foreseeable future) better off economically with subsurface injection.

-- Wetlands treatment (along with varying degrees of pretreatment) can probably remove enough contaminants to effectively reduce specific components of many geothermal effluents to a level compatible with irrigation requirements. Some nonstandard agricultural practices increase the potential for this cascading resource use.

-- General cost comparisons of subsurface injection versus wetlands construction indicate that injection is favored for cases of high effluent flows plus zero discharge requirements. If variances for the zero discharge could be secured, wetlands constructions would become economically attractive. Benefits from wetlands construction tend to counterbalance some excess wetlands costs (compared with injection). Site-specific analysis will be required to determine the overall benefit/cost comparisons on a case-by-case basis.

-- Legal and institutional constraints to wetlands utilization of geothermal effluents appear to be state-specific and relate to permitting activities. Section 318 (FWPCA), the aquaculture section, may provide an appropriate way to permit wetlands disposal.

-- The need for injection of spent geothermal fluid for resource conservation depends on the characteristics of the geothermal reservoir. Volcanic and vapor reservoirs are generally least dependent on injection for replenishment.

-- The potential for subsidence due to lack of fluid injection back to the reservoir depends on the geologic structure of the reservoir itself. The volcano-tectonic model and an all-liquid reservoir have the greatest potential for subsidence.

-- Local topography, evapotranspiration, and characteristics of the zone of aeration (i.e., thickness, composition, permeability, and soil type) can substantially alter the chemical composition of the effluent during downward migration from the surface to the aquifer.

-- Three areas appear to hold high potential for development of a geothermal wetlands. Several additional sites not in the initial screening may also prove practical, where commercial development is currently taking place.

-- Geothermal effluents and prospective wetlands sites (and wetlands responses) are variable, and each combination is unique. Potential applications will require site-specific studies. However, the economic, wildlife, recreational, aesthetic, and other possible benefits indicate that wetlands treatment is a feasible alternative to other methods of disposal.

ACKNOWLEDGEMENT

We wish to thank the U.S. Fish and Wildlife Service and the U.S. Environmental Protection Agency for critical review and financial support of this study. Work performed under the auspices of the U.S. Environmental Protection Agency for the U.S. Department of the Interior, Fish and Wildlife Service.

LITERATURE CITED

Barnes, A.B. 1980. Pitt Marsh: Bio-improvement proposal. Ducks Unlimited (Canada), Vancouver, B.C. 15 pp.

CH2M HILL. 1980. Draft wastewater management environmental assessment. Incline Village General Improvement District. Redding, California.

CH2M HILL, 1981. Utilization of Geothermal Effluents to Create Waterfowl Wetlands. Prepared for U.S. Fish and Wildlife Service, Fort Collins, Colorado.

Demgen, Francesca C. and J. Warren Nute. 1979. Wetlands aeration using secondary treated wetlands. Pages 727-739, *in* proceedings, Water Reuse Symposium, Vol. 1, March 25-30, 1979, Washington, D.C.

SOCIO-POLITICAL CONFLICTS
IN WILDLIFE MANAGEMENT ON COAL
LEASES IN NORTHEASTERN WYOMING

Mark S. Boyce
    Department of Zoology and
    Physiology,
    University of Wyoming,
    Laramie, Wyoming 82071

ABSTRACT

   Conflicting demands for land and wildlife resources have created several difficult management problems in areas experiencing energy resource development. We review three case studies. First, sagebrush reclamation on mined surfaces is highly beneficial to wildlife but may reduce agricultural productivity. Second, industry may desire high densities of highly visible pronghorn antelope because of improved public relations. Yet, the Wyoming Game and Fish Department may manage to sustain the herd at much lower densities because pronghorns create agricultural damage which must be reimbursed to the landowner from Game and Fish funds. Third, recreational uses of wildlife for hunting, fishing, camping, etc. may suggest much different reclamation strategies than an ecologically-oriented attempt to restore mined surfaces to former conditions. Reclamation for recreational users will greatly increase the public visibility of reclamation efforts but may not meet the objective of restoring the ecological integrity of mined surfaces. Some of these conflicts cannot be reconciled, but hopefully others may be resolved by compromises entailing management for "multiple-use" objectives.

Since 1977, we have been involved in research on pronghorn antelope (Antilocapra americana) and sage grouse (Centrocercus urophasianus) ecology on ARCO Coal Company coal leases in Campbell County, Wyoming. The principal focus of our studies has been to formulate mitigation procedures to minimize the effects of surface mining for coal on sage grouse and pronghorns, and to develop recommendations for effective reclamation of mined surfaces for these wildlife species. In the course of these studies, we have encountered socio-political issues which will probably result in compromising efforts to manage mine surfaces most effectively for wildlife. In this paper, we review three such socio-political conflicts of significance throughout the coal-rich Powder River Basin in northeastern Wyoming.

CASE STUDIES

Sagebrush Reclamation

Habitats in northeastern Wyoming are characterized by rolling sagebrush grasslands dominated by big sagebrush (Artemisia tridentata). Big sagebrush is extremely important habitat component for several native species of wildlife, particularly pronghorn antelope and sage grouse. Sagebrush has a higher nutrient content than most other available forage during winter and is seasonally utilized extensively by pronghorns and sage grouse.

Several studies (e.g., Bayless 1969; Beale and Smith 1970) have demonstrated that winter forage of pronghorn antelope is composed largely of shrubs and especially sagebrush. Autenrieth (1978) notes that browse plants, such as sagebrush, are essential to ensure the survival of pronghorns during severe winters. Martinka (1967) observed that winter deaths due to malnutrition were common on grasslands and that survivorship was higher on sagebrush ranges.

Sage grouse are extremely dependent upon sagebrush for forage and cover. Winter food habits of sage grouse usually consist of 100% Artemisia tridentata (Patterson 1952). Boyce (1981) demonstrates that areas without sagebrush are not utilized by sage grouse during any season of the year. In fact, the extensive declines in sage grouse populations among several western states have been attributed directly to increased agriculturization and concomitant sagebrush reduction practices (Braun et al. 1977).

Colenso et al. (1980) have argued that effective reclamation of sage grouse habitat can only occur if sagebrush is restored. Similarly, reestablishing sagebrush on mined surfaces will enhance the winter survival of pronghorns and probably mule deer. Technologies for reclamation of sagebrush are currently being developed. Although the procedure is rather expensive, Native Plants, Inc. (Salt Lake City, Utah) will guarantee at least 50% survival of their big sagebrush seedlings. Some mines in Campbell County, Wyoming are planning to transplant sagebrush plants to reclaimed surfaces from lands scheduled for future mining.

Unfortunately, ranchers in northeastern Wyoming frequently find sagebrush reclamation very offensive. Sagebrush clearly competes with grasses and forbs for water and soil nutrients, (Boyce 1981) and due to high content of essential oils, it is not usable as forage by domestic livestock (Barber 1968). Removing sagebrush by burning, chaining, digging, or spraying with an herbicide enhances grass production and soil moisture retention. Generally the more intensive the agricultural development, the greater the probability that sagebrush will have been removed or controlled.

The conflict is a clear one. Wildlife interests unquestionably favor the reestablishment of sagebrush on mined surfaces to hasten the reclamation of habitat preferred by native wildlife. Ranching interests do not favor the planting of sagebrush since it is largely viewed as a weed or pest species. On public lands, restoration of wildlife habitat is presently required by the Wyoming Department of Environmental Quality. On private lands, e.g., those owned by the coal company, industry must determine whose interests are to be best served by their reclamation effort.

Is a compromise management strategy possible? Many wildlife biologists argue that sagebrush can actually benefit domestic livestock range and that utilization of range for livestock and wildlife is certainly possible and even desirable. Sagebrush is argued to benefit livestock range because: (1) sagebrush can create snow drifts which increase growing season soil moisture and thus enhance forage growth, (Sturges 1975) and (2) sagebrush protects grasses from snow-pack and thus forage growing among sagebrush plants stays upright during winter and is thus more available to foraging livestock. Unfortunately the merits of sagebrush on livestock range are not always clear. The advantage offered by catching snow is partly offset by the increased transpiration of soil moisture by sagebrush plants which are not eaten by livestock (Sturges 1980). Although sagebrush may make grasses more available during winter, many ranchers only range livestock seasonally and provide winter hay.

Advantages to ranchers from sagebrush reclamation only become apparent when monetary returns from wildlife are realized. Scott and Terrel (1978) have reviewed an intensive pronghorn management scheme which could yield as high a profit from an animal unit of pronghorns as from cattle. Many view fee hunting as an undesirable alternative, but if fee hunting on private lands results in the restoration and maintenance of quality wildlife habitat, it seems to me worthy of consideration. Management for pronghorn habitat will also provide habitat for many other game and non-game species of wildlife.

Pronghorn Population Management

In general, all organisms experience some upper limit to population growth. Pronghorn antelope habitat can only support a maximum of K pronghorns. A simple model which describes density dependent population growth is the familiar logistic,

$$\dot{N} = rN(1-r/K)$$

where r is the intrinsic rate of increase, N is population size and $\dot{N}$ is the first derivative of N with respect to time (dN/dt). The logistic is invariably an over-simplification of the real world. We expect that rather than $\dot{N}/N(N)$ linear, more realistically $\dot{N}/N$ is some concave function of N (see Pielou 1977). Also, stochastic fluctuations in fertility and survivorship certainly occur (Boyce and Tate 1979), and such effects considerably complicate the dynamics of the system (Boyce and Daley 1980). Yet, the simplicity of the logistic makes it a valuable heuristic tool for understanding the general consequences of harvest policies (Clark 1976).

To model a harvest of pronghorns we consider a constant harvest rate, E, yielding the following model:

$$\dot{N} = rN(1-\frac{N}{K}) - EN$$

An equilibrium exists wherever $\dot{N} = 0$ and we find that for any E < r, equilibria exist along a parabola defined by

$$N' = K(1-E/r)$$

where the sustainable yield or harvest is

$$EN = KE(1-E/r).$$

From Figure 1 we see that an equilibrium sustainable yield exists where EN intersects the parabola of $\dot{N}$. Clearly maximum sustained yield (M.S.Y.) occurs at K/2. But, for any N from 0 to K, a harvest rate, E, exists whereby an equilibrium harvest and population can be maintained.

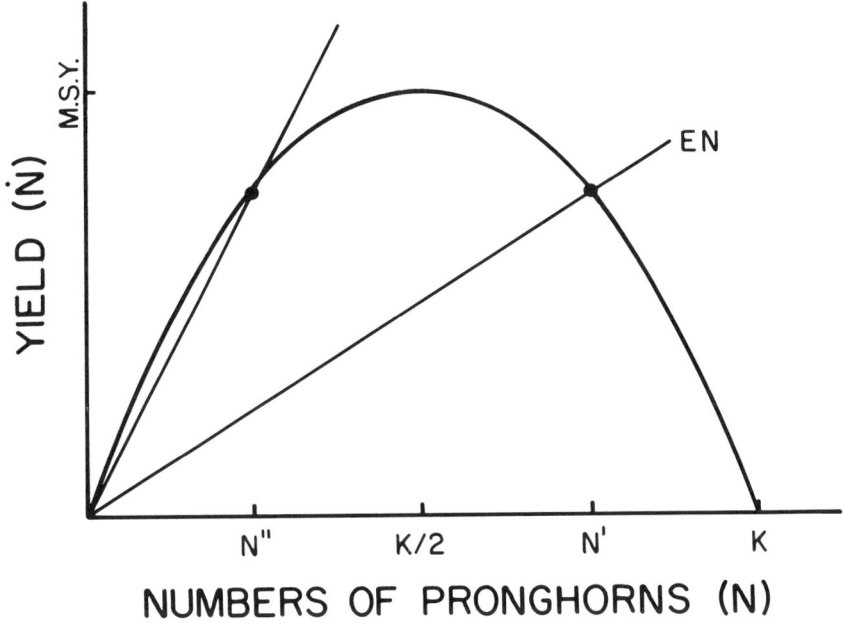

Figure 1.

Logistic population growth with two harvest rates, E. Maximum sustained yield occurs at N = K/2.

If yield of animals, say for meat, was the sole objective of a management policy, clearly M.S.Y. is the optimum. However, other criteria than yield usually shape the "optimum sustained yield" (O.S.Y.) which may be either greater or less than M.S.Y. For example, industry may have O.S.Y. density considerably greater than that for M.S.Y. because pronghorns are highly visible wildlife with considerable public relations value. Visitors to a mine site where pronghorns are abundant are immediately given the impression of minimal environmental impact. Agricultural interests, in contrast, may have an O.S.Y. density considerably below that for M.S.Y. since pronghorns compete for forage and water with domestic livestock, even though overlap in food habits may be quite small. Thus, with reference to Figure 1, mining industry may have an O.S.Y. density at N´ whereas agricultural interests may have an O.S.Y. density at N´´ or less.

Ranching is a dominant industry in the state of Wyoming and as a consequence, there exist state regulations promoting agricultural interests which inadvertently shape some of the policy and management decision-making of the Wyoming Game and Fish Department. Most importantly, the regulations which influence pronghorn management policy relate to (1) agricultural wildlife damage claims against the Wyoming Game and Fish Department's damage fund, and (2) landowner coupons which are designed to reimburse landowners for the use of their range by wildlife. Damage claims are paid by the Wyoming Game and Fish Department from a special fund created from the sale of hunting licenses. Landowner coupons are attached to every hunting license and are to be given to the landowner by the hunter upon killing an animal on his/her land. The landowner can then collect $8.00 for each pronghorn taken on his/her property.

The Powder River Basin coal seam falls mostly within pronghorn hunt unit #24 and within the larger Pumpkin Buttes herd. The combined damage payments from landowner coupons and claims for the period 1974-1980 from the Pumpkin Buttes herd exceeded those from any other area in the state. Total payments 1974-1980 were $121,726 for the Pumpkin Buttes herd. The principal reasons that the Pumpkin Buttes herd generates such large damage costs are (1) high population density of the herd (approximately one pronghorn per 15 acres (Scott and Terrel 1976)), and (2) approximately 80% private land within the herd unit.

Table I.

Statewide pronghorn antelope damage claim payments made by the Wyoming Game and Fish Department, 1974-1980.

Fiscal Year	Damage Claim Payments
1974	$ 960.
1975	4,912.
1976	4,586.
1977	4,615.
1978	8,999.
1979	17,335.
1980	21,927.

To reduce the damage claim costs incurred by the Wyoming Game and Fish Department, the optimal strategy is clearly to maintain an O.S.Y. density below that for M.S.Y. In a news release earlier this year, the stated intention of the Wyoming Game and Fish Department was to reduce the herd size by one third to reduce damage claims. The principal vehicle for achieving such an increased harvest is to increase hunting permit quotas. A one-third herd reduction seems quite unlikely in the immediate future because with increased license fees initiated in 1980, non-resident license sales have decreased rather dramatically. Note in Figure 2, that in 1980 the total kill is considerably less than the number of permits offered for sale. This is mostly due to poor sales since hunter success rates have remained very high (Figure 3).

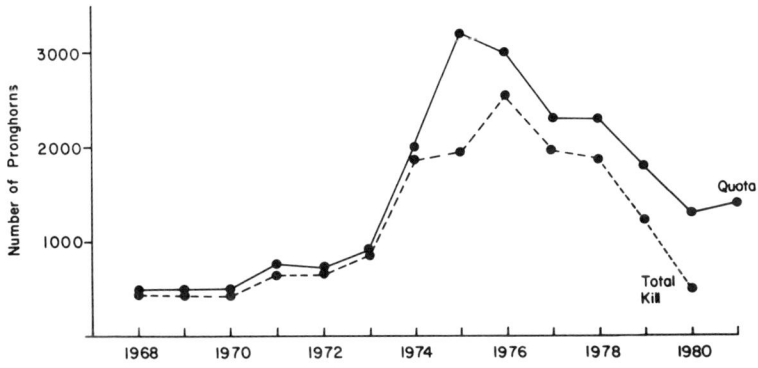

Figure 2.

Pronghorn antelope hunting permit quotas and harvests from hunt units 24 and 101 in Campbell County, Wyoming.

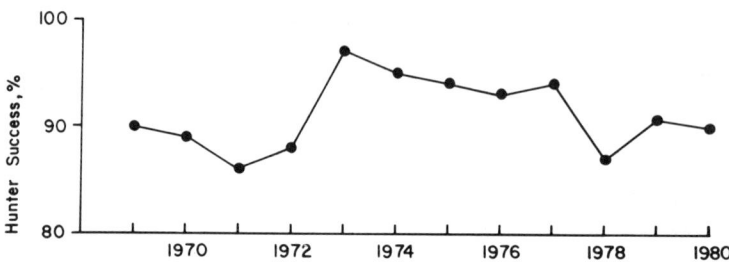

Figure 3.

Pronghorn hunter success in hunt units
24 and 101 in Campbell County, Wyoming.

From Figure 2 it is apparent that the Wyoming Game and
Fish Department's harvest policy in Campbell County has not
yet stabilized although the intention appears to be an increased exploitation rate. This may be undesirable from an
industry perspective. High pronghorn density creates good
public relations, but in addition, mining companies are expected to collect population trend statistics according to
the Wyoming Department of Environmental Quality's Wildlife
Guidelines No. 5. Explicitly these guidelines suggest that
"baseline data on wildlife use of the permit area will provide a valuable reference for development of the reclamation
plan and for evaluation of reclamation". Unfortunately, if
harvest policies are altered dramatically, it will be impossible to ascertain the consequences of development or reclamation. Therefore, it seems difficult to justify that industry
maintain pronghorn population monitoring programs when data
may be virtually uninterpretable.

Clearly, agricultural interests influence management
decisions of the Wyoming Game and Fish Department. In general,
the Wyoming Game and Fish Department would prefer to be

divorced from the issue of agricultural damage (Douglas Crowe, pers. comm.). Frequently bad feelings are created when damage claims are processed--due in part to the rigorous procedure which must be followed by a landowner to obtain compensation. Legislation is presently under consideration which would shift landowner payments for rangeland use by wildlife to a county-level agricultural board rather than the responsibility of the Wyoming Game and Fish Department (Douglas Crowe, pers. comm.). Such a system for all agricultural damage claims seems highly desirable since removal of this responsibility from the Wyoming Game and Fish Department should help to ensure more equitable policy based on the priorities of all people interested in wildlife.

Recreational versus Ecological Priorities in Reclamation.

Government policy on reclamation frequently implies that the objective should be to restore mined areas as closely as possible to premining conditions. If the area was principally rolling sagebrush steppe, postmining reclamation should attempt to restore a rolling sagebrush steppe. Many wildlife biologists feel that we could do much better. By selective planting and management, wildlife productivity and recreational use of the area may be substantially enhanced.

First I will develop arguments in favor of ecological priorities for reclamation. Our unfortunate ignorance of ecological systems and processes is the greatest justification for strict restoration of premining habitats. We cannot be certain that a tract of land is not critical habitat for some species. What consequences might occur in surrounding communities if a long-term disturbance occurs? How rare are some species about which we may have little data, e.g., sagebrush voles (<u>Lagurus</u> <u>curtatus</u>), which may occur on a mine site? How can one improve upon the natural vegetation and ecology of an area? Should ecologists be in a position to decide which species are "better"; and is "farming" of wildlife an ecologically sound strategy? The safest strategy for reclamation is to restore an area to previously existing habitats to ensure that some of these unknowns do not result in a disruption of ecological "stability" in an area.

In contrast to the "environmental" perspective, one can develop a strong case for management of wildlife to create maximum recreational opportunities. Driving through Campbell County, Wyoming, one is usually impressed by the expansiveness of the natural sagebrush steppe. Seemingly endless rolling hills are covered with seemingly homogenous wildlife, consumptive use is largely restricted to brief hunting seasons for pronghorn and sage grouse with occasional hunters pursuing mule deer (Odocoileus hemionus), and rabbits (Lepus spp. or Sylvilagus spp.). An option exists for mine reclamation to create areas attractive to a variety of wildlife. Many options exist including creation of waterfowl habitats on reservoirs, planting windbreaks of Russian olive (Eleagnus angustifolium), leaving mine highwalls for raptor habitat, and introducing exotic species of wildlife or wildlife food plants which could increase the recreational use of reclaimed surfaces. Although attempts to restore natural vegetation may restore populations of sagebrush voles and other native species, these taxa are of little if any recreational value.

Restoration to premining conditions will be attractive to ecologists, environmental groups, and perhaps O.S.M. bureaucrats, but unfortunately these people usually will never see the area. Local recreational users of wildlife will have more immediate interest in reclamation sites as potential places for camping, hunting, fishing, photography, etc.

I have no insight into the resolution of this conflict. As an ecologist I must emphasize that the safest ecological strategy is to restore natural conditions. But yet, some "farming" of wildlife is probably justified. Hopefully, compromises can be achieved depending upon the reclamation opportunities which a particular site might present.

CONCLUSIONS

Each of these cases represents socio-political conflict in wildlife management at different levels of organization, but in both cases, industry is faced with the ultimate management decisions. In the case of sagebrush reclamation, the conflict is a familiar one between agriculture and wildlife. The pronghorn harvest example is a complex political issue, but yet, underlying the policy of the Wyoming Game and Fish Department are agricultural demands in the form of wildlife damage claims. Recreational versus ecological priorities for wildlife reclamation constitutes a difficult philosophical issue among environmentalists and resource managers.

These conflicts cannot be completely resolved. Wildlife resources require land which may also have many other uses which benefit society. Similarly people may envisage different uses for wildlife; unfortunately, all interests cannot be satisfied on the same plot of land. These conflicts create very difficult management decisions for industry. Hopefully, compromise strategies can be employed.

ACKNOWLEDGMENTS

I thank Doug Crowe of the Wyoming Game and Fish Department for useful discussion and details on damage claims. James Tate, Jr., Bonnie C. Postovit, William Hepworth, Archie Reeve, Mike Naus, Steve Amstrup, and Dennis Rothenmaier assisted in various ways. Funding for my research has been provided by the ARCO Coal Company.

LITERATURE CITED

Autenrieth, R. (ed.). 1978. Guidelines for the management of pronghorn antelope. Proc. 8th Pronghorn Antelope Workshop, Jasper, Alberta. pp. 473-526.

Barber, T.A. 1968. Function of the cecal microflora in sage grouse nutrition. M.S. Thesis, Colorado State University, Fort Collins. 60 pp.

Bayless, S.R. 1969. Winter food habits, range use and home range of antelope in Montana. J. Wildl. Manage. 33: 538-551.

Beale, D.M., and A.D. Smith. 1970. Forage use, water consumption and productivity of pronghorn antelope in western Utah. J. Wildl. Manage. 34: 570-582.

Boyce, M.S. 1981. Robust canonical correlation of sage grouse habitat. D. Capen (ed.), Proc. Workshop on the Use of Multivariate Stat. in Studies of Wildlife Habitat (in press).

Boyce, M.S., and D.J. Daley. 1980. Population tracking of fluctuating environments and natural selection for tracking ability. Am. Nat. 115: 480-491.

Boyce, M.S., and J. Tate, Jr. 1979. Pronghorn (Antilocapra americana) demography and hunting quotas in the Powder River Basin, Wyoming. Proc. Int. Wildl. Congr. 14: (in press).

Braun, C.E., T. Britt, and R.O. Wallestad. 1977. Guidelines for maintenance of sage grouse habitats. Wildl. Soc. Bull. 5: 99-106.

Clark, C.W. 1976. Mathematical bioeconomics. J. Wiley, New York. 352 pp.

Colenso, B.E., M.S. Boyce, and J. Tate, Jr. 1980. Developing criteria for reclamation of sage grouse habitat on a surface coal mine in northeastern Wyoming. Proc. Symp. on Surface Mining Hydrology, Sedimentology and Reclamation (University of Kentucky, Lexington, Dec. 1-5, 1980) pp. 27-31.

Martinka, C.J. 1967. Mortality of northern Montana pronghorns in a severe winter. J. Wildl. Manage. 31: 159-164.

Patterson, R.L. 1952. The sage grouse in Wyoming. Sage Books, Denver, CO. 341 pp.

Pielou, E.C. 1977. Mathematical ecology. J. Wiley, New York. 385 pp.

Scott, M.D., and T.L. Terrel. 1976. Pronghorn antelope management potential on mining industry lands. Proc. Biennial Pronghorn Antelope Workshop 7: 135-151.

Sturges, D.L. 1975. Hydrologic relations on undisturbed and converted big sagebrush lands: the status of our knowledge. USDA Forest Serv. Res. Pap. RM-140. 23 pp.

Sturges, D.L. 1980. Soil water withdrawal and root distribution under grubbed, sprayed, and undisturbed big sagebrush vegetation. Great Basin Natur. 40: 157-164.

ENVIRONMENTAL SCIENCE AND THE
REGULATORY PROCESS: A CASE
STUDY WHERE DIALOGUE LED TO
PROBLEM SOLVING

D. F. Soule
   University of Southern California, Los Angeles
M. Oguri
   University of Southern California, Los Angeles
J. Lindstedt-Siva
   Atlantic-Richfield Co., Los Angeles
J.P. Ray
   Shell Oil Co., Houston

ABSTRACT

Environmental management depends upon an adequate information base for decision-making. Lack of such information and/or mechanisms for resolving permitting problems can cause delays, escalating costs or cancellation of projects. A case study in Long Beach, California involved Coastal Commission dredging and ocean disposal permits for a new ARCO Transportation Co. tanker terminal.

Initially, standard bioassay-accumulation tests showed uptake of cadmium in the clam, Macoma sp., which could result in permit denial and elimination of the project. Instead, ARCO, Shell and the University of Southern California designed a program in conference with EPA, the Corps of Engineers and the Port of Long Beach, to obtain information which each entity needed to make management decisions.

Results demonstrated that ambient cadmium levels in the dredge material were lower than those at the ocean dumpsite; statistically significant uptake of cadmium did not occur in multiple bioaccumulation tests with larger samples; methods and intervals of depuration could affect body-burden and cadmium was located primarily on external surfaces or in the gut. Agencies judged that dumping the dredged materials would not be deleterious, hence the project proceeded.

INTRODUCTION

In the decade of the 1970s, industry came to recognize the existence of the environmental regulatory process, but exhibited varying degrees of willingness to comply. The realities of regulation have led to more or less routine incorporation of the costs of the permitting process within project planning, even though costs at times have been difficult to relate to social and/or environmental benefits. However, the problems of "regulatory impasse", due to overlapping jurisdictions, and especially, to provisons which allow the attachment of future conditions to potential projects, can defeat the most well-intentioned of applicants. Mechanisms for resolution of such impasses may be non-existent, or they may be obstructed by restrictive individual interpretations of regulatory procedures and standards; however, solutions may sometimes be developed in situations where consultation rather than confrontation is practiced. A case history which illustrates this thesis is the Atlantic Richfield-Pier E Tanker Terminal Relocation Project in the Port of Long Beach, California.

PROJECT HISTORY

The Pier E project was a joint effort of ARCO Transportation Company and Shell Oil Company. Originally, the project objective was to streamline the pipeline connection between Pier E, Berth 118, Port of Long Beach, and tankage at Carson for both Shell and ARCO Petroleum Products refineries. Prior to the project, both refineries received oil via smaller pipelines and rather circuitous routes. Shell Oil Company in 1972 proposed to construct a new 107 cm (42-inch) pipeline from pier E to the Carson storage tanks, and filed an Environmental Impact Assessment document with the California Coastal Commission in 1976 (DMJM, 1976). When the Coastal Commission considered the project permit, a major concern was the potential increase in large tanker traffic at Berth 118, because large tankers obstructed navigation in the Back Channel to some extent. A condition was attached to the Shell permit to the effect that the ARCO marine terminal at Berth 118 would have to be relocated, out of the main channel. ARCO Transportation and the Port of Long Beach studied the feasibility of relocation and finally proposed that the terminal be located at Berth 121-122 (Figure 1), an adjacent small embayment then occupied by the Howard Hughes flying boat, known as the "Spruce Goose". By providing more vessel clearance and maneuvering room in the Back Channel, traffic conflicts between the present Berth 118 operations and vessels passing to and

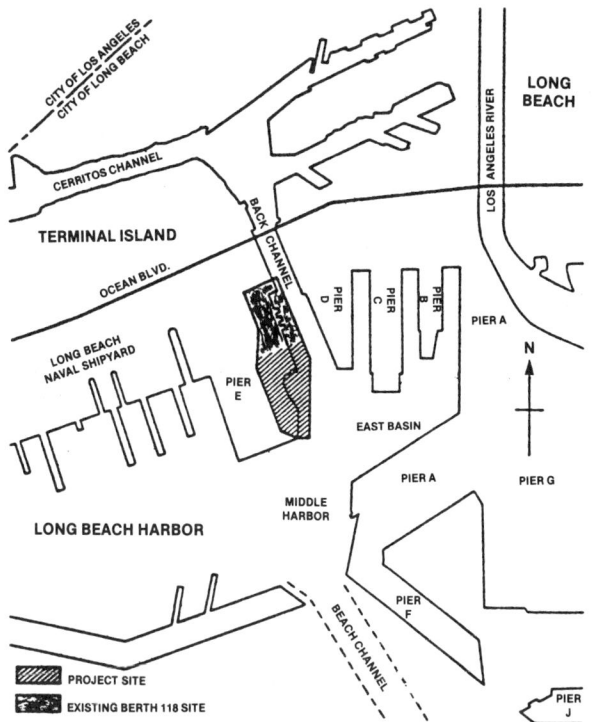

Figure 1. Arco Pier E Tanker Terminal, Long Beach. Note narrowing of Back Channel at Berth 118.

from the Inner Harbor would be eliminated. The safety of the Back Channel would be increased (ERG, 1979). Shell Oil would no longer be limited to 10.7 meter (35 ft) draft vessels at their terminal in Los Angeles Harbor, and lightering could be discontinued. A Draft Environmental Impact Report for the ARCO Terminal project was completed in September 1979 (ERG, 1979) and the Final EIR was filed in November (Port of Long Beach, 1979). Five major permits were required for the Pier E relocation project. They were from the following: 1), California Coastal Commission; 2), California Regional Water Quality Control Board; 3), South Coast Air Quality Management District; 4), U.S. Army Corps of Engineers and 5), U.S. Environmental Protection Agency. The last two permits are issued concurrently, based on identical applications; one of the federal agencies cannot give approval without the other.

On each of these permits, the Port of Long Beach was the applicant, on behalf of ARCO Transportation, the leasee. The first three permits were obtained within a few months after the application (Table 1), although approximately two years of effort had preceeded the South Coast Air Quality Management District approval.

Table I. Major Permits, Pier E Project

Permit	Application	Received
Calif. Coastal Commission (CCC)	Oct. 1979	Dec. 1979
Calif. Regional Water Quality Control Board (WQCB)	Oct. 1979	Dec. 1979
So. Coast Air Quality Management District (AQMD)	Mar. 1979	Apr. 1979
Army Corps of Engineers (C/E, Los Angeles District)	Oct. 1979	Mar. 1981
Environmental Protection Agency (EPA Region IX)	Oct. 1979	Mar. 1981

The Corps of Engineers/EPA permits took a bit longer. The COE permit was needed in order to dredge and transport dredge materials for the marine terminal at the new site, which was a shallow cul-de-sac and shoaling area of the channel. An EPA permit is required to dump the dredged material transported at an open ocean dumpsite, LA-2, off Point Fermin. To obtain the two joint permits, bioassays were required, following the procedures described in the Environmental Protection Acency/Corps of Engineers (1977) Manual, "Ecological Evaluation of Proposed Discharge of Dredged Material into Ocean Waters". Interpretations and modifications of bioassay procedures in the manual were the responsibility of the Corps of Engineers Los Angeles district office. Under interagency agreements, the Environmental Protection Agency must concur with those decisions before the permits can be issued. EPA Region IX in San Francisco thus had jurisdiction over the project. Harbors Environmental Projects (HEP) of the University of Southern California had been carrying out experimental bioassay projects related to sediments for a number of years and served as the contractor.

For the initial tests (HEP, 1979a,b) it was agreed that bioassays would be carried out for liquid phase sediments,

suspended phase and solid phase. Clams (_Macoma_, sp.) from
five replicates of the required 10 animals each in the solid
phase tests were then to be pooled for the 10 day bioaccumulation tests. Pooling of tissue was necessary because of
the small size of the individual animals. Results of the
chemical analyses carried out showed that levels of all
parameters examined in _Macoma_ sp. were lower than levels
in the test sediments (Table II).

The Corps of Engineers and EPA decided that, since statistical analysis could not be made on the basis of the
pooled samples, bioaccumulation tests should be repeated
using three replicates of 40 animals each, for statistical
analysis and to provide sufficient tissue for chemical
analysis.

Results of these tests (HEP, 1979b) indicated that
there were statistically significant differences in accumulation of cadmium and iron. However, it was pointed
out that, while the differences in cadmium concentrations
between reference (control) and test tissues were about
four-fold, the differences were about seven-fold
between reference and test sediments (Table III).
Reference sediments were taken from the outer harbor,
away from the project site, in supposedly cleaner waters.
It is important to note that the COE/EPA Manual does not
require chemical analysis of the test and reference site
water; however, the data demonstrated that there were
differences in bioaccumulation as compared with ambient
levels. Neither does the Manual require analysis of ambient levels of pollutants in the pool of test organisms
prior to initiation of tests, or of the sediments at the
conclusion of tests.

The test sediments used in the initial studies at Pier
E had been taken from the surface at only three stations
within the shoaling area of the project site (Figure 2)
and they contained a mean of 15.2 mg/kg of cadmium. The
assumption made by the agencies was that this was typical
of the entire project area, including the well-scoured
main channel, and at dredging depths up to several meters.

Although the COE/EPA Manual (p. F10) clearly indicates
that a statistically significant effect in a bioassay does
not necessarily imply an ecologically important impact in
the field, and that this must be kept in mind when interpreting results, the COE and EPA were reluctant to issue permits, especially since litigation was underway in another
district over bioassay results. Thus it appeared that the
project might be stopped, in spite of the obvious benefits

Table II. Bioaccumulation. Trace Metals Analysis. Values shown are in ppm for tissues and sediments and in mg/l for water samples. ND = none detected. (From HEP, 1979a, Appendix A)

	As	Cd	Cu	Cr	Fe	Pb	Hg	Mn	Ni	Zn	Oil & Grease
Macoma Reference Tissue	ND	ND	255	178.4	5380	ND	1.42	70.83	ND	486	2.335%
Macoma Test Tissue	ND	ND	242	90.3	1450	ND	ND	ND	ND	532	2.475%
Reference Sediment	570	33.8	165.6	144.6	45710	27.6	1.47	10300	147.2	17572	0.124%
Test Sediment	1106	1.61	203.7	140.8	42815	18.6	ND	4200	349.2	151.3	0.158%
House Water	0.65	0.045	ND	ND	0.5	0.44	0.035	ND	1.5	0.7	84mg/l
Site Water	0.65	0.057	ND	ND	0.5	0.50	0.035	ND	1.5	ND	153mg/l

Table III. Trace Metals Analysis. Values shown are ppm for tissues and sediment samples, mg/l for water samples and percent for oil and grease. (Values preceded by < indicate that none of the material was detected and the value shown is the detection limit). (From HEP, 1979b, Appendix B)

	As	Cd	Cu	Cr	Fe	Pb	Mn	Hg	Ni	Zn	Oil & Grease
Macoma ref. Tissue 1	77.6	3.10	373	77.6	6090	<16	31.0	3.10	<16	304	14.1%
Macoma ref. Tissue 2	42.4	4.24	226	70.6	7060	<16	70.6	2.82	<16	424	10.9%
Macoma ref. Tissue 3	90.9	1.52	273	90.9	6060	<22	45.5	<2	<22	303	10.1%
Macoma test Tissue 1	72.4	7.24	318	72.4	8680	21.7	57.9	1.5	<15	289	9.8%
Macoma test Tissue 2	121.0	17.2	379	103	12100	17.2	138	6.89	<15	3790	12.6%
Macoma test Tissue 3	14.4	14.4	260	101.0	8660	<15	57.8	5.78	<15	866	8.4%
Ref. Sediment	25.9	2.2	65	75.6	30200	10.8	432	0.43	<20	216	0.3%
Test Sediment	47.5	15.2	38	76.0	26600	5.7	475	0.38	285	228	0.2%
House Water	0.40	0.5	<0.25	0.1	<50	0.4	0.6	0.01	<5	21	17.0mg/l
Site Water	0.25	0.4	<0.25	0.1	<50	0.4	0.1	0.01	<5	1	20.0mg/l

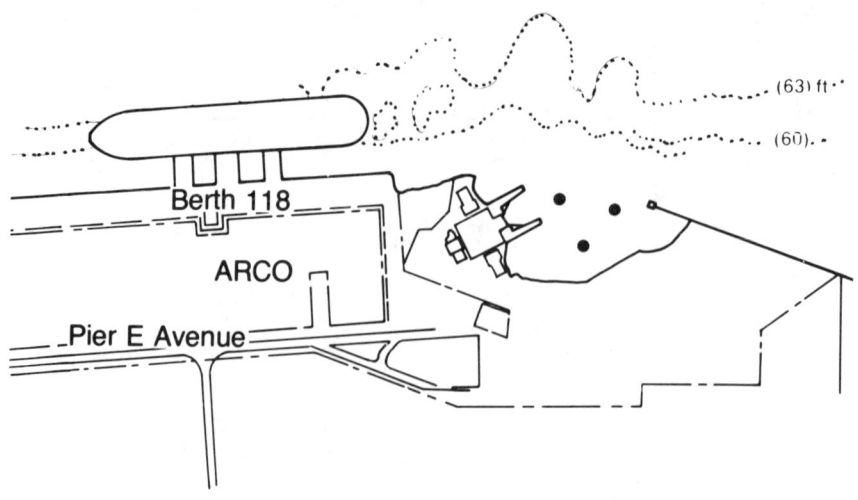

Figure 2. ARCO-Pier E, 1979. Sediment Collection Sites

of removing navigational hazards in the Port.

THE SEARCH FOR SOLUTIONS

The alternatives suggested by the Port of Long Beach were: 1), to seek an economic waiver; 2), to locate a fill site elsewhere in the Los Angeles-Long Beach Harbor; or 3), to find a land disposal site outside the harbor.

All of these alternatives would have meant, at a minimum, delays sufficient to cause some of the permits already granted to expire. In addition, alternatives 2 and 3 would have increased the project costs to the point that the feasibility of carrying out the project at all would have had to be recalculated. A harbor fill site was not available, although the Port of Los Angeles has an extensive dredge and fill project underway by the Corps of Engineers. It is questionable as to whether the Pier E material would have met almost identical criteria for a COE dumping permit within the harbor, based on the initial bioassay results. Before a landfill would accept the dredged material, a shoreline site and a Water Quality Control Board permit would have been required to de-water the material, even if a landfill could have been located and material trucked there. As for the first suggestion, obtaining an economic waiver from COE/EPA headquarters in Washington has never been tested; it would have been time consuming at best, and, most probably, litigation would have carried the matter into the courts because precedents would be set.

Conferences, not Confrontations

ARCO Transportation operates the present terminal facility at Berth 118, and Shell Oil Company shares the facility; in turn, Shell Oil Company owns the pipeline to the refineries, operated by ARCO. ARCO and Shell personnel met with representatives of the concerned agencies and consulted with Harbors Environmental Projects at the University of Southern California in efforts to resolve the impasse. There was never any question that the procedures specified in the Manual had not been carried out satisfactorily, but the tests represented the minimal available information on which to make judgments of regulatory compliance. Abandoning for the time being an enforcement mode, the participants, including the present authors, posed questions as to what data the regulators might wish to have in order to provide a better background for informed decision-making. Since cadmium was the pollutant of principal concern, these questions included:

1. Were the cadmium concentrations in the three surface sites, sampled in initial tests, representative of all the dredge material to be dumped?

2. Were the concentrations in the dredge material higher or lower than those a) at the EPA dump site, and b) at locations adjacent to the dumping area?

3. Were cadmium concentrations related to a) grain size, b) total organic carbon (related to residual oil), c) depth, d) distance from shore?

4. Were the evaluations of bioaccumulation skewed by a) small sample size, or b) by techniques of pre- or post-test depuration, or c) by other techniques in handling?

5. Was the cadmium primarily a) adhering to the external surfaces, b) transported in the digestive tract, or c) accumulated in reproductive organs?

In short, in a series of discussions, a program was developed that would test the validity of the assumptions made previously about cadmium distribution and about bioaccumulation test protocols.

Project Site Sediments

To answer the first question about the distribution of cadmium at Pier E, a stratified sampling program was proposed, since earlier data (HEP, 1975) had shown the variability of cadmium concentrations over small areas. Sediment samples were collected at 32 sites within the Pier E project area of approximately 41,000m_2 (440,000 ft_2), extending from the ship channel to the irregular shoreline, and to core depths of up to 2.44m (8 ft). The estimated amount of dredge material to be disposed is 500,000 cubic meters (650,000 cu. yds). A hydraulically driven Vibra-corer with a 3.1m (10 ft) barrel, and 9cm (3.5 in) diameter, was deployed from the USC research vessel Sea Watch. Duplicate cores were taken and the plastic core liners were capped. Sediment was then sampled with plastic instruments at the surface, at 1m (midcore) and at the bottom, regardless of depth of penetration. Samples were placed in plastic containers, chilled on board to 4°C (39°F), and transported to the USC Environmental Engineering laboratory for chemical analysis by Dr. K.Y. Chen and his staff.

Plots of the cadmium concentrations at the surface, mid-core and bottom (Figures 3,4,5) showed that the most polluted sediments were at the surface and in the shallower area which had been sampled during the previous studies (HEP, 1979a,b; 1980) almost along a transect of stations 20-23. The highest concentration of 6.53 mg/kg was, however, found at station 28 in the bottom of the core. This contrasts with the 15.2 mg/kg found earlier in the three pooled surface samples. The project area has received storm drain runoff and industrial wastes in earlier years prior to effluent control. The configuration of the small basin also created a depositional area when traffic in the channel caused stirring of the bottom sediments. Cadmium concentrations in the harbor as a whole have decreased two-to fourfold between 1973-74 (HEP, 1975) and 1978 (HEP, 1980). The computer map of mean cadmium concentrations (Figure 6) clearly indicates the variability in distribution, with statistically significant differences, so that the three sites (Figure 2) sampled previously were clearly not representative of the dredge material.

The EPA Dumpsite

Although the regulatory agencies are not required to consider concentrations of pollutants at the existing dumpsites, and in fact the COE, New York District was in litigation over a permit issued based on a disposal field study (POLB, 1979), information on dumpsite conditions could

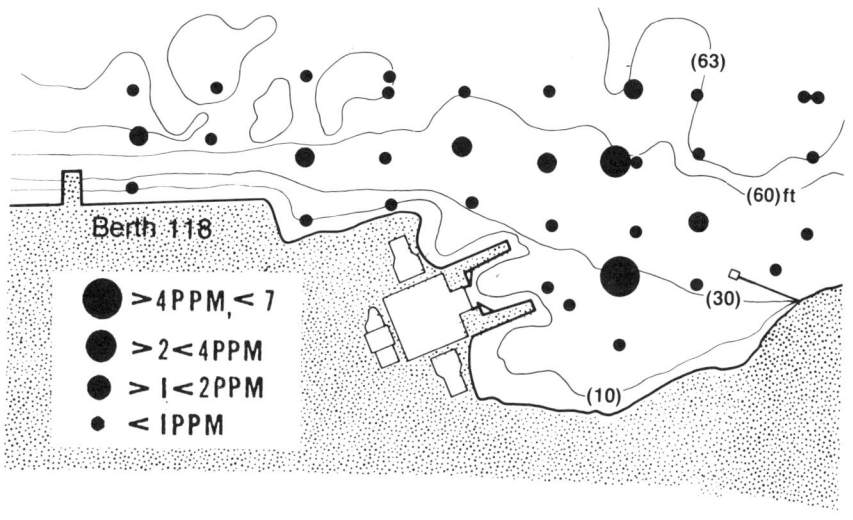

Figure 3. Pier E Vibra-Corer Stations. Cadmium Concentrations at Surface of Sediment Cores

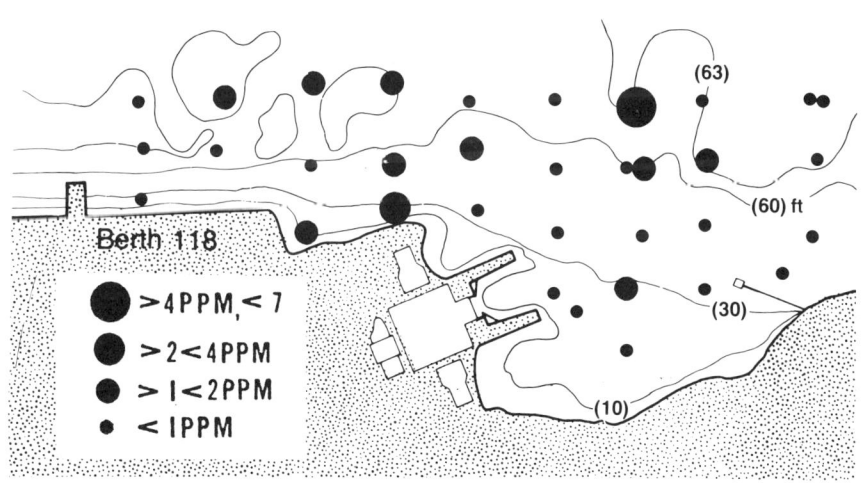

Figure 4. Pier E Vibra-Corer Stations, Cadmium Concentrations at One Meter Depth in Sediment Cores

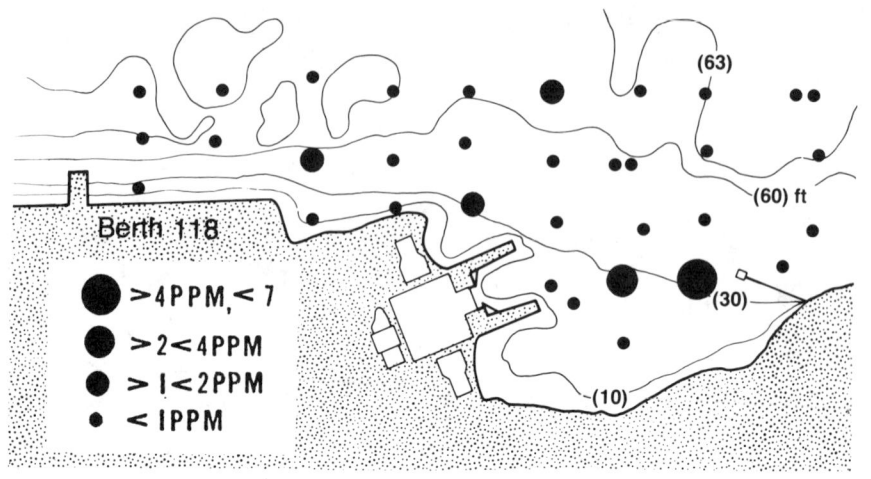

Figure 5. Pier E Vibra-Corer Stations. Cadmium Concentrations at Bottom of Sediment Cores.

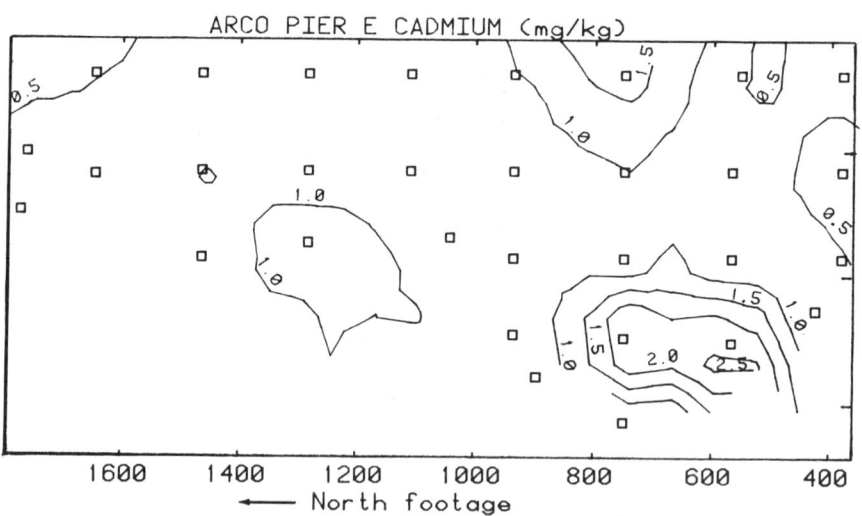

Figure 6. Computer Plot of Mean Cadmium Concentrations at ARCO Pier E Site (HEP, 1981)

provide useful background for decisionmaking. Therefore, a program to determine cadmium levels in the vicinity of the EPA dumpsite LA$_2$ was established; included were some sites that HEP had sampled in 1972-73 (Chen and Lu, 1974). Boxcores were taken from the RV VELERO IV in a transect toward Pt. Fermin (Figure 7). The EPA dumpsite is supposed to be at the 100 fm (183 m) contour at 118°17'24: W x 33°37'06"N, but the concentrations found indicated that the coordinates were probably being ignored, and dumping was taking place at the 100 fm (183 m) area closer to shore. There is also a large county sanitation discharge at Whites Point, to the west of Pt. Fermin, which would be a source of metals. Earlier studies indicated that higher levels of metals were found in the deeper basins and canyons of the San Pedro Channel and the dumpsite south of the head of a submarine canyon, San Pedro Valley, which is at station 8 (Figure 7).

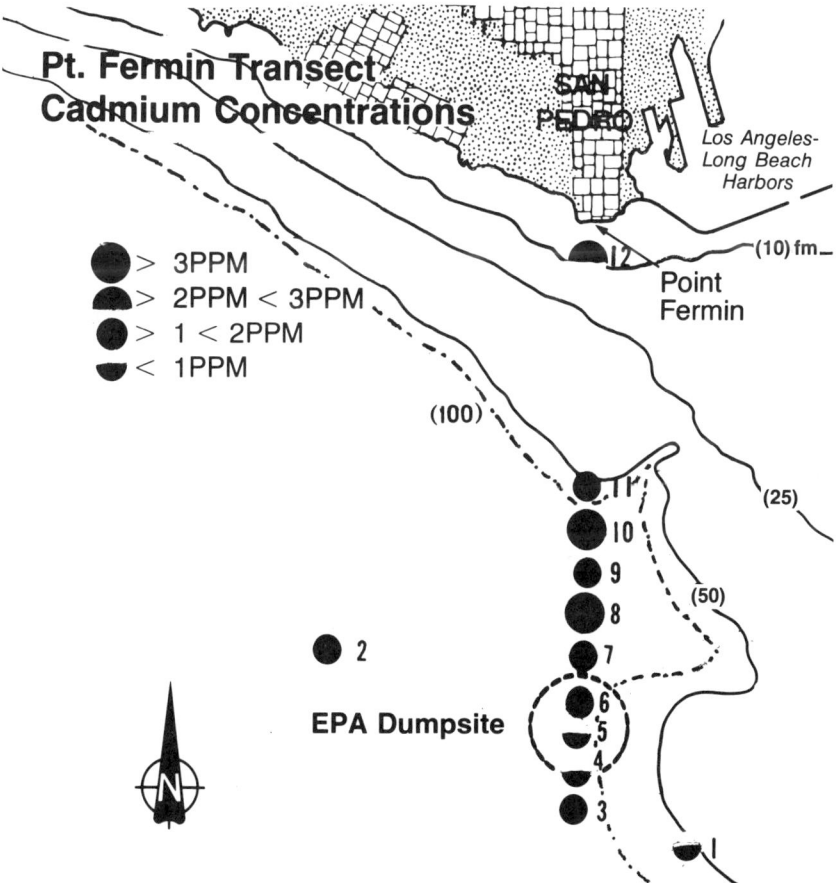

Figure 7. Cadmium Concentrations in Sediments Sampled by Box Corer, 1981.

Results of the present study showed the highest cadmium concentrations to be on the surface at stations 8 and 10. The range was from 0.22 mg/kg to 5.92 mg/kg, a higher minimum but lower maximum than the Pier E samples contained. Cadmium levels at Pier E were significantly lower statistically than those in the transect, in spite of the wide range. Thus, it could be concluded that the dredge material would pose no additional burden upon the dumpsite in regard to levels of cadmium.

Correlations of Cadmium

The question of the relationship of cadmium to other physical parameters in the marine environment is a complex one (Phillips, 1980). Principal Component Analysis (PCA) which translates statistical measurements graphically into vectors, produced somewhat surprising results, since it has been suggested that cadmium might be complexed to the finest-grained sediments. Instead, cadmium levels and Total Organic Carbon (TOC) levels were highly correlated and there was a strong inverse relationship between cadmium and depth of the sample within the Pier E cores themselves (Fig. 8). There was a slight trend, in dumpsite data, for increases in surface cadmium concentrations with increasing depth of the water column, but there were many exceptions (Fig.9). Chen and Lu (1974) found the highest level contour of 3.4 mg/kg in the deep basin northward of Santa Catalina Island; TOC values were highest in the same general area beyond to 450 fm (823 m) contour. Grain size did not appear to be directly related. Concentrations of cadmium were in part inversely related to distance from shore, but this apparent relationship does not follow the contours found by Chen and Lu (1974) in a more widely-spread sampling pattern.

BIOACCUMULATION PROCEDURES

Bioassay and bioaccumulation procedures which are specified in the COE/EPA Manual (1977) have been subjected to testing as well as to modification by local jurisdictions in order to reflect more adequately the local conditions. For example, in the ARCO bioaccumulation tests (HEP, 1981) an ambient water temperature of 11.5°C was used for the clam Macoma carlottensis, collected from off Pt. Fermin, but the polychaete worm Neanthes arenaceodentata, obtained from D.J. Reish's laboratory cultures, showed only 27.5% to 62.5% survival overall at the 11.5°C + 1° (53°F) temperature. A parallel test was immediately begun using 15.5°C + 1° (60°F) when the worms became lethargic and lay on the surface of the sediments in the containers. Survival in the latter tests averaged better than 90%. Permission

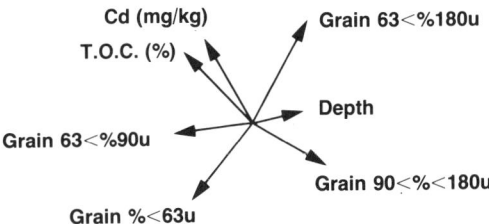

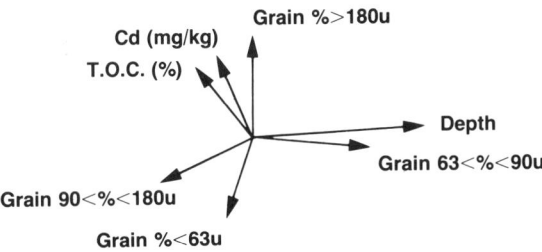

Figure 8. Relationships of Factors, from Principal Component Analysis, Pier E Sediment Samples

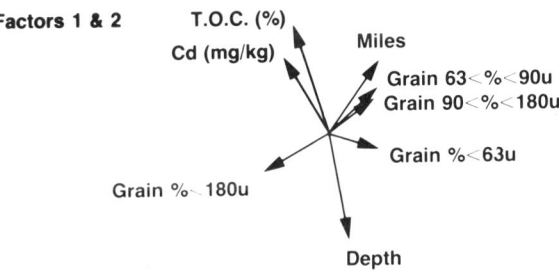

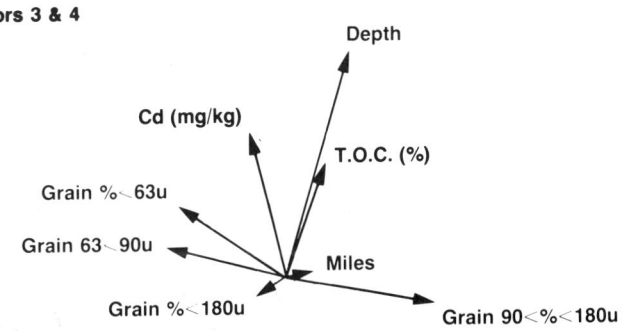

Figure 9. Relationships of Factors from Principal Component Analysis, Dumpsite Sediment Samples

had to be obtained by HEP from the Corps, the Port and ARCO to substitute the latter test group for chemical and statistical analyses. Since bioassay and bioaccumulation testing represents at least a considerable modification of the natural environment, research and experience are needed in order to recognize and to minimize such testing problems.

Sample Size

Accurate chemical analysis requires at least one gram of tissue; consequently, 10 specimens of the tiny invertebrates may form an inadequate sample. Clam tissue, which must be dissected from shells with glass knives, is therefore pooled and weighed to provide this. It is also desirable to use a larger number of organisms in view of the Manual requirement of 90% survival in controls. Either wild or cultured organisms can be damaged or senescent at the time of collection and testing, but may not exhibit obvious symptoms initially. With a control group of only ten organisms, a test could be invalidated by deaths unrelated to the test.

The University of Southern California has a large bioassay laboratory at Fish Harbor (in Los Angeles Harbor) equipped with an 86,000 gpd flowing seawater supply, with sand and 5 micron cartridge filters, U-V sterilization and temperature controls. Water temperatures are maintained in multiple aquaria in water tables. Plastic tents to reduce possible airborne contamination are placed over aquarium racks.

Survival

Survival is the important criterion of the 96-hour toxicity test, but is not a factor in measuring bioaccumulation, provided that an adequate tissue sample is available for analysis. Nevertheless, it is interesting to note that in the Pier E bioaccumulation investigations (HEP, 1981), survival in both test and reference (control) aquaria exceeded 90%. Insofar as is possible, survival in the reference group is considered to indicate that no significant stress was imposed other than the test conditions. In the standard bioaccumulation procedures, clams and worms are allowed to depurate (void gut contents) while on racks in flowing seawater for 48 hours prior to initiating the tests. The standard bioaccumulation period is 10 days, although longer periods for both toxicity and bioaccumulation tests might be more indicative of stress, particularly if sublethal criteria were developed. A 48-hour post-test depuration period is also standard.

In the final Pier E bioaccumulation tests, with five replicates of test and reference sediments each using Macoma sp. and Neanthes sp., there were no statistically significant differences in bioaccumulation, based on methods specified in the Manual. Table IV shows the mean values for cadmium in test and reference (control) specimens in the standard Pier E tests in the first column.

Depuration Methods

While the standard two-day pre-test and post-test depuration periods have been accepted as sufficient time for organisms to void their digestive tracts of possible ambient pollutants, the COE/EPA Manual (1977) does not provide for initial tissue analysis to demonstrate this or determine the ambient body burden. Clams, when stressed, are likely to close up valves and cease to feed or void. Alternatively, clams and polychaete worms may re-ingest voided feces, and also may exude mucus, which becomes colonized by bacteria and may be ingested. Thus, pollutants which have been voided may be reingested to some extent during the test or in the post-depuration period, skewing test results.

In the Pier E investigations, HEP (1981) analyzed Macoma tissues from the pre-test depuration of two days (Table IV), and the standard post-test depuration of two days. Depuration was continued, however, for testing after 4, 6 and 8 days for comparison (Table IV). This, of course, required large numbers of extra animals beyond the needs of routine bioassays.

Pre-test data on cadmium concentration showed 1.92 mg/kg wet weight of cadmium at the end of the standard 2 day depuration. This increased to 2.96 mg/kg at 4 days and then dropped to 2.17 mg/kg at 6 days at depuration.

Statistically significant differences were found in cadmium concentration after 2, 4, 6 and 8 days of post-test depuration, shown in Table IV, with an initial reduction at 4 days, but with an unexpected increase in cadmium at 6 days and a decrease at 8 days. Comparison with cadmium concentration in predepuration tissue is included in Table IV, and data on test water and sediment cadmium are included in Table V.

HEP hypothesized that reingestion of contaminated mucus or feces was occurring, even though the clams were removed from the sediments, placed on plastic grids of $1 cm^2$ mesh, and flushed continuously with flowing seawater. Bacteria multiplying on the feces, on shells or in mucus may take

Table IV. Mean (X) Cadmium Levels (mg/kg wet wt.) in Macoma (clam) Tissue after Differing Depuration Times and Neanthes (worm) in Standard Test Time.

Procedure	Depuration Time			
	Standard + 2da	+ 4da	+ 6da	+ 8da
Macoma				
(pre-dep)	1.92	2.96	2.17	
Test $\bar{X}$	2.074	1.446	2.528*	2.23
Ref $\bar{X}$				
Neanthes				
Test $\bar{X}$	0.062			
Ref $\bar{X}$	0.076			

* Differences are statistically significant.

Table V. Cadmium Levels in Water and Sediments Used in Bioaccumulation Tests.

Procedure	Test	Reference (control)
Water	0.2 mg/l	0.17 mg/l
Macoma test sediment	0.69 mg/kg	0.69 mg/kg
Neanthes test sediment	1.2 mg/kg	0.44 mg/kg

up cadmium themselves and may result in increased bioaccumulation. In subsequent experiments with different procedures, smaller grids were used which were further elevated, a jet of seawater was flushed beneath the grids, and grids were relocated in the water tables daily. This still resulted in a rise in cadmium after the test, but the peak was at 2 days with a gradual decrease thereafter to the 8-day post-depuration period (Soule and Oguri, unpublished data). Thus the standard 2-day depuration period may be insufficient time to achieve the evacuation of pollutants and obtain genuine body burden analyses.

Tissue Wash

Another possibility for inducing variation in test results lies in the handling of tissues during dissection. Glass instruments are used to separate tissue from shells, but no other treatment is used on tissues to reduce possible contamination of the tissue by material adhering to or adsorbed on the shells, prior to weighing and chemical analysis. To test this, two techniques of washing tissues after removal of shells were tested. In Wash I (Table VI), tissues were rinsed with 10 ml of isotonic saline solution after dissection. In Wash II, the pH of the isotonic saline rinse was lowered slightly with 10% hydrochloric acid. Both procedures resulted in lower cadmium values, although the differences were not statistically significant.

Thus, it can be seen that changes in the timing and in mechanics of testing can strongly influence test results. Because COE and EPA cannot require permittees to carry out fundamental research, decisions may be made on less than adequate data. Further studies should be carried out for developing and refining techniques, as well as for investigating alternatives to bioassays as they are presently conceived.

Site of Cadmium Uptake

Prior to the Pier E study, the senior author carried out dent research on uptake of heavy metals in the marine invertebrate bryozoans (Soule and Soule, 1981) using the scanning electron microscope (SEM) and electron beam probe analysis. It was decided to investigate the use of this technique to determine whether the cadmium uptake found in Macoma whole tissue tests represented material adsorbed or adhered to the external surface.
Samples of Macoma were taken from the bioassay tests and prepared in standard histological techniques of paraffin embedding and sectioning to 15 um thickness. Sections

Table VI. Mean Cadmium Levels in <u>Macoma</u> (clam) Tissue (mg/kg wt wt) in Different Washing Techniques After Ten-Day Bioaccumulation Tests*

Reference	Standard Test Test	Wash I (saline rinse)	Wash II (saline and acid pH)
1.47	2.074	1.636	1.336

*Differences not statistically significant

were mounted and coated with gold for SEM. In the event that normal concentrations would not be sufficient to demonstrate bioaccumulation with the electron probe, a series of cadmium "spikes" of $CdCl_2$ up to 50 ppm were administered to the live clams for 24 hours. The sensitivity of the Tracor Northern Analyzer coupled to a Cambridge SEM used in the USC Materials Science laboratory has a sensitivity range that would have shown bioamplifications such as those cited in literature for some fish species (Chen and Eichelberger, 1976). However, the electron probe did not reveal cadmium in the Macoma sections. Therefore, similar material was prepared and sectioned at 5 um, gold-coated, and examined with a Cameca Mass Spectrometer at Charles Evans & Associates in Menlo Park, California. The mass numbers and relative abundances of cadmium are shown at the bottom of Figure 10. The figure summarizes the ion probe data, obtained from the 50 ppm $CdCl_2$ spike; the levels in standard tests remained well below the limits of detection, probably below 5 ppm. Concentrations of 50 ppm are lethal (LD_{50}) in 24 hours for Macoma.

Results demonstrated that the cadmium was located primarily on the external gills and in the digestive tract. Levels in the reproductive organs and blood were generally lower, near the limits of detection. This information suggests that bioaccumulation in Macoma may be transitory.

CONCLUSIONS

An alternative title to this paper could be "Regulation, Research and Resolution," three R's of a different sort. There are several lessons to be learned from this case history, some of which are specific only to the Pier E project and others which offer more general insight to the regulatory process, to research, and the relationships of both to industry. With regard to Pier E, based on the test results and other research, HEP personnel were of the opinion that dumping the dredge materials at the EPA dumpsite would not be likely to degrade the present environment or result in significant bioaccumulation. Development of additional data on the project site and dumpsite aided all concerned to make informed decisions. Permits for the new facility were issued by COE and EPA as soon as the HEP (1981) report was reviewed. The project is now underway.

Of wider application, however, are several conclusions reached due to this project. For example, additional standard bioassay replicates, as well as additional chemical analyses at the sites, reduced the probability that results were anomalous rather than representative. Examination of

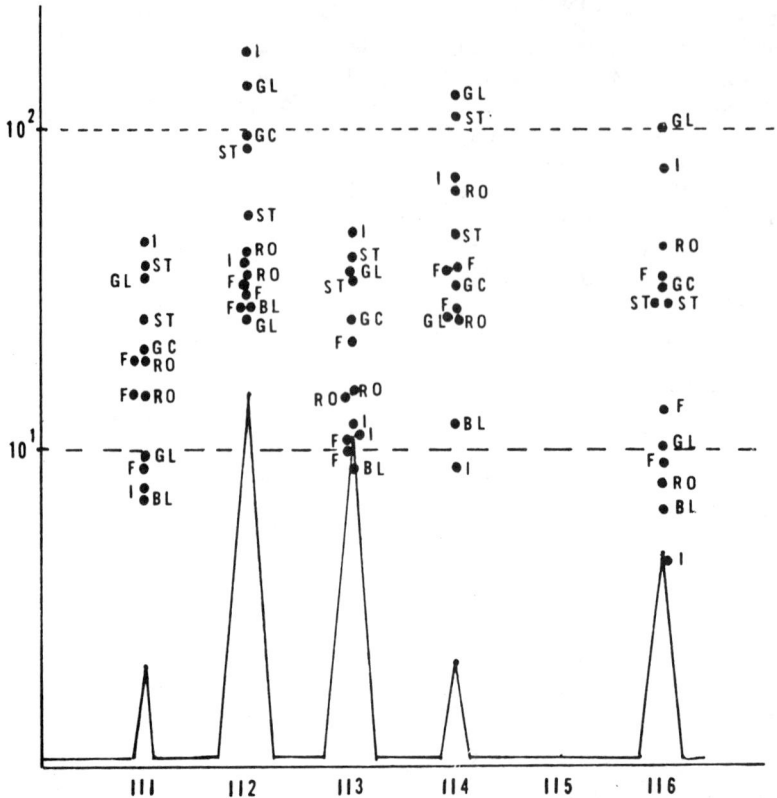

Figure 10. Comparison of Cadmium levels in <u>Macoma</u> organs; BL = blood; F = foot; GC = gut content; GL = gill; I = intestine; RO = reproductive organ; ST = stomach. Solid line = background level of SI chip; log scale: 10^1 = probable level of 5 ppm; 10^2 = probable level of cadmium "spike" 50 ppm.

---

techniques for bioassay demonstrated that the potential exists for unintentionally affecting the results substantially, which can and should be corrected.

It is important to recognize that no additional data would have been obtained, and none of the experimental methods examined, without the willingness of Corps of Engineers and EPA personnel to carry on a dialogue with representatives of ARCO, SHELL, the Port of Long Beach and Harbors Environmental Projects.

Willingness, however, is contingent upon the existence of latitude for dialogue and interpretation. There are numerous instances in which permit applicants and consultants complain about the lack of highly specific procedures for obtaining permits; a + b + c = d is convenient, but it is a two-edged sword. With highly structured rituals prescribed, there would be no latitude for deciding what questions should be asked, what scientific tasks should be done, and how the entire process could be better directed toward environmental protection without total prohibition of human activity. While latitude for decision-making sometimes results in indecision, the lack of latitude would eliminate dialogue and make the permitting ritual an end in itself without regard to the purpose of the permit or the project.

ACKNOWLEDGMENTS

The authors of the present paper participated in negotiations for the scope of investigations reported. We wish to thank other participants for input and cooperation; Lee Childres, of ARCO Four Corners Pipeline Company who was the Pier E project manager, Russell Bellmer of the U.S. Army Corps of Engineers, Los Angeles District, Chris Vais of EPA Region IX, Leland R. Hill, director of Port Planning, Port of Long Beach, and Richard Sandell, then with the Port of Long Beach.

LITERATURE CITED

Chen, K.Y. and B.A. Eichenberger. 1976. Concentrations of trace elements and chlorinated hydrocarbons in marine fish: a bibliography. In Marine Studies of San Pedro Bay, California Part 11 (D. Soule and M. Oguri, eds.). Harbors Environ. Projects, University of Southern California, pp. 283-298.

Chen, K.Y. and J.C.S. Lu. 1974. Sediment compositions in Los Angeles--Long Beach Harbors and San Pedro Basin. Marine Studies of San Pedro Bay, California, Part 7. (D. Soule and M. Oguri, eds.). Univ. Southern California, 177 pp.

Daniel, Mann, Johnson & Mendenhall (DMJM), 1976. Environmental impact assessment, Shell Oil Company pipeline, Pier E, Berth 118 to Wilmington Refinery. Report for the Port of Long Beach Environmental Management Division, November 1976.

Environmental Resources Group (ERG). 1979. Draft environmental impact report on the Atlantic Richfield Company marine tanker terminal relocation to Berths 121-122. Report prepared for Port of Long Beach, Sept. 1979.

Environmental Protection Agency/Corps of Engineers. 1977. Ecological evaluation of proposed discharge of dredged material into ocean waters. Implementation Manual. Environ. Effects Lab., U.S. Army Eng. Waterways Exper. Station, Vicksburg, Miss.

Harbors Environmental Projects. 1975. Report to the U.S. Army Corps of Engineers on environmental investigations and analyses, Los Angeles Harbor 1973-1975. Allan Hancock Foundation, Univ. Southern California. 552 pp (Draft).

Harbors Environmental Projects. 1979a. Bioassay investigations of sediments for the Pier E Marine Tanker Terminal and Appendix A, Bioaccumulation. Report to the Port of Long Beach, Office of Port Planning. Univ. of Southern California.

Harbors Environmental Projects. 1979b. Bioassay investigations of sediments for the Pier E Marine Tanker Terminal and Appendix B, Reassessment Bioaccumulation. Report to the Port of Long Beach, Office of Port Planning. Univ. of Southern California.

Harbors Environmental Projects. 1980. The marine environment in Los Angeles and Long Beach Harbors during 1978. Marine Studies of San Pedro Bay California, Part 17. (D. Soule and M. Oguri, eds.) Univ. S. Calif. 667 pp.

Harbors Environmental Projects. 1981. Bioaccumulation investigations, ARCO-Pier E Tanker Terminal. Report to the Port of Long Beach, Office of Port Planning. Univ. of Southern California. 93 pp.

Phillips, D.J.H. 1980. Toxicity and accumulation of cadmium in Marine and estuarine biota. In Cadmium in the Environment Part 1 Ecological Cycling (J.O.Nriagu, ed.) Wiley-Interscience, New York. pp. 425-569.

Port of Long Beach (POLB). 1979. Final environmental impact report Atlantic Richfield Co. marine tanker terminal relocation to Berth 121. Prepared by Port of Long Beach, Environmental Management Div., Nov. 1979.

Port of Long Beach (POLE). 1980. Letter to Mr. Lee Childress Four Corners Pipeline Co., from Leland R. Hill, Director of Port Planning. Oct. 6, 1980.

Soule, J. D. & D. F. Soule. 1981. Heavy metals uptake in Bryozoans. In Proc. of the 5th Internat. Conf. on Bryozoans. Olsen & Olsen, Denmark (in press).

SECTION 2

EFFECTS ON THE ENVIRONMENT

A NEW APPROACH TO OILED BIRD REHABILITATION
AFTER OIL SPILLS ON THE EAST COAST

L. S. Frink
Tri-State Bird Rescue,
Delaware Audubon Society, Inc.

Abstract. Tri-State Bird Rescue has successfully pooled the resources of area organizations to provide effective rehabilitation teams to clean oiled birds and to initiate research programs to improve treatment methods for contaminated wildlife.

Oil has been demonstrated to have a negative impact on waterbirds both as a result of contamination from chronic oil pollution and from major oil spill incidents on waterways around the world. Croxall [1] agrees with estimates that one million birds die annually just in the North Atlantic as a result of oil pollution.

An added concern is that the species most commonly affected are those birds which are not prolific breeders and whose populations are not easily monitored in the wild.

When an oil spill occurs from pipeline leakage or on waterways near land masses, oiled birds are the most visible result of an oil spill and cause the loudest public outcry.

The Delaware River-Bay system is the second-largest oil-importing waterway in the United States, with over 2,100 round-trips by barge and tanker annually, and an average of 2.1 oil spills per day. Five major oil spills occurred here between December 1973 and December 1976 killing thousands of waterfowl.

Oil destroys the water-proofing and insulating properties of plumage. The bird suffers from hypothermia and has difficulty swimming and flying. Ingestion of all grades of oil has toxic consequences, and aromatic components in the more highly refined oils can also be absorbed through the skin. Clinical signs of stress are almost always present.

In oiled bird rehabilitation efforts following the five 1973-1976 spills, mortality was in excess of 81% [2], due to physiologic and psychologic stresses, including overhandling by untrained workers, delays in treatment, improper cleaning and feeding techniques, poor housing conditions and overcrowding.

Tri-State organized as a result of intense public pressure following these spills with the initial goal of providing an effective, humane response team to deal with stricken birds. Tri-State is comprised of professionals and volunteers from 8 area organizations: the Delaware, Valley Forge and independent New Jersey Audubon Societies, the Philadelphia Zoological Gardens, the University of Pennsylvania School of Veterinary Medicine, the Delaware SPCA, and federal and state Fish and Wildlife agencies.

In our efforts to provide equipped, trained workers, we hold annual seminars which cover such basics as the proper handling of wild birds (to insure worker safety and reduce stress to the birds). We teach some physiology, such as the distinct differences between the trachea and the esophagus (the passageway utilized in tube-feeding birds). Workers receive hands-on training in cleaning birds and doing physical examinations.

During actual spill situations, we immediately control access to the facility, minimizing stress to the birds by admitting only trained workers who rotate shifts throughout the entire oil spill response. Our equipment ranges from the mundane to the exotic - from popsicle sticks to hematocrit centrifuges. We began with a simple goal of providing trained, equipped workers. The Delaware Bay and River Cooperative is a coalition of 13 industries which use the Delaware River system for transportation of materials. The goal of the coalition is to prevent oil spills and to minimize environmental damage from any oil spills which do occur. The Cooperative came forward with funding for Tri-State which has enabled us to expand our perspectives. With this significant financial support, Tri-State is now actively instituting research which will:
1) enhance the survival ability of oiled birds; and,
2) develop medical and rehabilitative procedures which are geared to have a broad applicability to many problems facing wild birds.

We are fortunate to have this opportunity to do research in the laboratory as well as during actual field situations.

During oil spills, then, a strong emphasis is placed on keeping accurate records on every bird, from the moment it enters the facility until it dies or is successfully released into the wild.

The oiled bird is given a basic medical examination when it enters the facility; its body weight and temperature are

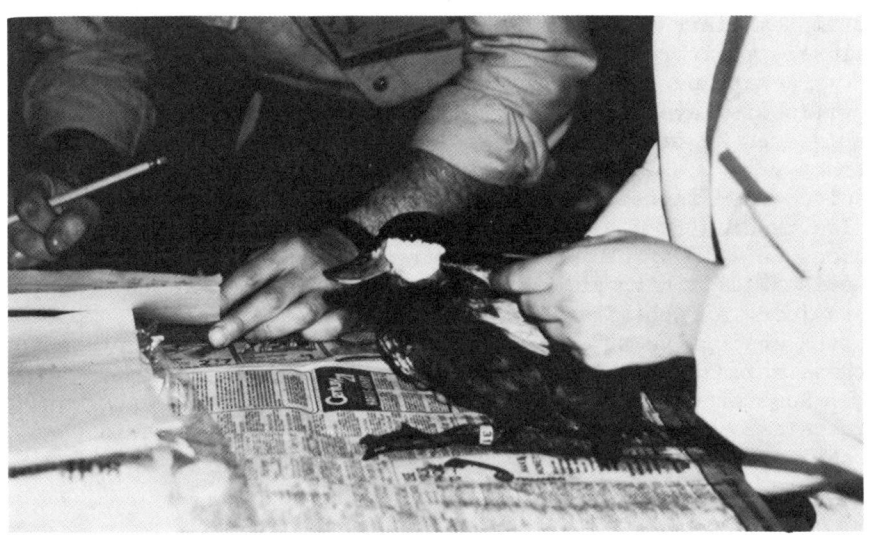

Figure 1. Detailed records are kept on each bird through every step of the rehabilitation process.

recorded, the degree of oiling, appearance of lacerations or fractures are noted, and, if diagnostic procedures indicate, fluids are given to counteract dehydration (Fig. 1).

The bird is washed in a series of detergent baths, with minimal feather disruption; the type of cleaning technique is dictated by the grade and amount of oil. The bird is then rinsed throughly and dried with forced hot air or ambient warm air from heating lamps (Fig. 2).

All birds brought in dead or dying at the facility are necropsied; gross observation is followed by histopathologic examination.

The findings from these post-mortem examinations provide a framework for our research. It is generally thought, for example, that oiled birds have lowered resistance to infection, and may require prophylactic or reactive treatment with antibiotics. We found no useful baseline data on the use of broad-spectrum antibiotics in ducks available to us; with cooperation from the University of Pennsylvania and the Philadelphia Zoo, my associate Dr. F. Joshua Dein geared one research project to learning the behavior of the broad-spectrum antibiotic chloramphenicol in ducks. The drug was administered by intramuscular and intravenous injection and by oral dose. Serum samples, collected by in-dwelling catheters measured the amount of antibiotic utilized by each bird - determining the half-lives of the drug. One important finding was that the chloramphenicol given by oral dose behaved erratically and unpredictably, suggesting that with oral administration of the drug clinical efficacy should be closely monitored [3].

Seventy percent of the birds necropsied after three 1980 spills displayed gross signs of hemorrhagic enteritis, as evidenced by fresh and clotted blood present in the entire tract from the duodenum to the cloaca. This enteritis resulted from ingestion of oil through preening, but enteritis also results from a number of diseases and infectious conditions in wild birds. We determined to initiate research to devise a diet high in easily absorbable nutrients. By thus providing a rich diet which does not tax the already compromised absorptive surfaces, we can increase nutrient uptake in these animals.

Some of our research involves surmounting mechanical and technical problems encountered in holding waterbirds in captivity. We had to provide housing which was inexpensive, easy to construct, assemble, and store.

Our 8' x 4' plywood pens provide an elevated 4' x 4' foam padded, carpet-covered dry area; the padding protects diving ducks from breast feather abrasion. The bottom of the pen is a 4' x 4' pool with constant overflow drainage. The birds are thus given free access to the pool, allowing them the natural sequence of swimming and preening which will accomplish realignment of feathers and restoration of waterproofing. Through the pen-pool design, and by curtaining each pen off, we reduce physical and visual stress to a minimum.

Birds which are considered suitably water-proofed, seem free from disease, and weigh-in close to average for their sex/species, are banded with permanent Fish and Wildlife Service bands by licensed master banders and are released in suitable habitat. When possible, the birds are monitored for two-to-four weeks after release.

Our data from these spills provide us with information we can use in subsequent treatments of injured or contaminated waterbirds. Our statistics for 3 1980 spills, for example, show a strong correlation between body weight/temperature and mortality. Diving ducks with a body weight or temperature at or above average for their species treatment group achieved a 100% release rate. Females which did not meet this standard showed an 80% mortality, males a 56% mortality.

We have devised standardized records forms for all the organizations we have trained along the Atlantic Coast, and will hopefully be able to analyze data from larger sample sizes.

As I mentioned, in the 5 oil spills between December 1973 and December 1976, over 81% of the birds brought in for treatment died. Tri-State responded to 3 spills in 1980, with a release rate in excess of 80% for treated birds.

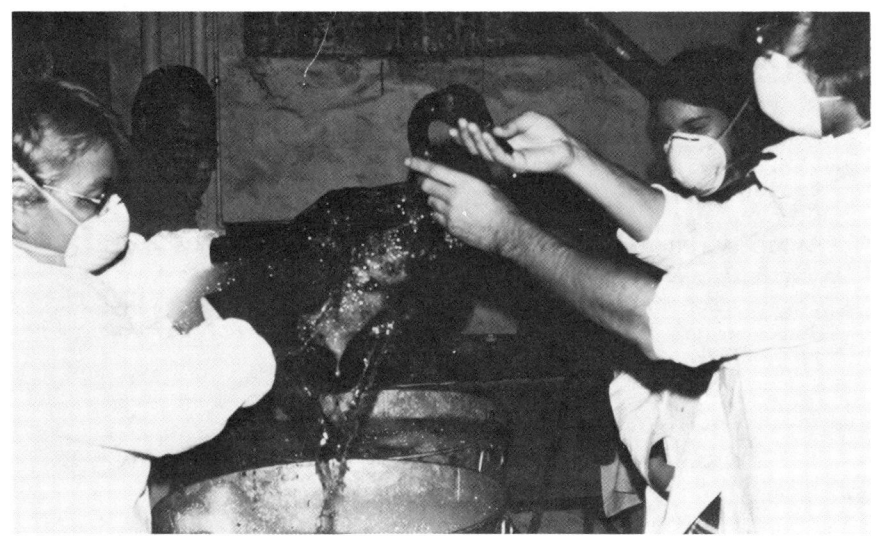

Figure 2. Cleaning techniques are determined by the type and amount of contaminant.

Following all three oil spill incidents, private organizations, industries and government agencies were able to benefit from previously rehearsed mobilization plans to perform quickly and effectively.

In conclusion, we feel Tri-State has accomplished two goals:

1) Tri-State has achieved its primary objective of providing trained and equipped workers to deal with oiled birds. We feel that our high release rates are the result of giving equal consideration to each step of oiled bird response - rapid retrieval of oiled birds, immediate medical attention, and effective cleaning techniques and rehabilitative care.

2) Tri-State has taken advantage of an issue of great public interest to acquire the funding and manpower needed to design and implement research, much of it in the field during environmental crises, and we have purposefully geared this research to benefit not only oiled birds but a number of indigenous avian species with a variety of health problems.

LITERATURE CITED

1 Croxall, J. (1979) Birds and oil pollution. Cooper, J.T., Eley, J.E. (Editors) *First Aid and Care of Wild Birds*, David & Charles, Ltd., Great Britain.
2 Perry, M.C., Ferrigno, F., Settle, F.H. (1978) Rehabilitation of birds oiled on two mid-Atlantic estuaries. *Proceedings of the Annual Conference of the Southeast Association of Fish and Wildlife Agencies*, 32: 318-325.
3 Dein, F.J., Monard, D.F., Kowalczyk, D.F. (1980) Pharmacokinetics of chloramphenicol in Chinese spot-billed ducks (*Anas poecilorhyncha*). *Journal of Veterinary Pharmacology and Therapeutics*, 3: 161-168.

METABOLIC AND TEMPERATURE
RESPONSES OF POLAR BEARS
TO CRUDE OIL

R. J. Hurst
   Department of Biology,
   Univ. of Ottawa, Ottawa,
   Ontario, Canada

N. A. Øritsland
   Institute of Zoophysiology,
   Univ. of Oslo, Oslo, Norway

P. D. Watts
   Institute of Arctic Eco-
   physiology, Churchill,
   Manitoba, Canada

ABSTRACT

    The metabolic and temperature responses of three sub-adult polar bears (*Ursus maritimus*) were monitored before and after exposure to a 1 cm. slick of Midale crude oil. Body and skin temperatures were measured by surgically implanted radio transmitters and metabolism by open circuit respirometry.
    Resting metabolic rate increased by 27% to 86% and averaged over 50% higher following oil fouling. Activity metabolism also increased in the one animal tested after oil contact, to 24% above the pre-oil level. Body temperature increased slightly in 2 of the animals and dropped in the 3rd polar bear within 24 hours of oil contact. Skin temperatures of all animals remained abnormally high for approximately one week following oiling.
    The metabolic changes induced by oil, and the potential energetic consequences, are discussed in relation to a possible metabolic compensation for a reduction in fur insulation; a change in the minimum level of energy turnover of body tissues; and a skin reaction to the physical oil coating.

INTRODUCTION

The polar bear (*Ursus maritimus*) is a circumpolar species adapted to a semi-marine existence in a cold arctic environment. Extensive travel over sea ice and across open-water leads distinguishes the polar bear as a uniquely amphibious arctic mammal(Stirling and Kiliaan, 1980; Larsen, 1971; Øritsland, 1969). It also assures that the probability of contact with an oil spill is roughly analogous to that of marine mammals such as seals.

Although offshore petro-chemical exploration has recently advanced at an exponential rate, the potential effect of oil on marine mammals and polar bears remains virtually unknown (Geraci and St. Aubin, 1980; Geraci and Smith, 1977). The suggestion has been voiced that the physiological impact of oil on polar bears could be substantial(Stirling, 1980; Birchard et al, 1978; and others).

A critical consideration in evaluating the potential of oil contact has been the intimate association of polar bears with their primary food source, the ringed seal (*Phoca hispida*) (Stirling and McEwan, 1975). Muller-wille(1974) reported ringed seals in Hudson Bay completely covered with oil. Moreover, ringed seals have demonstrated rapid absorption of hydrocarbons into body tissue through oil exposure by both immersion and ingestion(Engelhardt, 1978; Engelhardt et al, 1977).

Also of considerable concern is the potential vulnerability to physical oil contact during seasonal concentrations of polar bears at limited areas of ice-free water(Stirling and Latour, 1978; Smith and Stirling, 1975). Polar bears hunt primarily in moving pack ice and areas of ice-bound open water, termed 'polynyas'. The apparently higher productivity, due to upwellings of surface nutrients or other factors(Stirling, 1980), tends to concentrate many marine species in these areas. The result is to effectively 'shrink' the area of critical habitat and make some species particularly vulnerable to marine perturbations.

Exposure of marine mammals, particularly seals, to crude oil spills is well documented(Baker, 1976; Davis and Anderson, 1976; Anon, 1970). The severity, or in fact manifestation, of physiological impacts from such free-ranging contact remains poorly substantiated(See Geraci and Smith, 1977; Brownell and LeBoeuf, 1971). More germane to our understanding of the impact of oil pollution on marine or amphibious mammals are a limited number of controlled studies. These include investigations into the effects of oil on ringed seals(*Phoca hispida*) and harp seals(*Phoca groenlandicus*) (Engelhardt, 1978; Engelhardt et al, 1977; Geraci and Smith, 1976); sea otters(*Enhydra lutris*) (Costa and Kooyman, 1980; Williams, 1978); fur seals (*Callorhinus ursinus*) (Kooyman et al, 1976); and muskrats (*Ondatra zibethica*) (McEwan and Koeling, 1973). Also of relevance are studies on the effects of oil on the insulative value of fur (Kooyman et al, 1977; Øritsland and Smith, 1975).

Marine or amphibious mammals encounter particularly severe and unique thermal challenges in their diurnal or seasonal transitions from air to water. It has been suggested (Geraci and St. Aubin, 1980; Kooyman et al, 1977) that mammals which rely on fur to prevent excessive heat loss to the environment would experience more severe oil-induced changes in thermoregulation than those which rely primarily on blubber. Polar bears, although possessing extensive layers of subcutaneous fat (Øritsland, 1970), rely principally on fur to contend with thermal gradients of as much as 90°C (Scholander et al, 1950a;b).

Crude oil could alter the thermal integrity of the fur itself and jeopardize the thermal balance of an exposed individual. The present study documents some of the thermal and energetic responses of 3 polar bears following controlled exposure to a simulated oil spill.

MATERIALS AND METHODS

Experimental Animals

Three sub-adult polar bears were captured as 'nuisance bears' in the area of Churchill, Manitoba and held at the Churchill Polar Bear Project facilities for 3-4 months prior to experiments. During this time they were fed a diet of beef scraps and ringed seal. Water and/or snow were supplied ad libitum. The animals were of comparable age (2-3 years) and size (110-165 kg). When handled for surgery and for movement to and from the respiration chamber, they were immobilized with a jab stick injection of ketaset (ketamine hydrochloride) and rompun (Xylazine hydrochloride) at a dose of 5.5 mg·kg^{-1}.

Experimental procedure

The three animals were studied over a 6 week period; each individual staggered through an essentially identical sequence of measurements. The order of experimentation was bear 2 followed by bear 4 and bear 3. Animals acted as their own controls in the general measurement sequence of: resting metabolism, activity metabolism, and wind exposure, followed by oil contact. Measurements were then repeated in the order of wind exposure, resting metabolism and activity metabolism.

Oil Contact

The bears were exposed to a simulated oil slick in a pool measuring 3.7 x 1.5 x 1.5 m and containing 6,800 l of sea water. Oil was introduced to the pool surface in an amount necessary to create a 1 cm slick (57 l) (See Wadhams, 1976),

36 hours prior to the first bears' contact. Following exposure, the remaining surface volume was calculated and more oil was added to reconstitute the 1 cm thickness. The oil used was a medium viscosity Midale crude with a specific gravity at 15.6 °C of 0.89, a pour point of 10°C, and a viscosity of 7.43 and 10.24 centistokes at 50°C and 40°C respectively (Energy, Mines and Resources Canada, 1981). The physical characteristics are generally comparable to Prudhoe Bay crude oil (Beak, 1974). Animals were enticed to the pool edge and a sliding door was closed, forcing the animal into the pool. After a designated time (15 minutes for bear 2, 30 minutes for bear 4, and 53 minutes for bear 3), the entrance door was opened allowing the bear to exit the pool and re-enter the wind cage.

Metabolism

An index of resting and activity metabolism was determined using open circuit chambers and calculating the volume of oxygen consumed per unit weight and time. The flow of expired air from the chambers was established by means of centrifugal pumps and determined by a 350 l gasmeter (Collins Inc.). Flow rates were maintained at approximately 200 $l \cdot min^{-1}$ and 600 $l \cdot min^{-1}$ for resting and activity measurements respectively. Oxygen concentration was measured by use of a zirconium oxide analyzer (Applied Electronics Inc.). Ambient air outside the chamber was recorded before and after each sample period while oxygen concentration of expired air was recorded every minute. All values were corrected to S.T.P.

Resting Metabolism

The respiration chamber was a metal cylinder with a functional volume of 2,300 l which was air-sealed except for intake and air collection openings installed at opposite ends of the chamber. 4 weight sensitive load cells (W.C. Dillon Co.), installed under the floor of the chamber and connected to a microvoltmeter, relayed any movement of the animal. Sampling was of 2 hour duration on each of 2-4 successive nights when disturbance factors were minimal. Excessive movement predicated a delay of measurements or rejection of the results.

As the absolute conditions for measuring 'basal metabolic rate' (Kleiber, 1975) can rarely be attained (Pauls, 1981; Bligh and Johnson, 1973), the term 'resting metabolic rate' (RMR) has been used in this study.

Activity Metabolism

The treadmill used in this study has been described previously (Best et al, 1981; Øritsland et al, 1976). It was hydraulically powered and equipped with a variable speed control. A steel chamber (2.8 x 1.3 x 1.1 m) enclosing the treadmill tract served as an open circuit respiration chamber. Internal air was circulated to avoid stratification. Each animal, as control, was run for 3 or 4 days at no more than 2 runs per day. All runs were of 30 minute duration at a walking speed of 1.5 m·s^{-1} (5.4 km.hr^{-1}).

Temperatures

Body temperatues were determined with temperature sensitive radio transmitters (J. Stuart Enterprises) surgically implanted 2-3 weeks prior to oil contact. Deep body transmitters were implanted mid-ventrally, 20 cm posterior to the sternum and adjacent to the peritoneum. Skin or sub-cutaneous transmitters were placed just under the skin of the neck. Periodic pulses, convertible to temperature ($\pm 0.01^{\circ}C$), were received and displayed in milliseconds on a digital processor (Telonics). Body temperature was monitored at 15 min intervals during the wind/observation periods, 5 minute intervals during determination of resting metabolism and 2 minute intervals during the testing of activity metabolism.

Ambient air and water temperatures were monitored with standard tempscribe recorders (Bachard Inst.). Thermister probes (Y.S.I. Electronics) were used to confirm ambient temperatures and to monitor respiration chamber air temperature. The pool water temperature was maintained at $0^{\circ}C \pm 2^{\circ}C$ to best simulate the water temperatures to which a polar bear would normally be exposed (Weller, 1976).

RESULTS

Oil Contact

In the pool, oil rapidly contaminated all fur coming in contact with the water-oil surface. The oil coating was of a substantial but unquantifiable amount. An initially uneven coating of oil became more generally distributed over and into the fur in the days following exposure. Much oil was lost through shaking and grooming and also to sawdust within the respiration chamber.

Metabolism

Resting Metabolism

Control values of resting metabolism, determined from the average of minimum daily oxygen consumption rates, were generally comparable with the weight specific basal metabolic rate predicted by Kleiber (1975). The mean values of metabolism, using 25 measurements per day, were higher (Table I). In all cases there was a substantial increase in oxygen consumption following oil fouling. See Figure 1.

Table I. Resting metabolic rate of 3 polar bears expressed in watts and followed by standard deviation. N (in brackets) refers to number of nights for minimum MR, and number of measurements taken at 10 minute intervals for mean MR. * denotes statistical significance at p ≤ 0.05.

#	Min. Metabolic Rate (W)			Mean Metabolic Rate (W)		
	Before Oil	After Oil	Change	Before Oil	After Oil	Change
2	138±34 (4)	206±6 (2)	1.49	189±32 (100)	240±16 (50)	1.27*
4	133±10 (4)	243±15 (3)	1.82*	160±12 (100)	299±51 (75)	1.86*
3	210±6 (2)	261±35 (4)	1.24	263±28 (50)	365±70 (100)	1.38*

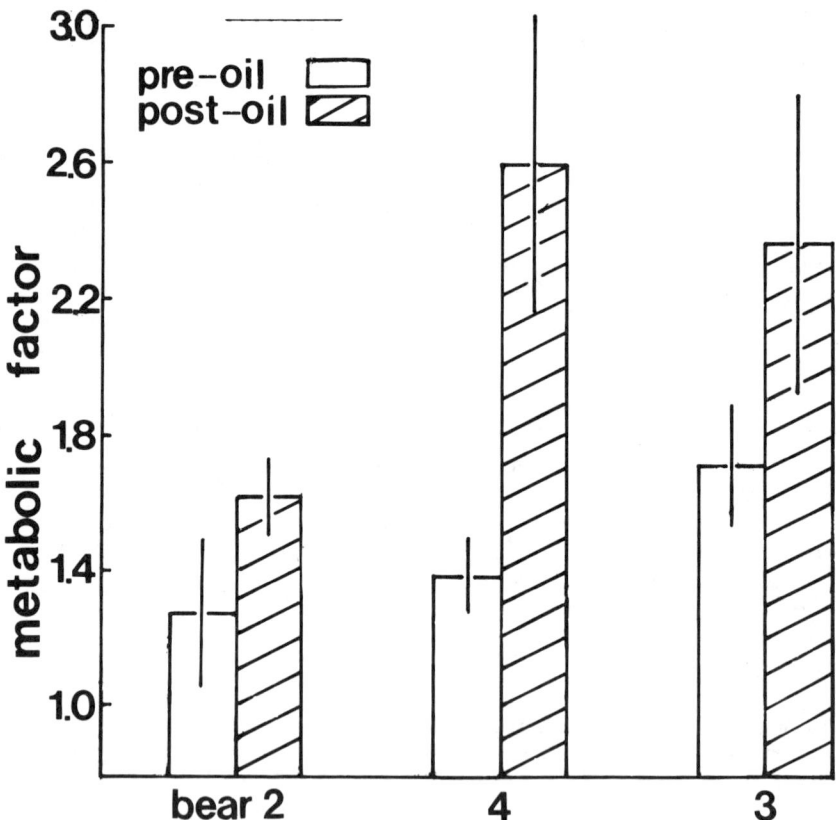

Figure 1. Resting metabolic rate expressed as a factor of predicted basal metabolic rate. Vertical bars denote standard deviations of the mean.

Activity Metabolism

Comparative results of exercise metabolism, before and after oiling were obtained for bear 2 only. Mean oxygen consumption, as measured in the final 12 minutes of each test, were 1.68 ± 0.2 $mlO_2.g^{-1}.hr^{-1}$ verses 2.10 $mlO_2.g^{-1}.hr^{-1}$ pre and post-oil respectively. This constitutes an average increase in the energetic cost of walking of 24%. In the initial post-oil treadmill trial of bears 4 and 3 the animals were unwilling or incapable of continuing the walk. Mechanical failure of the treadmill prevented further exercise testing of bears 4 and 3. See Figure 2.

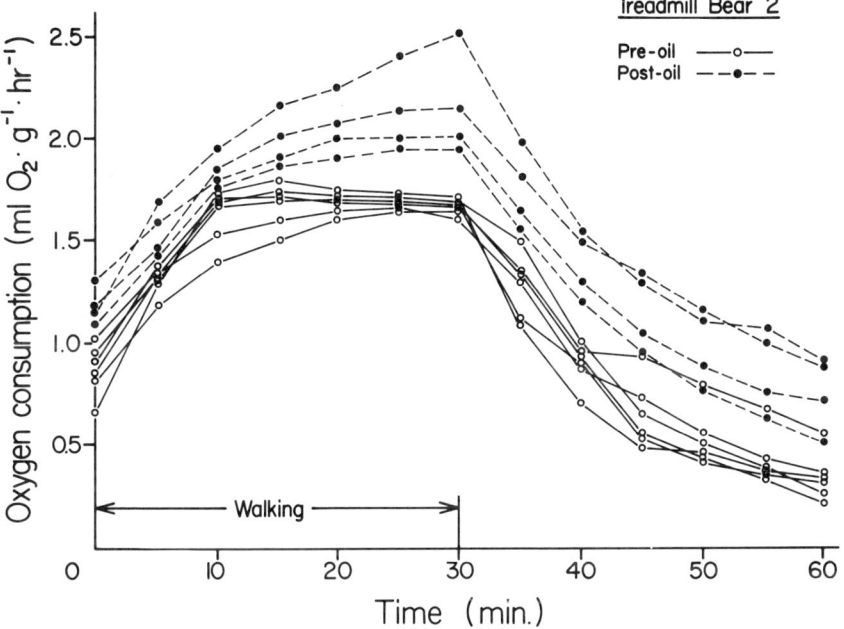

Figure 2. O_2 consumption at a walking speed of 5.4 $km.hr^{-1}$.

Temperatures

Resting Temperatures

Respiration chamber temperatures ranged from $-16-+3°C$. Deep body temperatures of bears 2 and 4 increased following exposure, whereas bear 3 showed a significant drop in T_b within 24 hours of oil exposure (Figure 3) which continued throughout the observation period. Skin temperatures increased in all cases following oiling although the rise was less apparent in bear 3.
See Table II.

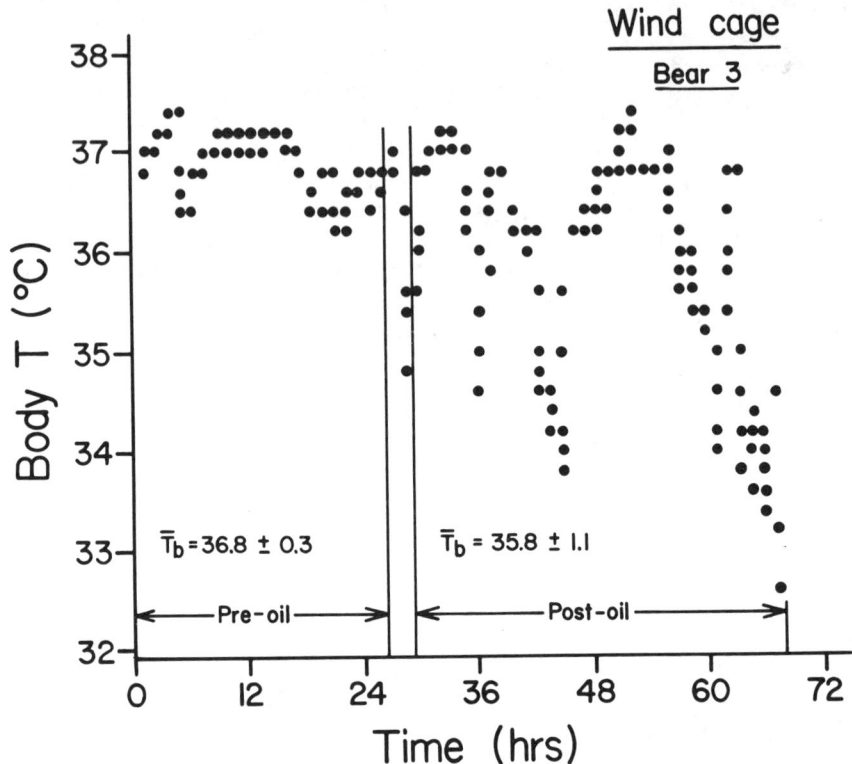

Figure 3. T_b of bear 3 for a 68-hour period before, during and after oil contact. Wind of 5 m.s^{-1} was generated in alternating periods of 4 hours on, 4 hours off, etc. for the duration of the wind cage period.

Table II. T_{body} and T_{skin} of polar bears during rest. Mean temperature is followed by standard deviation (°C). N (in brackets) refers to the number of measurements taken at 10 minute intervals. * denotes significance at $p \leq 0.05$

#	Body Temperature (T_b)			Skin Temperature (T_s)		
	Before oil	After oil	Change	Before oil	After oil	Change
2	36.5±0.17 (75)	36.9±0.14 (50)	+0.43*	35.5±0.32 (75)	37.3±0.16 (50)	+1.83*
4	35.8±0.78 (50)	36.6±0.61 (100)	+0.80*	34.5±1.50 (100)	35.7±0.71 (75)	+1.24*
3	36.8±0.50	36.1±0.94	-0.70*	36.3±0.09	36.7±0.20	+0.37*
$\bar{X}$	36.2±0.57 (225)	36.4±0.78 (225)	+0.19	35.2±0.90 (225)	36.4±0.58 (150)	+1.29*

Activity Temperatures

Ambient temperatures ranged from -22--$-13°C$ during activity testing. Deep body temperatures in bear 2 prior to exercise averaged slightly higher in the trials after oil contamination. The increment as a consequence of exercise was dramatically higher at $1.92°C$ post versus $1.34°C$ pre-oil (Figure 4).

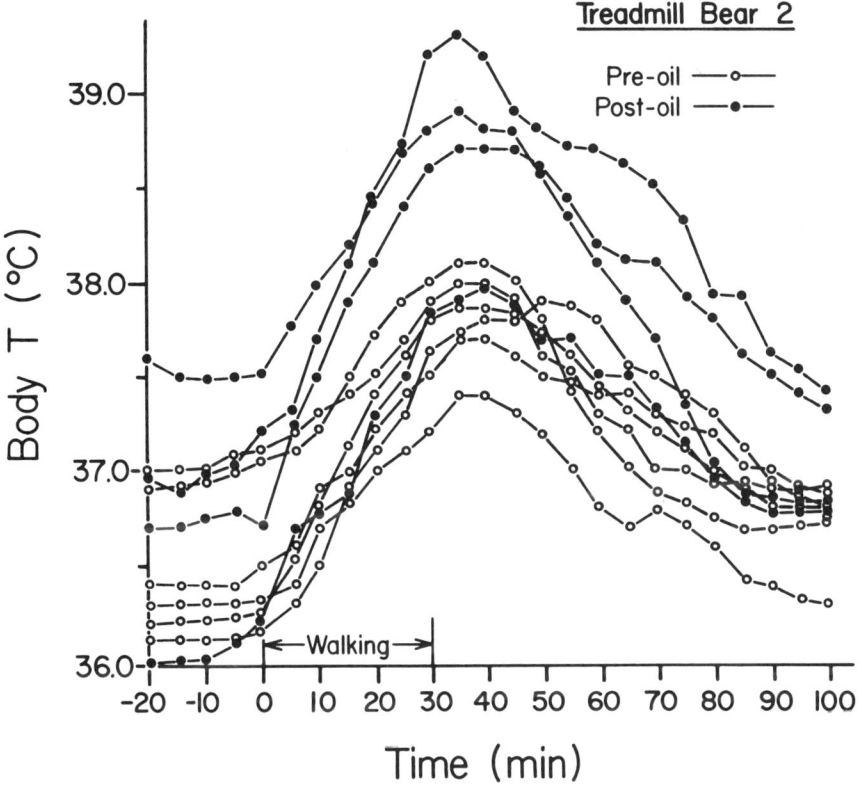

Figure 4. Deep body temperature of bear 2 before, during, and after 30 minutes of exercise at a walking speed of 5.4 km.hr^{-1}.

DISCUSSION

Oil Contact

Oil adhered to the polar bear fur on contact, suggesting that physical fouling of fur would occur rapidly. Oil distribution was uneven and depended on the point of contact with the water-oil surface. It did, however, become more evenly distributed with time as it was worked over and into the fur surface.

The results suggest that a polar bear crossing an oil-fouled lead would become physically contaminated by a substantial amount of oil. The experiment was not designed to explore the behaviour of the animals under conditions of free choice, and thus we can not address the question of probability of contact with any certainty.

Metabolism

Resting Metabolism

Due to the logistical limitations and the intractable nature of the animal, the resting metabolic rate of adult or sub-adult polar bears has not been previously reported. The basal metabolism of polar bear cubs of 12 kg or less has been documented (Blix and Lentfer, 1980; Scholander et al, 1950b). The control results of this study show mean resting metabolic values slightly higher than predicted (Fig.1). The mean minimal RMR, however, was as generally predicted (Kleiber, 1975). Although some semi-aquatic mammals may have acquired metabolic rates higher than those of similar sized terrestrial mammals as a compensation for high heat loss (Grant and Dawson, 1978; Iversen, 1972; Hart, 1962), there is no support for such a strategy in polar bears.

The increase in resting metabolic rate of the polar bears following oil fouling was consistent and substantial(Fig. 1). In vitro tests (Øritsland et al, 1981) illustrated a dramatic reduction in insulation, or increase in thermal conductance of polar bear fur, as a consequence of oil fouling. The evidence for thermal stress is further substantiated by the severe, and often violent, shivering observed in the period after oiling.

The response of homeotherms contacting crude oil strongly supports the probability of metabolic compensation in response to a reduction in insulation. McKewan and Koeling (1973) showed an oil-induced shift in the lower critical temperature of ducks from $12°C$ to $25°C$. Depending on the extent of oiling, the MR in air increased to 1.7 and 2 times the resting value in mallards (*Anas platyrhynchos*) and scaup (*Aythya sp.*) (McKewan and Koelink, 1973) and 2 to 2.4 RMR in dabbling ducks (Hartung, 1967). Holmes and Cronshaw (1977) suggested that dramatic changes in metabolism of oil-soaked birds were due entirely to physical changes of the plumage and not due to systemic toxicity or irritation of the skin. McKewan et al (1974) demonstrated a comparable response in muskrats, wherein the metabolic rate was dependent on the degree of oiling and increased by 120% following heavy fouling of the fur.

Davis and Anderson (1976) reported a lower peak weight in naturally oiled harbour seal pups(*Phoca vitulina*) but

cautioned that the difference may have been partly attributable to the greater experimental disturbance suffered by oiled pups. Kenyon(1974) noted that malnutrition was common in oil-contaminated fur seals. Costa and Kooyman(1980) found that oiling of 13% of the fur surface caused a 70% increase in metabolic rate of sea otters in 15°C water. Similarly, Kooyman et al(1976) showed that light oiling of 30% of the fur surface resulted in a 1.5x increase in MR of fur seals immersed in water.
The crude oil also may have caused a skin or peripheral tissue reaction or a more general malfunction of the thermoregulatory system.

Rice et al(1977) speculated that an increased breathing rate documented in fish was coupled with an increased energy demand for detoxification and elimination of hydrocarbons from the body tissue. Since the polar bears took up considerable quantities of oil (Engelhardt, 1981), the above explanation can not be rejected.

Activity Metabolism

The pre-oil cost of walking was higher than that predicted from general expressions scaling the energy cost of locomotion (Fedak and Seeherman, 1979; Taylor et al, 1970). Such an unpredictably high cost, also documented in other species (Chassin et al, 1976), has been discussed elsewhere in relation to polar bears (Øritsland et al, 1981).

The walking heat production prior to oiling was more than 9 times BMR. From a thermoregulatory viewpoint, the work load would generate excesss heat. Thus, a post-oil metabolic increase to counteract cold induced by changes in the insulative value of the fur was entirely unexpected. The degeneration of motor co-ordination, particularly evident in the later weeks of observation and suggested by the unwillingness or inability of bears 4 and 3 to complete their single post-oil trials, implies a possible change in the mechanical efficiency of walking. Although this may have accounted for the 24% increase in oxygen consumption, a stress-or toxicity-induced metabolic elevation should not be dismissed.

Stirling(1974) found that free-ranging polar bears under observation in the high arctic spent approximately 29% of their time walking. An increase in the cost of walking as a consequence of oiling could lead to changes in hunting strategies and/or an acceleration of the exhaustion of body energy stores.

Temperatures

Body Temperatures

The respiration chamber results show that a 38% rise in the metabolic rate was insufficient in bear 3 to sustain T_b at the normal pre-oil level. The different thermal strategies between oil-exposed individuals (Fig. 3; Table II) may reflect a difference in the degree of oiling, the energetic condition of the animals or other factors. It must be emphasized that the maintenance of normal T_b in oil-fouled animals, particularly arctic homeotherms, may occur concurrently with severe thermal stress. This may be accomplished by utilizing peripheral body tissues as insulation, although there is the potential of freezing injuries. The alternative response, an obligatory rise in metabolism as noted in this and other studies, is 'paid for' by an acceleration of the exhaustion of energy stores.

The pre-oil elevation of deep body temperature during exercise is in general agreement with previous observations of walking polar bears (Best et al, 1981; Best, 1976). The general elevation in T_b and the greater rise in T_b during post-oil exercise (Figure 4) was not predictable from an insulative viewpoint, but may conform to a fever response (Øritsland et al, 1981). Fever is produced by an upward displacement of a central 'reference point' temperature controlling the rate of metabolism (Stitt et al, 1974; Dubois, 1948). The general consequences of fever are an increase in heat production accompanied by an elevation of T_b and often T_s (Palmes and Park, 1965). The greater rise in T_b during post-oil exercise conforms with findings of fever in humans wherein there was a greater rise in T_b during the rising phase of fever than during the afebrile state (See Bligh, 1973 p235).

The documented rise in T_b in all animals was, however, less than that of a true pyrogen-induced fever. Exercise tests, post-oil, were also limited to one polar bear. Although the experimental evidence for oil acting as a pyrogen is limited (See also Øritsland et al, 1981), the subject deserves consideration in that normally acceptable techniques of polar bear capture by helicopter could cause dangerous overheating in oil-stressed individuals.

Skin Temperatures

Skin temperatures rose to a statistically higher level following oil-exposure and remained elevated for a period of approximately one week in all 3 bears (Table II). Normally healthy polar bears would compensate for a reduction in fur insulation by reducing their peripheral blood circulation, causing a corresponding increase in tissue insulation and decrease in skin or sub-cutaneous temperatures.

Engelhardt(1981), in a comprehensive assessment of the accumulation and clearance of the Midale oil fractions, documented extensive hydrocarbon loading of polar bear body fluids and tissues. The author suggested that pathological exposure likely did occur through inhalation and skin absorption. The dynamics of the uptake were, however, masked by the extensive hydrocarbon loading through ingestion by grooming. Juck(in Øritsland et al, 1981) recorded, during necropsy of 2 of the study animals, an oil covered epidermis which had a dry, crusty and swollen appearance that was most evident in a large, almost completely hairless patch along the mid back of bear 3. This confirms the contention of an extreme skin reaction to the application of oil.

Laboratory mammals have demonstrated an inflamatory skin reaction to topical application of petroleum hydrocarbons (Blood et al, 1979; Nau et al, 1966). There is little precedent in the area of marine mammals. Van Haaften(1973) found harbour seals with patches of oil on their fur, below which were inflamed areas of skin and in a few cases skin lesions. Engelhardt et al(1977) reported rapid absorption of hydrocarbons from Norman Wells crude oil into body fluids and tissues of ringed seals when exposed by both immersion and ingestion. Costa and Kooyman(1980) demonstrated that the subcutaneous temperature beneath the oiled fur of sea otters dropped $5-10°C$. This localized reduction in peripheral circulation was followed in time, however, by an unexpected increase in subcutaneous temperature to that approaching body temperature.

The thermal implications are serious. Thermoregulatory demands generally play a significant role in the distribution of blood flow between the body core and periphery, with periodic reduced temperature of the skin and extremities recognized as an important, if not essential, prerequisite of mammalian life in a cold northern environment (See Henshaw, 1978; Hart, 1962; Irving and Krog, 1955). The apparent elevation of skin temperature documented in this study compares with that of some tropical species at cold temperatures (Baust and Brown, 1980). Such animals must, as a consequence, pay the price of tremendous heat loss to the environment. Such a costly circulatory response could compound the effects of increased fur conductance and thereby augment an already potentially serious thermal imbalance.

The metabolic changes following oil contamination may be due to: 1) a normal metabolic compensation for a reduction in fur insulation; 2) a direct effect on the minimum level of energy turnover of the tissues; 3) a skin reaction in combination with the effects of reduced insulation or, finally; 4) a combination of all three mechanisms. The results of this study suggest that following oil exposure the thermoregulatory balance of a free ranging polar bear would be stressed. This could lead to serious alterations in activity patterns, hunting strategies, and ultimately survival.

ACKNOWLEDGEMENTS

Research was conducted at the Institute of Arctic Ecophysiology(formerly the Polar Bear Project), Churchill, Manitoba. Funding was provided by a grant from the Department of Indian and Northern Affairs Canada to N.A.Ø. The project was further supported by the Norwegian MAB project and an NSERC grant to F.R. Engelhardt. We would like to thank Dr. Engelhardt and the Federal Government, the Manitoba Dept. of Renewable Resources, and the N.W.T. Government for their valuable and continued support. A personal thankyou is extended to the citizens of Churchill and to the dedicated members of the Polar Bear Project who participated in the project including: D. Belme, C. Cuyler, S. Hanson, M. Leonard, P. Ratson and L. Stegenga.

LITERATURE CITED

Anon. 1970. Ecological effects of oil-Chedabucto Bay. Report of the Task Force, Operation Oil. Vol II. p.43-50.

Baker, J.M. 1976. Ecological changes in Milford Haven during its history as an oil port. IN: Marine Ecology and Oil Pollution. (J.M. Baker, ed) J. Wiley and Sons, N.Y. p 55-66.

Baust, J.G. and R.T. Brown. 1980. Heterothermy and cold acclimation in the arctic ground squirrel (Citellus undulatus). Comp. Bioch. Physiol. 67:447-452.

Beak Consultants Limited. 1974. Contingency Planning Considerations, Northern Oil and Pipelines. Environmental-Social Committee, Northern Pipelines Report No. 74-23.

Best, R.C., K. Ronald and N.A. Øritsland. 1981. Physiological indices of activity and metabolism in the polar bear. Comp. Biochem. Physiol. 69:177-185.

Best, R.C. 1976. Ecological energetics of the polar bear (Ursus maritimus Phipps 1774), M.Sc. thesis. University of Guelph, Guelph, Ontario. 136 pp.

Birchard, E.C., S.A.M. Conover, G. Greene, and A.S. Telford. 1978. Assessment of the ecological effects of an oil spill in an offshore subarctic environment. Proc. Conf. on Assess. Ecolog. Impacts of Oil Spills, Keystone, Col., p 835-855.

Bligh, J. 1973. Temperature Regulation in Mammals and Other Vertebrates. North-Holland Publ. Co., London. 436 pp.

Bligh, J. and K.G. Johnston. 1973. Glossary of terms for thermal physiology. J. Appl. Physiol. 35:941-961.

Blix, A.S. and J.W. Lentfer, 1979. Modes of thermal protection in polar bear cubs- at birth and on emergence from the den. Am. J. Physiol. 236:R67-R74.

Blood, D.C., J.A. Henderson and O.M. Radostits. 1979. Oil and petroleum product poisoning. In: Veterinary Medicine, 5th Ed. Bailliere-Tindall, London. p985-986.

Brownell, R.L. and B.J. LeBoeuf. 1971. California sea lion mortality:Natural or artifact? In: Biological and Oceanographical Survey of the Santa Barbara Channel Oil Spill 1969-1970. Biology and Bacteriology (D. Straughan, ed.), Vol. 1. Allan Hancock Foundation, U. of S. Calif., Los Angeles. p287-306.

Chassin, P.E., C.R. Taylor, N.C. Heglund and H.J. Seeherman. 1976. Locomotion in lions:energetic cost and maximum aerobic capacity. Zoology 49:1-10.

Costa, D.P. and G.L. Kooyman. 1980. Effects of oil contamination in the sea otter, Enhydra lutris. Report, Outer Cont. Shelf. Envir. Assess. Progr. NOAA, Alaska.

Davis, J.E. and S.S. Anderson. 1976. Effects of oil pollution on breeding grey seals. Mar. Poll. Bull. 7:115-118.

Dubois, E.F. 1948. Fever and the regulation of body temperature. Springfield:Thomas.

Engelhardt, F.R. 1981. Oil pollution in polar bears:exposure and clinical effects. Proc. 4th Arctic Mar. Oil Spill Progr. Tech. Sem., Ed. Alta., Envir. Can., Envir. Protec. Serv. p 139-179.

Engelhardt, F.R. 1978. Petroleum hydrocarbons in arctic ringed seals, Phoca hispida, following experimental oil exposure. Proc. Conf. Assess. Ecolog. Impacts Oil Spills, Am. Inst. Biol. Sci. p 614-628.

Engelhardt, F.R., J.R. Geraci, and T.G. Smith. 1977. Uptake and clearance of petroleum hydrocarbons in the ringed seal(Phoca hispida). J. Fish. Res. Brd. Can. 34:1143-1147.

Fedak, M.A. and H.J. Seeherman. 1979. Reappraisal of energetics of locomotion shows identical cost in bipeds and quadrupeds including ostrich and horse. Nature 282:713-716.

Geraci, J.R. and T.G. Smith. 1977. Consequences of oil fouling on marine mammals. In:Effects of Petroleum on Arctic and Sub-Arctic Marine Environments and Organsims. Vol.II. (D.C. Malins Ed.) Academic Press, N.Y. p399-410.

Geraci, J.R. and T.G. Smith. 1976. Direct and indirect effects of oil on ringed seals(Phoca hispida) of the Beaufort Sea. J.Fish. Res. Brd. Can. 33:1976-1984.

Geraci, J.R. and D.J. St. Aubin. 1980. Offshore petroleum resources development and marine mammals:a review and research recommendation. Mar. Fish. Rev. 42:1-12.

Grant, T.R. and T.J. Dawson. 1978. Temperature regulation in the platypus, Ornithorhynchus anatinus. Physiol. Zool. 51: 1-6.

Hart, J.S. 1962. Mammalian cold acclimation. In: Comparative Physiology of Thermoregulation. Vol.II. Academic Press, N.Y. p1-149.

Hartung, R. 1967. Energy metabolism in oil-covered ducks. J. Wildl. Manage. 31:798-804.

Henshaw, R.E. 1978. Peripheral thermoregulation:haematologic or vascular adaptations. J.Therm. Biol. 3:31-37.

Holmes, W.N. and J. Cronshaw. 1977. Biological effects of petroleum on marine birds. In: Effects of Petroleum on Arctic, Sub-arctic Marine Environments and Organisms. Vol.II. (D.C. Malins, ed.) Academic Press, N.Y. p 359-398.

Irving, L. and J. Krog. 1955. Temperature of the skin in the arctic as a regulator of heat. J.Appl. Physiol. 7:355-364.

Iversen, J.A. 1972. Basal energy metabolism of mustelids. J. Comp. Physiol. 81:341-344.

Kenyon, K.W. 1974. The effect of oil pollution on marine mammals. Dept. Int. 102, Statement Task Force B, Task Force on Alaska Oil Development. U.S. Govt. Proc. Report.

Kleiber, M. 1975. The Fire of Life. 2nd ed. R.E. Kreiger, Publ. Co., N.Y. 456 pp.

Kooyman, G.L., R.W. Davis, and M.A. Castellini. 1977. Thermal conductance of immersed pinniped and sea otter pelts before and after oiling with Prudhoe Bay crude oil. In: Fate and Effects of Petroleum Hydrocarbons in Marine Ecosystems and Organisms. (D.A. Wolfe, ed.) Permagon Press, N.Y. p 151-157.

Larsen, T. 1971. Capturing, handling and marking polar bears in Svalbard. J.Wildl. Man. 35:27-36.

McEwan, E.H. and A.F.C. Koelink. 1973. The heat production of oiled mallards and scaup. Can. J. Zool. 51:27-31.

McEwan, E.H., N.Aitchison and P.E. Whitehead. 1974. Energy metabolism of oiled muskrats. Can.J. Zool. 52:1057-1062.

Muller-wille, L. 1974. How effective is oil pollution legislation in arctic waters?: an example from Repulse Bay, N.W.T. Musk-ox 14:56-57.

Nau, C.A., J. Neal and M. Thornton. 1966. C_9-C_{12} fractions obtained from petroleum distillates: an evaluation of their potential toxicity. Arch. Environ. Hlth. 12:382-392.

Øritsland, N.A. 1970. Temperature regulation of the polar bear (Thalarctos maritimus). Comp. Biochem. Physiol. 37:225-233.

Øritsland, N.A. 1969. Deep body temperatures of swimming and walking polar bear cubs. J.Mammal. 50:380-382.

Øritsland, N.A. and T.G. Smith, 1975. Heat budget and insulation in marine mammals: Effect of crude oil on ringed seal pelts. Report to Beaufort Sea Project. Arctic Biol. Station, St. Anne de Bellevue, P.Q., Canada.

Øritsland, N.A, F.R. Engelhardt, F.A. Juck, R.J. Hurst and P.D. Watts. 1981. Effect of crude oil on polar bears. Dept. Indian Affairs and Northern Develop., Publ. No. QS-8283-020-EE-A1. (in press).

Øritsland, N.A., C. Jonkel and K. Ronald. 1976. A respiration chamber for exercising polar bears. Norw. J.Zool. 24:65-67.

Palmes, E.D. and C.R. Park. 1965. The regulation of body temperature during fever. Arch. Environ. Hlth. 11:749-759.

Pauls, R.W. 1981. Energetics of the red squirrel: a laboratory study of the effects of temperature, seasonal acclimatization, use of the nest and exercise. J.Therm. Biol. 6:79-86.

Rice, S.D., J.W. Short and J.F. Karinen. 1977. Comparative oil toxicity and comparative animal sensisity. In: Fate and Effects of Petroleum Hydrocarbons on Marine Ecosystems and Organisms (D.A.Wolfe, Ed) Permagon Press, N.Y. p 78-94.

Scholander, P.R., V.Walters, R. Hock and L. Irving. 1950a. Body insulation of some arctic and tropical mammals and birds. Biol. Bull. 99:225-236.

Scholander, P.F., R.Hock, V.Walters, F. Johnston and L.Irving. 1950b. Heat regulation in some arctic and tropical mammals and birds. Biol. Bull. 99:237-258.

Smith, T.G. and I. Stirling. 1975. The breeding habitat of the ringed seal (Phoca hispida): The birth lair and associated structures. Can.J. Zool. 53:1297-1305.

Stirling, I. 1980. The biological importance of polynyas in the Canadian Arctic. Arctic 33:303-315.

Stirling, I. 1974. Mid-summer observations on the behaviour of wild polar bears (Ursus maritimus). Can.J. Zool. 52:1191-1198.

Stirling, I. and H.P.L. Kiliaan. 1980. Population ecology studies of the polar bear(Ursus maritimus) in northern Labrador Can Wild. Serv. Occ. Pap. No. 42. 19pp.

Stirling, I. and P.B. Latour. 1978. Comparative hunting abilities of polar bear cubs of different ages. Can.J. Zool.56: 1768-1772.

Stirling, I. and E.H. McEwan. 1975. The caloric value of whole ringed seals(Phoca hispida) in relation to polar bear (Ursus maritimus) ecology and hunting behaviour. Can.J. Zool. 53: 1021-1027.

Stitt, J.T., J.D. Hardy and J.A. Stolwijk. 1974. PGE_1 fever: its effect on thermoregulation at different low ambient temperatures. Am.J.Physiol. 227:622-629.

Taylor, C.R., K. Schmidt-Nielsen and J.L. Raab. 1970. Scaling energetic cost of running to body weight of animals. Am. J. Physiol. 219:1104-1109.

Van Haaften, J.L. 1973. Die Bewirtschaftung von seehunden in den Niederlanden. Bietrage zur Jagd-und Wildforschung 8:345-349.

Wadhams, P. 1976. Sea ice topography in the Beaufort Sea and its effect on oil containment. Aidjex Bull. 33:1-52.

Weller, G. 1976. Outer continental shelf. Environmental assessment program in the Beaufort Sea. Arctic Bull. Vol. 2:125-142.

Williams, T.D. 1978. Chemical immobilization, baseline haematological parameters and oil contamination in the sea otter. Report no. MMC-77/06 to U.S. Marine Mammal Commission, Wash. D.C. 27pp.

EFFECTS OF PETROLEUM HYDRO-
CARBONS ON BRANCHIAL ADENOSINE
TRIPHOSPHATASES (ATPases) IN
FRESH AND SALT WATER ACCLIMATED
RAINBOW TROUT (SALMO GAIRDNERI)

M.P. Wong
University of Ottawa

F.R. Engelhardt
Department of Indian Affairs
and Northern Development

ABSTRACT

Adenosine triphosphatases (ATPases) are the enzymes generally associated with the transport of ions across the gill epithelium in teleost fish. The effects of particulate Norman Wells and Venezuelan crude oils and of the polycyclic aromatic hydrocarbon benz(a)anthracene on these enzymes were studied in vivo. The five enzymes, Na^+-K^+-, $Na^+-NH_4^+-$, HCO_3^--, $Mg^{++}-$, and $Ca^{++}-$activated ATPases assayed were microsomal preparations from the gills of fresh (FW) and salt water (SW) acclimated rainbow trout. Elevated Na^+-K^+, and depressed Mg^{++} ATPase activities were found in salt water fish when compared to FW fish. The petroleum hydrocarbons generally had disruptive effects on the enzymes. The enzymes from FW fish were more sensitive to the hydrocarbons than those from SW fish. This evidence provides a biochemical basis for the ionic imbalance found in oil-exposed fish.

INTRODUCTION

Acute and chronic effects have been found in organisms in the aquatic environment exposed to pollution from petroleum hydrocarbons (Malins, 1977). Fish captured in areas of oil spills have shown structural gill damage, suggesting that physiological functions may be impaired (Blanton and Robinson, 1972). Studies in this laboratory and in others have found that exposure to crude oils and their fractions resulted in alterations in gill morphology and iono-osmoregulatory dysfunction (Wong et al, 1981; Engelhardt et al,1981; McKeown and March, 1978; DiMichele and Taylor, 1978). The direct cause and effect relationships have yet to be established.

It has been shown that crude oils and aromatic hydrocarbons can affect enzyme activity in fish (Yarbrough et al, 1976; Payne, 1976; Varanasi and Malins, 1977; Sabo and Stegeman, 1977; Nava and Engelhardt, 1981). It is generally accepted that branchial adenosine triphosphatases (ATPases) are the enzymes associated with active transport of ions across the fish gills (Maetz, 1971; Jampol and Epstein, 1970; Epstein et al, 1967). It is possible that crude oils and their constituents may interfere with the activity of these enzymes, thereby promoting ionic disturbances. This has been shown in DDT-exposed fish (Janicki and Kinter, 1971; Leadem et al, 1974). The present study will provide insight into the effects of petroleum hydrocarbons on these enzymes in rainbow trout.

METHODS AND MATERIALS

Juvenile rainbow trout, Salmo gairdneri (Richardson), 100 to 200 g, were obtained from a commercial hatchery (Thistle Springs Trout Farm, Ashton, Ontario) and were kept under laboratory conditions for 4 weeks before experimentation. Groups were acclimated to freshwater (FW) or 21 parts per thousand salt water (SW), pH= 7.4-7.6, and held on a 12L-12D photoperiod regime for at least 3 weeks before use. Water quality parameters were monitored periodically during acclimation. Nitrogenous waste (ammonia) and chlorine levels were negligible. Oxygen concentrations were kept above 75% saturation.

The petroleum hydrocarbons used in the exposure tests were three-day weathered Venezuelan, Norman Wells crude oils, or benz(a)anthracene (30 µg/L). The exposure regimes were the same as those described previously (Wong et al,1981; Engelhardt et al,1981). Following the 7 day exposures, the fish were removed from the aquaria, stunned, weighed and exsanguinated. This was performed at the same time of day to preclude any possible diurnal changes in the enzymes (Zaugg and Wagner, 1973). Extraction and preparation of the gills for ATPase activity determinations were similar to those described by Watson and Beamish (1980) and are summarized in Table I.

TABLE I. Preparation of ATPase Enzymes from the Gills of Rainbow Trout

---

Gills dissected into 0.9% NaCl ($0-4°C$)
↓
Filaments removed, weighed, and placed in 0.25M sucrose, 5mM EDTA, 40 mM Tris-HCl medium (10% $^w/_v$; $0-4°C$; pH= 7.4)
↓
Homogenized in Sorvall (Omni-mixer) 90 s
↓
Ground for 15 strokes (Teflon grinder), with 0.1% Na-deoxycholate added after 8 strokes
↓
Centrifuged (900x g) 15 min. ($0-4°C$) in a Sorvall RC-2B
↓
Supernatant centrifuged (10,000x g) 30 min. ($0-4°C$)
↓
Supernatant centrifuged (35,000x g) 2 hr. ($0-4°C$)
↓
Final supernatant discarded, microsomal pellet washed, recentrifuged (35,000x g) 30 min. ($0-4°C$)
↓
Final pellet resuspended in 3-6 ml of homogenizing medium ($0-4°C$) without deoxycholate
↓
0.1 ml of microsomal suspension placed in 1.8 ml of incubation media, incubated ($18°C$, 30-60 min.). Reaction started by addition of 0.1 ml 100 mM vanadium-free Na_2ATP
↓
Reaction stopped with 0.5 ml of 30% TCA ($0°C$)
↓
ATP hydrolysis of P_i determined by an automated modification of Fiske and Subbarow (1925) procedure (Technicon method, No. SE2-004F3)

The levels of substrate ions and pH employed were those that yield maximum specific activities (Watson and Beamish, 1980). The composition of the incubation media used in assessing the various ATPase activity is shown in Table II. In 1.8 ml of incubation medium, 0.1 ml of gill microsomal suspension was added. Enzyme reactions were initiated by the addition of 0.1 ml of 100 mM vanadium-free disodium adenosine triphosphate (ATP) (Sigma) after 15 min. pre-incubation at the assay temperature. The assays were performed at $18 \pm 1°C$ for 30-60 min. The reaction was terminated by placing the incubation tubes on ice and addition of 0.5 ml of 30% TCA. The suspensions were centrifuged (3000x g) for 10 min. and aliquots were removed for duplicate determinations of inorganic phosphate (P_i) by an automated modification of the Fiske and Subbarow (1925) procedure (Technicon method, No. SE2-004F3) using a Technicon Auto Analyzer II.

TABLE II. Branchial ATPase Incubation Media Composition

	Na^+-K^+	Na^+-NH_4^+	HCO_3^-	Mg^{++}	Ca^{++}
Tris-HCl	20 mM	20 mM	20 mM	20 mM	20 mM
pH	7.4	7.4	7.4	7.4	7.4
Mg^{++}	5 mM	5 mM	5 mM	5 mM	5 mM
Na^+	100 mM	100 mM	---	---	---
NH_4^+	---	40 mM	---	---	---
K^+	20 mM	---	---	---	---
HCO_3^-	---	---	50 mM	---	---
Ca^{++}	---	---	---	---	3 mM
ouabain	---	---	1 mM	1 mM	---
SCN^-	---	---	---	5 mM	---

Activities of Na^+-K^+-, and Na^+-NH_4^+-stimulated ATPase were calculated as the difference between activities in the presence and absence of the cardiac glycoside, ouabain (1 mM). Activity of HCO_3^- stimulated ATPase was calculated as the difference between activities in the presence and absence of (5 mM) KSCN. Magnesium ion and calcium ion dependent activities were estimated by the difference between blank determinations and those with the respective cations. All activities are expressed as the number of micromoles of inorganic phosphate liberated per milligram of protein per hour ($\mu M\ P_i \cdot mg^{-1} \cdot hr^{-1}$). Protein concentration was determined using bovine serum albumen as the standard (Lowry et al, 1951).

Analysis of variance was used for comparisons of the data. Significance attributed to statistical differences was accepted at values of p =0.05 or better.

RESULTS

It is apparent from Figure 1 that significant differences in gill ATPase activities were found between fresh and salt water adapted trout. The Na -K - stimulated ATPase activity was elevated in the salt water fish and magnesium dependent ATPase activity was higher in the freshwater fish. No significant changes were found for the HCO_3^- stimulated ATPase activity between the two salinities.

Petroleum hydrocarbons were found to affect the five gill enzymes measured, specifically Na -K -stimulated ATPase (Figure 2), $Na^+-NH_4^+$-stimulated ATPase (Figure 3), HCO_3^--stimulated ATPase (Figure 4), Mg^{++}-dependent ATPase (Figure 5), and Ca^{++}-dependent ATPase (Figure 6). In Venezuelan crude oil exposed fish, significant inhibition of Na^+-K^+-ATPase activity were found in both fresh and salt water conditions. The HCO_3^--ATPase activity decreased and increased in freshwater and salt water, respectively, while Mg^{++}-ATPase was stimulated in freshwater and showed no changes in salt water. Norman Wells crude oil inhibited Na^+-K^+-, HCO_3^--, $Na^+-NH_4^+$-, Ca^{++}-ATPases and stimulated Mg^{++}-ATPase in freshwater. In salt water the Mg^{++}- and $Na^+-NH_4^+$-dependent activities decreased and the Ca^{++}-stimulated. Benz(a)anthracene exposures did not produce significant changes in enzyme activity in freshwater but did inhibit Na^+-K^+-ATPase in salt water.

DISCUSSION

The present study has clearly indicated that branchial ATPase activities were altered *in vivo* in exposed fish, from both fresh and salt water. This suggests that disruptions in the activities of ionoregulatory enzymes may contribute to the detrimental effects of sublethal concentrations of petroleum hydrocarbons. This correlates well with the ionic imbalance found in fish using the same exposure.(Wong et al, 1981; Engelhardt et al, 1981). Other workers have found that fish ATPase activities of teleosts were affected by various xenobiotics, such as pesticides, heavy metals, and phenol, a constituent of refined coal effluent and a minor component of petroleum (Table III). It is somewhat surprising that benz(a)anthracene did not cause major disruptions in the enzymes since polycyclic aromatic hydrocarbons are known to be toxic (Neff, 1979). The benz(a)anthracene level used in the present study, although appreciably less than that used in other studies, was low but environmentally relevant. Further, because such hydrocarbons selectively accumulate in fish, they may ultimately have a deleterious impact of this nature.

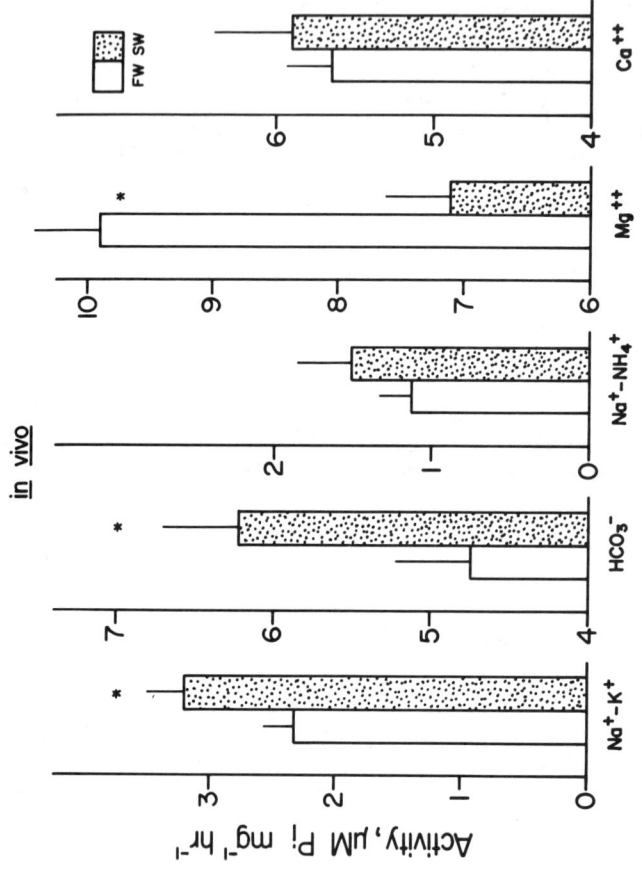

Figure 1: Gill ATPase activities in fresh and salt water acclimated trout. N= 6. (*) denotes statistical significance at the 0.05 level or better. (Errata, see page 296.)

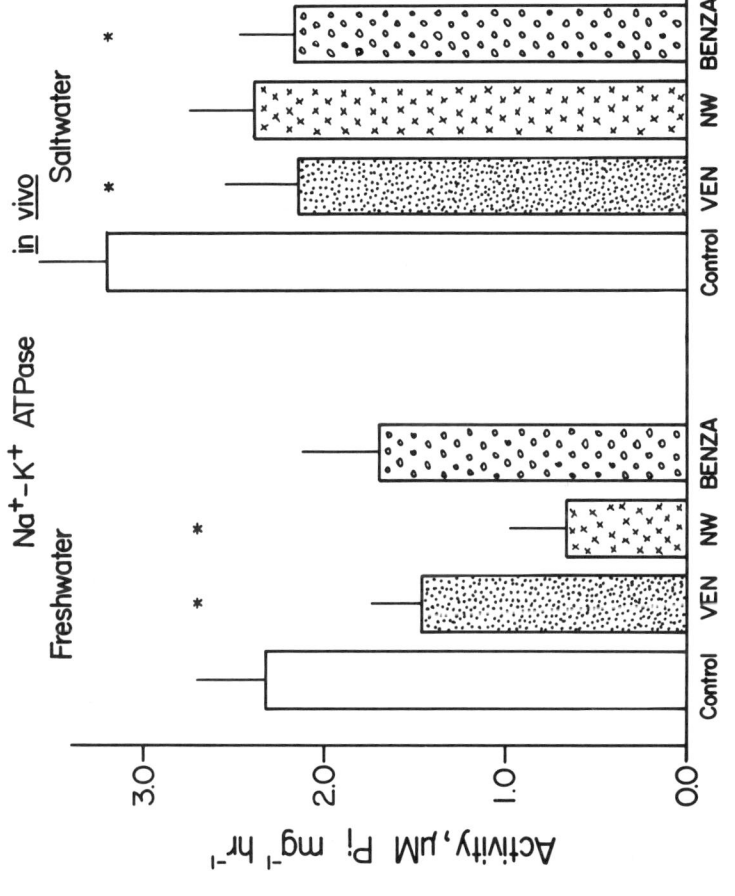

Figure 2: Sodium-potassium stimulated ATPase activities in gills of fish exposed to petroleum hydrocarbons in fresh and salt water. VEN= Venezuelan crude oil; NW= Norman Wells crude oil; BENZA= Benz(a)anthracene. N=6, except VEN=4. (*) denotes statistical significance at 0.05 or better.

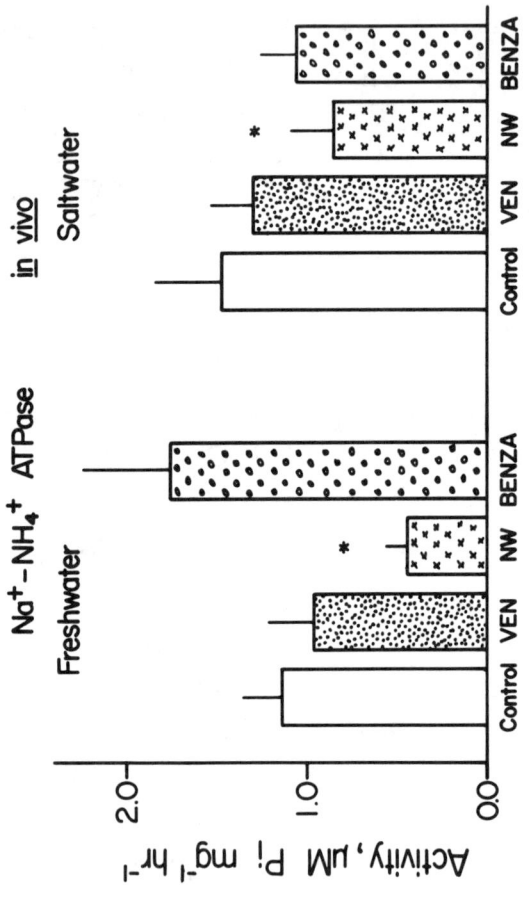

Figure 3: Sodium-ammonium stimulated ATPase activities in gills of fish exposed to petroleum hydrocarbons in fresh and salt water. VEN= Venezuelan crude oil; NW= Norman Wells crude oil; BENZA= Benz(a)anthracene. N=6, except VEN= 4. (*) denotes statistical significance at 0.05 or better.

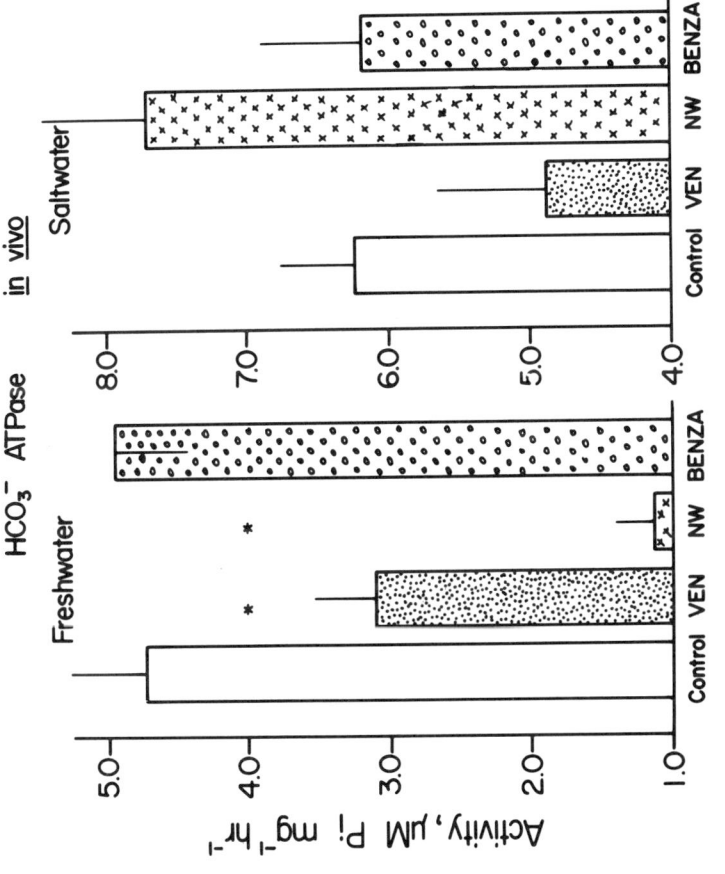

Figure 4: Bicarbonate stimulated ATPase activities in gills of fish exposed to petroleum hydrocarbons in fresh and salt water. VEN= Venezuelan crude oil; NW= Norman Wells crude oil; BENZA= Benz(a) anthracene. N=6, except VEN=4. (*) denotes statistical significance at 0.05 or better.

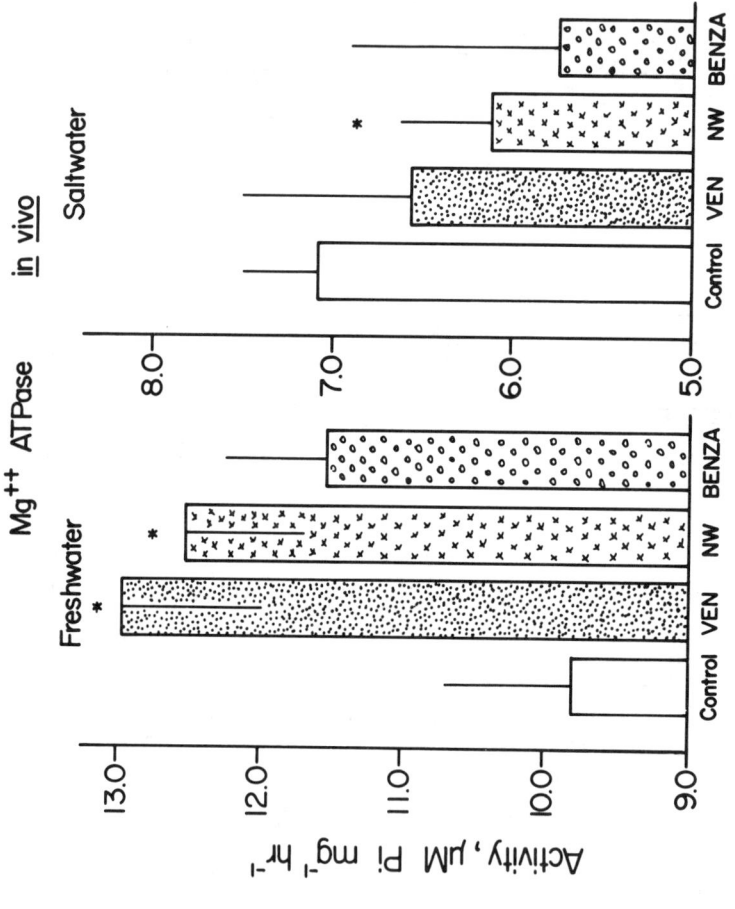

Figure 5: Magnesium-dependent ATPase activities in gills of fish exposed to petroleum hydrocarbons in fresh and salt water. Ven= Venezuelan crude oil; NW= Norman Wells crude oil; BENZA= Benz(a) anthracene. All N=6, except VEN= 4. (*) denotes statistical significance at 0.05 or better.

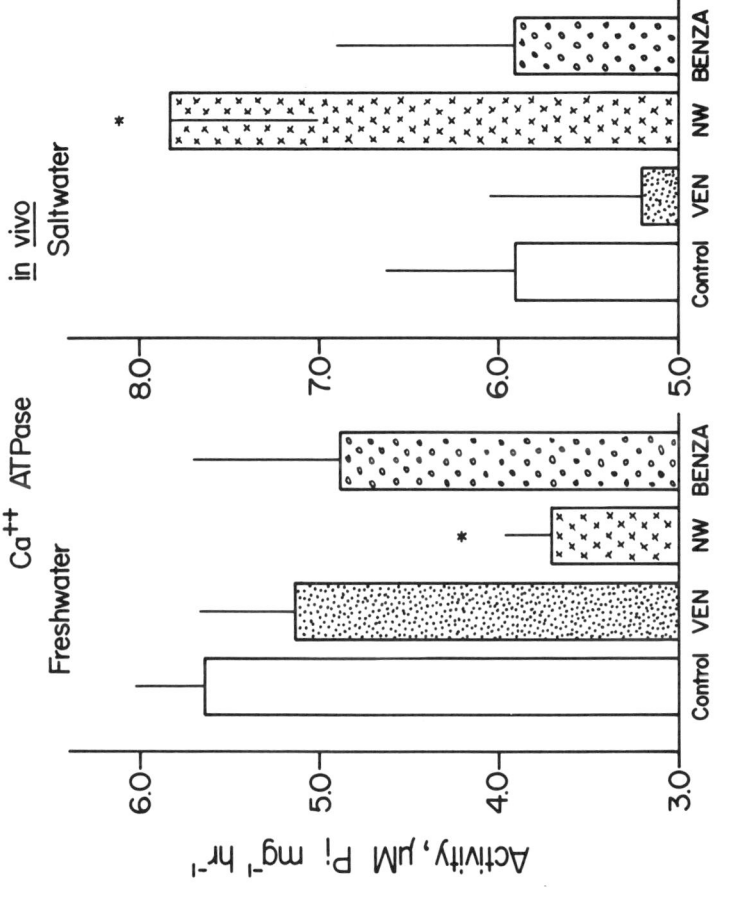

Figure 6: Calcium-dependent ATPase activities in gills of fish exposed to petroleum hydrocarbons in fresh and salt water. VEN= Venezuelan crude oil; NW= Norman Wells crude oil; BENZA= Benz(a) anthracene. N= 6, except Ven =4. (*) denotes statistical significance at 0.05 or better.

TABLE III. The Effects of Various Contaminants on Teleost Gill ATPase Activities, (+), stimulation and (-), inhibition.

SPECIES	Na^+-K^+	$Na^+-NH_4^+$	HCO_3^-	Mg^{++}	Ca^{++}	REFERENCE
Salmo gairdneri	DDT (-)	---	---	DDT (-)	---	(Davis and Wedemeyer, 1971)**
Salmo gairdneri	DDT (-)	---	---	DDT (-)	---	(Leadem et al, 1974)*
Oncorhynchus kisutch	Cu^{++} (-)	---	---	---	---	(Lorz and McPherson, 1976)*
Fundulus heteroclitus	DDT (-)	---	---	---	---	(Miller and Kinter, 1977)*
Oncorhynchus kisutch	Zn^{++} (-) Cu^{++} (-) Cd^{++} (-)	---	---	---	---	(Lorz et al, 1978)*
Rutilus rutilus	---	---	---	---	Cu^{++} Pb^{++} (-) Zn^{++} Hg^{++}	(Shepherd and Simkiss, 1978)**
Salmo gairdneri	phenol (-)	---	---	phenol (-)	---	(Poston, 1979)*
Salmo gairdneri	Zn^{++} (+)	Zn^{++} (+)	---	Zn^{++} (+)	Zn^{++} (+)	(Watson and Beamish, 1980)*
Salmo gairdneri	Zn^{++} (-)	Zn^{++} (-)	---	Zn^{++} (-)	Zn^{++} (-)	(Watson and Beamish, 1981)**

(*) – in vivo
(**) – in vitro

Other enzymes in fish have been studied as potential indices of environmental stress due to oil exposure. The β-glucuronidase activities in muscle and alkaline phosphatase activities in gills were elevated, while malic dehydrogenase was depressed in the livers of oil treated mullet, Mugil cephalus (Yarbrough et al, 1976). Sabo and Stegeman (1977) showed that lipid metabolism was altered in the livers of No. 2 fuel oil contaminated Fundulus heroclitus. Also, fish captured in oil polluted areas have higher levels of benzopyrene hydroxylase activity in the liver and gill tissue (Payne, 1976). Altered levels of mixed-function oxidases have often been found in various tissues from fish challenged by oil or its fractions (Varanasi and Malins, 1977). These findings further emphasize that petroleum hydrocarbons can induce disruptions in key physiological and biochemical processes in fish.

Alterations in ATPase activity and disturbances in osmoregulatory capacity may reduce the ability of fish to successfully adapt to varying salinities. Field studies have shown that Atlantic salmon, Salmo salar, will abort their upstream migration when they encounter copper-zinc mining effluents (Saunders and Sprague, 1967). Also, coho salmon, Oncorhynchus kisutch, exposed to heavy metals had depressed gill Na-K-ATPase activity. This resulted in reductions in the number of fish that were able to successfully migrate downstream to the sea. Further, it was evident these fish were unable to tolerate seawater (Lorz et al, 1978). Since many migrating fish, such as salmonids, anguillids, and clupeids represent valuable resources, the impact of oil contamination from spills and industrial discharges may reduce returning numbers as well as inducing mortalities due to ionic stress.

ACKNOWLEDGEMENTS

The authors gratefully acknowledge the financial support provided by Natural Sciences and Engineering Research Council of Canada and by Imperial Oil Ltd. to F.R. Engelhardt. We thank Dr. T.A. Watson of Simon Fraser University of Burnaby, British Columbia, for providing the enzyme assay procedures. Mr. M.E. Duey receives special thanks for his assistance. We appreciate the valuable and stimulating discussions with Dr. T.W. Moon during the course of this study. We also wish to thank Mr. G. Ben-Tchavtchavadze for his excellent work on the photographic plates.

LITERATURE CITED

Blanton, W.G. and M.C. Robinson. 1973. Some acute effects of low-boiling petroleum fractions on the cellular structures of fish gills under field conditions. In: D.G. Ahearn and S.P. Meyers (eds.), The Microbial Degradation of Oil Pollutants. Louisiana State University, Center for Wetland Resources Publ. LSU-SG-73-01. p.265-273.

Davis, P.W. and G.A. Wedemeyer. 1971. Na^+-K^+-activated ATPase inhibition in rainbow trout : a site for organochlorine pesticide toxicity? Comp. Biochem. Physiol. 40B., 823-827.

DiMichele, L. and M.H. Taylor. 1978. Histopathological and physiological responses of Fundulus heroclitus to naphthalene exposure. J. Fish. Res. Board Can. 35, 1060-1066.

Engelhardt, F.R., M.P. Wong, and M.E. Duey. 1981. Hydromineral balance and gill morphology in rainbow trout, Salmo gairdneri, acclimated to fresh and sea water, as affected by petroleum exposure. Aquat. Toxicol. (in press).

Epstein, F.H., A.I. Katz, and G.E. Pickford. 1967. Sodium and potassium activated adenosine triphosphatase of gills : role in adaptation of teleosts to salt water. Science, 156, 1245-1247.

Fiske, C.H. and Y. Subbarow. 1925. The colorimetric determination of phosphorus. J. Biol. Chem. 66, 375-400.

Jampol, L.M. and F.H. Epstein. 1970. Sodium-potassium activated adenosine triphosphatase and osmotic regulation by fishes. Am. J. Physiol. 218, 607-611.

Janicki, R.H. and W.B. Kinter. 1971. DDT: disrupted osmoregulatory events in the intestine of the eel, Anguilla rostrata adapted to sea water. Science, 173; 1146-1148.

Leadem, T.P., R.D. Campbell, and D.W. Johnson. 1974. Osmoregulatory response to DDT and varying salinities in Salmo gairdneri. I. Gill Na, K-ATPase. Comp. Biochem. Physiol. 49A. 197-204.

Lorz, H.W. and B.P. McPherson. 1976. Effects of copper or zinc in fresh water on the adaptation to sea water and ATPase activity and the effects of copper on migratory disposition of coho salmon, Oncorhynchus kisutch. J. Fish. Res. Board Can. 33, 2023-2030.

Lorz, H.W., R.H. Williams and C.A. Fustish. 1978. Effects of several metals on smolting of coho salmon. U.S. Environmental Protection Agency, EPA-600/3-78-090, 85 pp.

Lowry, O.H., N.J. Rosebrough, L.A. Farr and R.J. Randall. 1951. Protein measurements with the Folin phenol reagent. J. Biol. Chem. 193, 265-275.

Maetz, J. 1971. Fish gills : Mechanisms of salt transfer in freshwater and seawater. Phil. Trans. Roy. Soc. Lond. 262B. 209-249.

Malins, D.C. 1977. Effects of Petroleum on Arctic and Sub-Arctic Marine Environments and Organisms. Academic Press, N.Y. 500 p.

McKeown, B.A. and G.L. March. 1978. The acute effect of Bunker C oil and an oil dispersant on : serum glucose, serum sodium and gill morphology in both freshwater and seawater acclimated rainbow trout, Salmo gairdneri. Water Res. 12, 157-163.

Miller, D.S. and W.B. Kinter, 1977. DDT inhibits nutrient absorption and osmoregulatory function in Fundulus heteroclitus. In: Physiological Responses of Marine Biota to Pollutants. Academic Press, N.Y., pp. 63-74.

Nava, M.E. and F.R. Engelhardt. 1981. Induction of mixed function oxidases in the American eel, Anguilla rostrata. Arch. Environm. Contam. Toxicol. (in press).

Neff, J.M. 1979. Polycyclic Aromatic Hydrocarbons in the Aquatic Environment. Applied Science Publishers Ltd. London, 262 p.

Payne, J.F. 1976. Field evaluation of benzopyrene hydroxylase induction as a monitor for marine petroleum pollution. Science, 191, 945-946.

Poston, T.M. 1979. Effects of phenol on ATPase activities in crude gill homogenates of rainbow trout. Pacific Northwest Laboratory, U.S. Dept. of Energy, Richland, Washington, 52 p.

Sabo, D.J. and J.J. Stegeman. 1977. Some metabolic effects of petroleum hydrocarbons in marine fish. In: Physiological Responses of Marine Biota to Pollutants. (A. Calabrese and J.F. Vernberg, eds.) Academic Press, N.Y. p.279-287.

Saunders, R.L. and J.B. Sprague. 1967. Effects of copper-zinc mining pollution on a spawning migration of Atlantic salmon. Water Res. 1, 419-432.

Shephard, K. and K. Simkiss. 1978. The effects of heavy metal ions on Ca^{++}-ATPase extracted from fish gills. Comp. Biochem. Physiol. 61B, 69-72.

Varanasi, U. and D.C. Malins. 1977. Metabolism of petroleum hydrocarbons: accumulation and biotransformation in marine organisms. In: Effects of Petroleum on Arctic and Subarctic Marine Environments and Organisms. (D.C. Malins, ed.). Academic Press, N.Y. p.175-270.

Watson, T.A. and F.W.H. Beamish. 1980. Effects of zinc on branchial ATPase activity in vivo in rainbow trout, Salmo gairdneri. Comp. Biochem. Physiol. 66C, 77-82.

Watson, T.A. and F.W.H. Beamish. 1981. The effects of zinc on branchial adenosine triphosphatase enzymes in vitro from rainbow trout, Salmo gairdneri. Comp. Biochem. Physiol. 68C, 167-173.

Wong, M.P., M.E. Duey, and F.R. Engelhardt. 1981. Stress, osmoregulatory imbalance, and gill histopathology in Salmo gairdneri exposed to sublethal levels of fresh and weathered particulate hydrocarbons. Can. J. Zool. (submitted).

Yarbrough, J.D., J.R. Heitz, and J.E. Chambers. 1976. Physiological effects of crude oil exposure in the striped mullet, Mugil cephalus. Life Sci. 19, 755-760.

Zaugg, W.S. and H.H. Wagner. 1973. Gill ATPase activity related to parr-smolt transformation and migration in steelhead trout, Salmo gairdneri : influence of photoperiod and temperature. Comp. Biochem. Physiol. 45B, 955-965.

ERRATUM ADDED IN PRESS:

p. 286, Figure 1. No significant differences occurred in the HCO_3 stimulated ATPase activities from FW and SW fish. The values (mean ± Standard deviation) should be 4.72 ± 0.49 for FW and 5.24 ± 0.43 for SW.

EFFECTS OF PETROLEUM
HYDROCARBONS ON TROPHIC
FACTORS OF DAPHNIA PULEX

F. R. Engelhardt
    Indian Affairs & Northern
    Development, Canada
R. G. Trucco
    University of Ottawa, Canada
C. K. Wong
    University of Toronto,
    Erindale, Canada
J. R. Strickler
    Australian Institute of
    Marine Sciences

INTRODUCTION

The study summarized here outlines findings on the influence of petroleum hydrocarbons on the zooplankton trophic level, specifically those factors which may affect the food web by their dynamic role. The position of Daphnia pulex, the model experimental zooplankter, in the food web is shown in Figure 1. Factors such as zooplankton density and reproductive potential, metabolic rate and consequent food requirements, feeding behavior, susceptibility to predation, and bioaccumulation of petroleum hydrocarbons were examined as being of potential influence on food web dynamics.

Exposure to petroleum was carried out by several vectors, as oil-in-water emulsions, as water soluble fractions (WSF), or as present in algal food. Two crude oils, an inert paraffin oil, and selected aromatic hydrocarbons were tested over a range of concentrations to 100 ppm (μL oil/L water). The effects of weathered oil were differentiated from those of fresh oil emulsions.

RESULTS AND DISCUSSION

Crude Oil Characteristics

Norman Wells (NW) crude oil is a light arctic crude with an aromatic proportion of 15 to 20%. Venezuelan (VEN) crude oil is a medium weight crude with a larger component of aliphatic and aromatic fractions in the middle boiling range. Weathering by evaporation and dissolution resulted in major losses within hours of the low molecular weight fractions of

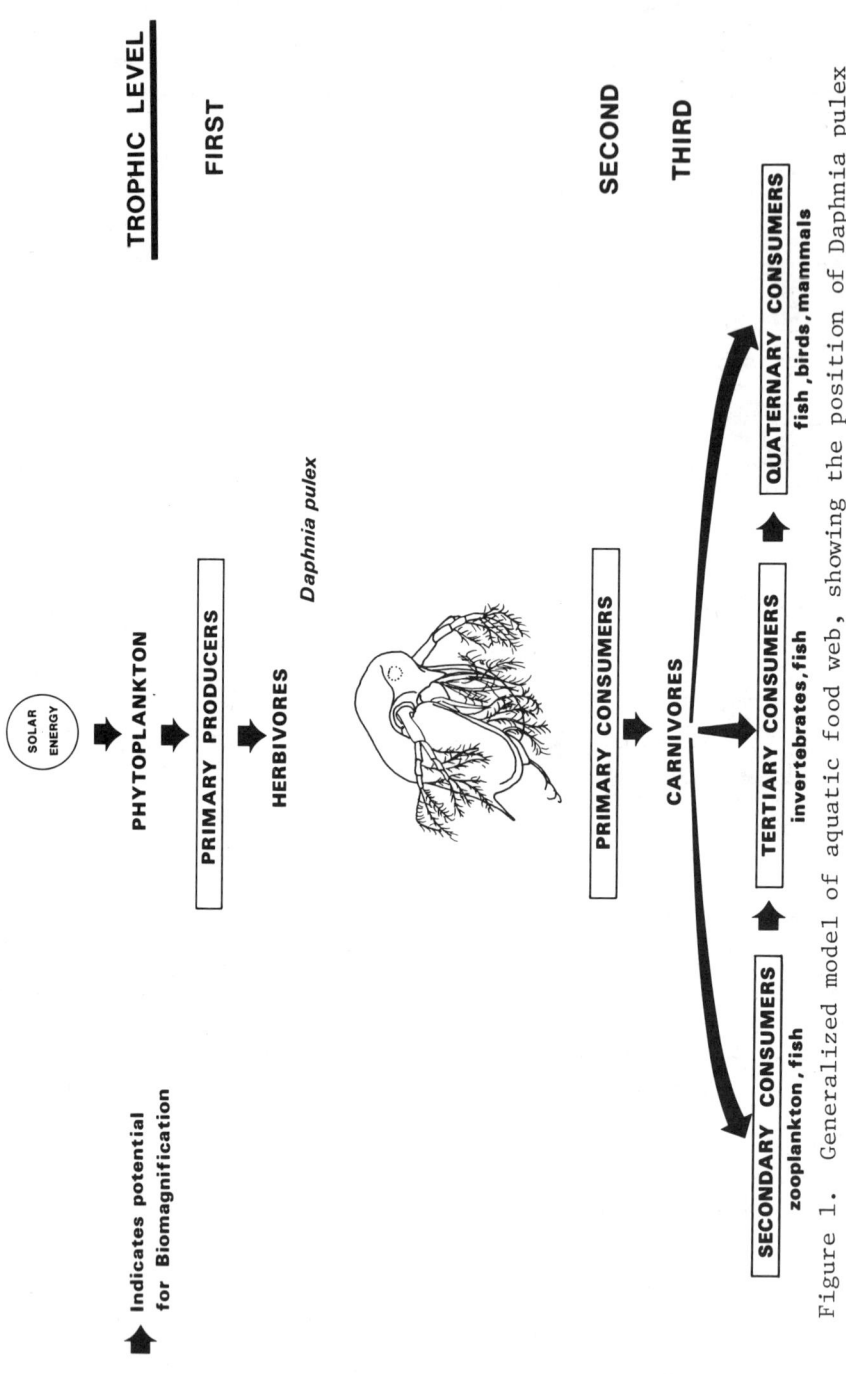

Figure 1. Generalized model of aquatic food web, showing the position of Daphnia pulex as a model herbivorous zooplankter.

NW crude. Emulsions were prepared for the testing sequences by homogenization (see Engelhardt et al., 1981, Aquatic Toxicology 1). Fresh crude and paraffin oils at 100 ppm gave a particle density of about 20 thousand particles (5 to 10 µm) per mm^3. Both particle number and concentration decreased to about one third during 3 days of weathering in a test solution.

Survival

Test populations of D. pulex were exposed to particulate (emulsion) fresh NW crude for up to 192 hours (emulsions were renewed every 72 hours) at nominal concentrations of 100 ppm or less (see Wong et al., 1981, Bull Environ. Contam. Toxicol. 26).

The toxic effect of oil emulsions increased with concentration. Neither paraffin oil nor crude oil up to 5 ppm had any survival effect. Total mortality resulted at 50 and 100 ppm. Weathered oil was less toxic, either because of a loss of the more acutely toxic low molecular weight fractions or because of a decrease in total concentration.

Fecundity

Reproductive potenial was tested as in the survival tests described above, and was found to be affected by oil exposure as low as 1 ppm. Both the percentage of young produced and the brood size were lower. The physical presence of oil particles contributed to the overall toxic effect in that paraffin particles also decreased brood size. A separate test showed that a minimum exposure time of 3 hours at 50 and 100 ppm was required to depress fecundity.

Both the decrease in zooplankton survival and their reduced fecundity would lower food availability for the next higher consumer level, that of carnivores such as fish and other zooplankter species.

Metabolism

Metabolic rate was assessed by measuring oxygen comsumption of D. pulex with a Gilson microrespirometer. Several modes of exposure were tested, differentiating the effects of NW from VEN oil, of concentration, of emulsions from WSF, and of duration of exposure. Control values averaging at 0.162 µL O_2/hour/D. pulex agree closely with literature values reported for daphnids of 1.5 to 1.8mm in length, as used throughout this study.

A depressed metabolic rate resulted from 50 and 100 ppm

doses which was likely an expression of the acute toxicity of fresh crude oil. Twenty-four hour doses at less that 10 ppm, however, caused a more than doubled oxygen consumption. An extrapolated consequence of this increase in metabolism would be a greater requirement for food (algae) by low dose impacted zooplankters. The increased energy needs may also explain the decreased fecundity noted in D. pulex exposed to low levels of oil contamination.

Behaviour

Feeding - Feeding behaviour changed immediately after the addition of 50 ppm of NW crude oil emulsion. The frequencies of thoracic appendage movement, mandibular beat, and swallowing decreased while the rates of rejection increased. Since paraffin oil emulsions had no effect, the response was one of chemical toxicity. The combined effect of these changes was a reduction in ingestion rate. This effect may be particularly relevant bioenergetically in consideration of the increases in metabolic rate noted at low oil doses.
Swimming - A separate experiment tested swimming behaviour after NW oil emulsion treatment at 50 and 100 ppm. Swimming activity became erratic and slower, and was nearly eliminated by 24 hours of exposure, but showed partial recovery in the survivors 48 hours after the end of exposure. Complete recovery in surviving daphnids occurred by 192 hours post-exposure. Decreased swimming effectiveness would make zooplankton more susceptible to predation, changing the dynamics of predator-prey interactions.

96-Hour LD_{50} for Aromatic Hydrocarbons

The aromatic component of crude oil has been most consistently identified with its chemical toxicity effects. Examination of selected major component aromatic hydrocarbons, increasing in molecular weight from benzene to benzopyrene, showed that their toxicity when dissolved in water increased with molecular weight, as indicated by decreasing LD_{50} values. The NW crude oil emulsion was less toxic, at a measured LD_{50} value of 40 ppm, than was benzene. A conclusion derived from the graded response is that while weathered oil has lost most of the volatile hydrocarbons such as benzene, the larger molecular weight aromatics still made a major contribution to overall oil toxicity.

Hydrocarbon Accumulation

Five different carbon-14 labelled aromatic hydrocarbons

were administered to _D. pulex_ by three exposure vectors:
1) dissolved in water; 2) in unicellular algae (_Ankistrodesmus falcatus_), water clean; 3) in both water and algae. As indicated by the accumulation factors, calculated on the basis of body levels after 24 hours of exposure, the different aromatics were unique. Benzene was the least accumulated, while naphthalene had the greatest uptake level, especially from algal food. Benzanthracene and benzopyrene, however, accumulated most from solution, presumably absorbed through the daphnid body surface. In general, _D. pulex_ showed a remarkable ability to accumulate aromatic hydrocarbons, above all naphthalene.

Hydrocarbon Depuration

Consistent with the high degree of accumulation of naphthalene during exposure periods was its low loss rate when zooplankters were held in clean medium. Only 17% of the accumulated load was lost over 72 hours of clearance. The other aromatics were lost more rapidly, as shown by their larger depuration factors. Whether clearance occurred by dissolution or by a detoxification metabolism cannot be established at this time, but it is of interest to note that dead or dying _D. pulex_ cleared benzopyrene less rapidly than did live daphnids.

The ability of _D. pulex_ to bioaccumulate aromatic hydrocarbons, especially naphthalene, suggests that the ingestion of zooplankton by secondary consumers will lead to a transfer of petroleum fractions throughout the food web. A biomagnification is also possible, and certainly occurred in the step between algae and _D. pulex_.

CONCLUSIONS

It has been demonstrated that petroleum, in the form of oil particles, its WSF or as individual aromatic hydrocarbons, can have a marked effect on several factors which influence trophic interactions. These effects may be seen at a level of a few ppm or less in ambient hydrocarbon concentration. The influences of the observed alterations on other trophic levels are inferred and remain to be quantitated.

ACKNOWLEDGEMENTS: This study was supported by grants from the Natural Sciences and Engineering Research Council of Canada, Strategic Grants Program.

OIL SPILLS AND MANGROVES: A REVIEW OF
THE LITERATURE, FIELD, AND LAB STUDIES

Charles D. Getter
   Research Planning
     Institute, Inc.
   925 Gervais Street
   Columbia, SC 29205

ABSTRACT

    The purpose of this paper is to synthesize available data concerning oiled mangroves, specifically to examine the nature and magnitude of a characteristic suite of stress responses and the distribution of these responses within an oiled forest. These observations were than organized to develop an Index of Mangrove Stress (IMS). Five levels of expression of stress responses are then described. Use of the IMS to calculate the area and extent of impact allows investigators a convenient means to quantify stress and to monitor recovery of oil-stressed mangrove forests.

INTRODUCTION

    In a field survey of five oil spills in Florida (Howard Star, Garbis, and Keys Marsh) and Puerto Rico (Peck Slip and Ensenada Honda), it was concluded that despite the size of the spill or the substance spilled, mangrove forests tended to express a characteristic suite of stress responses following oiling (1). Each spill differed mainly in the magnitude of expression of each stress response. Also differing between spills was the distribution within the forests of the magnitude of stress responses. This was attributed to geomorphic and topographic control of the movement of oil within forests.

    Three models were presented to describe the distribution of oil-induced stressed responses in mangrove forests:

    1) Outer fringe impact. This was described as an impact which is concentrated at a seaward berm and which usually comprises a continuous band of scattered

patches of defoliated mangroves along their coastal fringe.

2) Inner fringe impact. This impact also occurs in fringing forest types, but is characteristic within forests with a well-developed berm. Oil impact is concentrated against this berm which may lie well within the fringing forest. The result is a continuous band or a series of scattered patches of dead and defoliated mangroves along the berm well within the inner forest.

3) Inner basin impact. In cases where oil breaches an outer or inner berm, impact has been observed to be spread widely over a broad area, becoming diffused rather than concentrated against a berm. This results in numerous patches of partially defoliated trees, usually in a basin forest.

The purpose of this paper is to expand upon these findings to the worldwide literature base in order to summarize available observations concerning responses and their distribution. This will allow a comparison of major oil spills in mangroves, especially with regard to:

1) The appearance of a characteristic suite of mangrove stress responses.

2) The magnitude of their expression.

3) The distribution of stress responses within the forest.

This summary will include laboratory findings from two literature sources and some preliminary lab tests of our own as well. The application of this summary will be used to generate an Index of Mangrove Stress (IMS). This index will then be described according to the stress responses and their magnitude of expression. Applications of the index will be discussed with emphasis on field assessment of stress responses and monitoring of recovery of oiled mangroves.

SUMMARY OF LAB STUDIES

Three lab studies are known to have been conducted to date with mangroves and oil. These are:

1) Mathias (2) - Mathias exposed Avicennia seedlings to varying concentrations of water-accommodated fractions of diesel fuel and reported growth alterations and lethality.

2) Hannah and Phillips (3) - In an unpublished manuscript, they reported exposing black mangroves (Avicennia) to varying concentrations of four oil types, noting mortality and change in leaf numbers.

3) Our research group has exposed red mangrove seedlings (Rhizophora) to 38,600 ppm for 28 days in an oiled substrate with three oil types. Table I gives results for change in leaf number and total weight. Table II presents a summary all available lab data including this test.

SUMMARY OF FIELD OBSERVATIONS

A continuous literature search from 1978 to the present has produced nearly 70 references and personal communications which give maps, narratives, figures, tables, and/or photographs detailing stress responses by oiled mangroves in the Gulf of Mexico, Caribbean Sea, Nigeria, Saudi Arabia, and Indonesia. Certain of these references have created summaries of information (4, 5, 6). These and other primary references are summarized by stress response in Table III. A list of references (for completeness) other than those in Table which are pertinent to this summary follows:

1) Baker (7)
2) Birkeland et al. (8)
3) Diaz-Piferrer (9)
4) Getter et al. (10)
5) Gundlach et al. (11)
6) Lair et al. (12)
7) Lewis (13)
8) Lopez (14)
9) Lugo et al. (15)
10) Lugo and Snedaker (16)
11) Nadeau and Berquist (17)
12) Odum and Johannes (18)
13) Rutzler and Sterrer (19)

DEVELOPMENT OF THE INDEX OF MANGROVE STRESS (IMS)

First, from the summary, it was possible to expand the list of stress responses to include the magnitude of expression of each. Since the data base was collected with different methods, a broad categorization was required to discuss the magnitude of expression of stress responses:

1) Intensely stressed - This is reported where a stress response is expressed at the 80-100 percent level within a compartment examined (e.g., 80-100% mortality of trees, roots, seedlings, epiphytes, snails, crabs, and infauna). In most cases, this constitutes a nearly total elimination of the compartment affected.

2) Stress greater than natural variability - This is reported where stress responses are not intense, but

TABLE I. Red mangrove seedling change in total weight (g) and change in number of leaves, exposed to three types of oil for 28 days. D.F. = degrees of freedom; M.S. = mean square; F.R. = F-ratio; E.C.B. = estimate of exact change probability. *Significant at p less than 0.05.

CHANGE IN TOTAL WEIGHT

ANOVA Summary Table

SOURCE	D.F.	M.S.	F.R.	E.C.B.
Total	14	3.9557		
Oil Types	2	21.1940	19.58	0.0006*
Error	12	1.0827		

Means and Standard Deviations

OIL TYPES	N	MEAN	S.D.
Kuwait Crude	5	3.86	0.83
Bunker C	5	-0.24	0.72
No. 2	5	1.48	1.42
Grand Total	15	1.70	1.99

CHANGE IN NUMBER OF LEAVES

ANOVA Summary Table

SOURCE	D.F.	M.S.	F.R.	E.C.B.
Total	14	1.7810		
Oil Types	2	6.0667	5.69	0.0362*
Error	12	1.0667		

Means and Standard Deviations

OIL TYPES	N	MEAN	S.D.
Kuwait Crude	5	2.00	1.41
Bunker C	5	1.80	1.10
No. 2	5	0.00	0.00
Grand Total	15	1.27	1.33

TABLE II. Concentrations of different types of petroleum hydrocarbons and observed toxic effects from various references of laboratory investigations.

CONCENTRATIONS (in ppm)	FUEL TYPES	TOXIC EFFECTS	REFERENCE
100	Diesel	Growth alterations	2
300	No. 2	91.7% mortality; survivors with no new leaves	3
300	Bunker C	23.9% mortality; survivors with fewer new leaves	3
300	South Louisiana crude	20.0% mortality; survivors with fewer new leaves	3
300	Kuwait crude	No mortalities; increased number of new leaves	3
1,000	Diesel	Growth deformities	2
10,000	Diesel	Lethal	2
38,600	Bunker C No. 2 Kuwait crude	Fewer new leaves, depressed weight gain	This paper
100,000	Diesel	Lethal	2

TABLE III. Summary of field-observed stress responses of mangroves with primary references.

STRESS RESPONSES	PRIMARY REFERENCES
1) Tree mortality	4, 20
2) Defoliation of canopy	1, 20, 21
3) Root mortality	1, 15, 21
4) Bark fissuring	6, 10
5) Seedling mortality	5
6) Epithelial scarring	5
7) Lenticel expansion	20
8) Adventitious pneumatophores	22
9) Leaf deformities/chlorosis	1, 5, 20
10) Propagule stunting/bending	6, 10
11) Epiphytic mortality	1, 20
12) Leaf stunting	1
13) Reduction in leaf number	10
14) Tree snail density	1, 4
15) Tree crab density	1, 4
16) Number of lenticels	6
17) Changes in infaunal community	23

where they are measurably greater than those found at control sites due to natural variability.

3) No responses observed - This is reported if no data are available for the compartment or if the level of response is less than or equal to that determined for comparison sites due to natural variability.

Secondly, it is possible to categorize the methods of field survey necessary in order to observe or measure each of the suite of stress responses. The purpose of this categorization is to generate an index which will be simply applied during field surveys. These were divided into three types:

1) Overflight-observed responses - These are stress responses which are obvious to an observer in an aircraft flying at low altitude (less than 200 ft) and at low speeds (less than 100 mph) over impacted forests such as extensively denuded canopies and patches of defoliation.

2) Ground-obvious responses - These are responses which are obvious to the trained observer and which are measured during ground stations in oil-impacted mangrove forests, such as seedling mortality, snail and crab population reductions, and propagule stunting or bending.

3) Ground-statistical responses - These are stress responses which are not obvious to a trained observer during ground stations, but which require field measurements at impacted and comparison sites and subsequent sorting and/or statistical analyses in order to confirm stress responses. These include responses such as change in leaf size and weight and infaunal community changes.

The list of responses, their relative magnitude of expression, and field means of observations and monitoring were combined to produce the Index of Mangrove Stress (IMS) (Table IV). A discussion of each index value follows.

DESCRIPTION OF THE INDEX OF MANGROVE STRESS (IMS)

Extensively Denuded Canopy (IMS=IV)

These areas are readily visible from plane, boat, or ground stations and comprise areas of extensive tree mortality and total canopy defoliation. The inner fringe impact at Machos Creek, Puerto Rico, reported by Getter et al. (1), is approximately one-half acre of IMS=IV (Figure 1).

TABLE IV. Stress responses and their magnitude of expression combined to generate an index of mangrove stress (IMS). The index categories are as follows: extensively denuded canopy (IMS=IV); patches of partial defoliation (IMS=III); numerous, visible stress responses (IMS=II); statistically observed stress responses (IMS=I); and no stress responses (IMS=0).

STRESS RESPONSE	IMS=IV	IMS=III	IMS=II	IMS=I	IMS=0
1) Tree mortality	●	●	•		
2) Defoliation	●	●	•	•	
3) Root mortality	●	•	•		
4) Bark fissuring	●	•	•		
5) Seedling mortality	●	•	•	•	
6) Epithelial scars	●	•			
7) Lenticel expansion	●	•	•		
8) Adventitious pneumatophores	●	•			
9) Leaf deformities	●	●	•		
10) Propagule size	●	●	•		
11) Epiphytes dead	●	●	•		
12) Leaf stunting	●	●	•	•	
13) Reduction in leaf number	●	●	•	•	
14) Tree snail density	●	●	•		
15) Tree crab density	●	●	•	•	
16) Number of lenticels	●	●	•		
17) Changes in infaunal community	●	●	•		

● Intensely stressed

• Response greater than natural variability but not intense

☐ No responses observed

An outer fringe impact at Ensenada Honda, Puerto Rico, is approximately one-half acre of IMS=IV as well. These areas are likely to have oil-laden substrate constituting a continuous source of oiling. Seedlings colonizing these areas are responsive to the toxicity of the oiled substrate. Initially, seedling mortality is complete. Subsequent recolonization is not expected until survival-level thresholds are reached through weathering or cleanup activities.

FIGURE 1. An area with an extensively denuded, red mangrove canopy (IMS=IV) in a Puerto Rican mangrove forest, three months after oiling from the Peck Slip barge in December 1978.

Obvious stress symptoms observed during ground stations at an IMS=IV include:

1) Massive periderm fissuring.

2) Lenticel expansion.

3) Bark peeling, exposed root cortex, and subsequent root death and disintegration.

4) Total mortality of seedlings present during and colonizing after the spill.

5) Onset of leaf chlorosis, formation of abscissic layers, and subsequent dropping of all leaves (within two months in the Peck Slip case).

6) Massive epithelial scarring and flaking of oil-covered periderm from trunk and branches.

7) Desiccation and mortality of all epiphytes and attached animals on root systems (oysters in IMS=IV area may shed from prop roots and die as observed following the spill of Howard Star).

8) Radical changes in the density and distribution of crabs and snails (85% reduction to the former, three months after Peck Slip).

Patches of Partial Defoliation (IMS=III)

These areas are visible from the air and are usually quite obvious to any observer conducting ground stations. They comprise patches of dead trees and either total or nearly total defoliation. They are distinctly visible on infrared film, appearing green or pink, where unstressed vegetation is normally deep red. Typically, the center of such a patch is a small area of extensively denuded canopy (IMS=IV), surrounded by gradually increasing percent leaf cover or partial defoliation (IMS=III). Tree mortality is typically 80-100 percent at the center of the patch and decreases, radiating from the center. Typically, leaf deformities and chlorosis are extensive and the number of crabs and snails are severely reduced. Lenticels on trunks and roots are typically expanded and in one case (Peck Slip), were extruding a waxy-orange substance. Infaunal communities at IMS=III sites may have significant alterations to the community structure, especially of crustaceans and bivalves (10).

The appearance of adventitious pneumatophores on black mangroves at these patches is known from Peck Slip and the Keys Marsh sites as reported by Getter et al. (1) (Figure 2). Snedaker et al. (22) described this phenomenon in some detail, based on observations at the Howard Star spill. A significant portion is dead, and most living seedlings have stunted or deformed leaves. Patches of partial defoliation are known at spill sites, including the Garbis (Florida), Howard Star (Florida), and Showa Maru (Indonesia).

Numerous Visible Stress Responses (IMS=II)

These areas are not readily visible from an airplane or a boat. However, stress symptoms are usually apparent to a trained observer while conducting ground stations. Typical of inner basin impacts at Peck Slip and Howard Star sites are a small but significant number of dead trees, some visible defoliation of the canopy, a higher incidence of root mortality, bark fissuring, seedling mortality, some lenticel expansion, and leaf deformity and stunting (Figure 3). Typical evidence at IMS=III sites are significant numbers of stunted

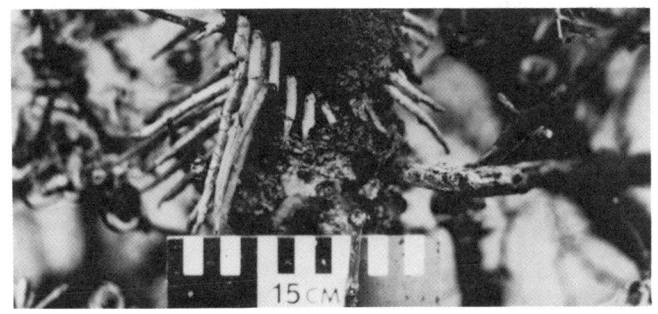

FIGURE 2. Anomolous, adventitious pneumatophores at the base of a black mangrove trunk in a patch of partially defoliated canopy (IMS=III) at Keys Marsh, Florida, in March 1980.

or dismorphic propagules (Figure 4). Tree snail and tree crab distributions and densities may be obviously changed.

Statistically Observed Stress Responses (IMS=I)

Typical of lightly oiled areas, certain of the suite of stress responses are cryptic, not being readily visible even to the trained observer. Precise field and lab measurements and their statistical analysis are necessary for demonstrating the presence of these responses which may include some defoliation, seedling mortality, leaf stunting, shape change, or reduction in leaf number. Changes may also be demonstrated at lightly oiled areas expressing IMS=I for tree-dwelling crabs and snails and also for infaunal communities.

No Stress Responses (IMS=0)

Where no obvious or measurable differences can be demonstrated between impact and comparison sites, no stress responses or IMS=0 are reported. Natural variability within compartments of mangrove forests may be very high, especially in areas which are already stressed by frost, hurricanes, high salinity, impoundment, or oil. The presence of existing stressors complicates all assessments, requiring a vigorous effort to separate mangrove stress responses due to oil from those due to ambient stressors.

APPLICATION OF THE INDEX

Observations of the recovery of oiled mangrove forests indicate that oiled forests with extensively denuded canopies (IMS=IV) will not fully recover within 2-5 years. Patches of

FIGURE 3. Leaf stunting and decreased number of leaves on oiled buttonwood mangroves (Conocarpus) from an area with numerous visible stress symptoms (IMS=II) in the Florida Keys (left and center) and a comparison or unoiled area (right).

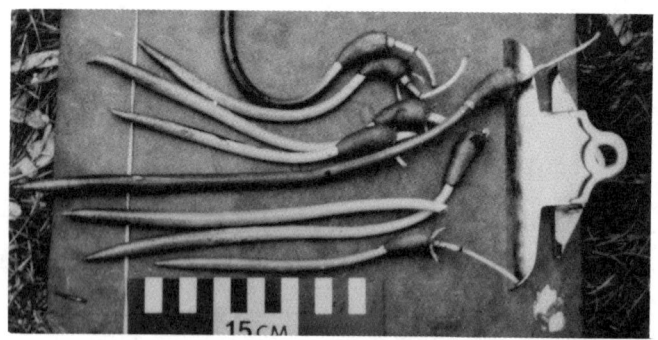

FIGURE 4. Curled and elongated red mangrove propagules from an area showing numerous observed stress symptoms (IMS=II) following oiling by the Peck Slip barge in Puerto Rico in December 1978.

partial defoliation (IMS=III) have been observed to begin re-colonizing immediately, but the process has been observed to take a number of years, depending on the extent of substrate oiling. Oiled forests ranked as IMS=III or IV, therefore, will have the potential for long-term expression of stress responses. These are areas which are candidates for assessment of oil-induced damages and for monitoring of recovery. Areas with numerous, visible stress responses (IMS=II) should be monitored as well, especially in areas subject to previous or subsequent stress. For example, it is possible that a frost occurring at Tampa Bay, Florida, could have more effect on oil-stressed mangroves (IMS=II) than on unoiled trees.

Such an effect could result in mortality of IMS=II areas which might have otherwise survived. Areas showing only cryptic responses (IMS=I) and those showing no response (IMS=0) after six months will almost certainly recover within a short time.

By applying the IMS to an oiled mangrove shoreline and by calculating the appearance and recovery of number of hectares of each IMS type over time, an investigator would have an objective manner of expressing change over time. Recommended are monthly flights and ground stations during the first year with larger time intervals following that.

ACKNOWLEDGMENTS

Field studies which formed the basis of the Gulf of Mexico and Caribbean Sea field work were sponsored by the NOAA Office of Marine Pollution Assessment (OMPA) and the Florida Department of Natural Resources. The literature search was sponsored by NOAA/OMPA. Applications of the IMS have been tested during field work in Florida, sponsored by the Florida Coastal Energy Impact Program.

Many thanks are extended to Dr. Jennifer M. Baker, Research Director of the Field Studies Council (Orielton Field Centre, Pembroke, Dyfed, Wales, U.K.), for providing literature on international spill studies. Dr. Samuel C. Snedaker and Mr. Melvin S. Brown of the University of Miami, Rosenstiel School of Marine and Atmospheric Science are acknowledged for their work with me in developing an earlier "mangrove stress index" specific to interpretation of stress responses during a study of the Howard Star oil spill at Tampa Bay, Florida.

RPI assistants, Thomas G. Ballou, E. Kevin Brophy, and Jeffrey A. Dahlin, assisted in the oil experiments with red mangroves reported here. Diana Gaines typed the manuscript.

LITERATURE CITED

1) Getter, C. D., G. I. Scott, and J. Michel, 1981, The effects of oil spills on mangrove forests: a comparison of five oil spill sites in the Gulf of Mexico and the Caribbean Sea: in Proc. 1981 Oil Spill Conf., API Publ. No. 4334, Amer. Petrol. Inst., Wash., D.C., pp. 535-540.

2) Mathias, J. A., 1976, The effect of oil on seedlings of the pioneer mangrove, Avicennia intermedia, in Malaysia (abs.): in Proc. of a Symp. on Ecology and Management of Some Tropical Shallow-water Communities, Jakarta. Results summarized in C. Thia-Eng and J. A. Mathias (eds.),

Coastal Resources of West Sabah: An Investigation into the Impact of Oil Spills, Penerbit Universiti Sains Malaysia, Pulau Pinang, 1978.

3) Hannah, R. P. and L. A. Phillips, 1975, The effects of selected petroleum refinery products and crude oils on the seeds and seedlings of the black mangrove, Avicennia germinans: Prog. Rept. No. 9A, unpubl.

4) Baker, J. M., I. M. Suryowinoto, P. Brooks, and S. Rowland, 1980, Tropical marine ecosystems and the oil industry; with a description of a post-oil survey in Indonesian mangroves: in Proc. PETROMAR '80, Graham and Trotman Ltd., London, England, pp. 617-643.

5) Chan, E. I., 1977, Oil pollution and tropical littoral communities: biological effects of the 1975 Florida Keys oil spill: in Proc. 1977 Oil Spill Conf., Amer. Petrol. Inst., Wash., D.C., pp. 187-192.

6) Getter, C. D., J. M. Nussman, E. R. Gundlach, and G. I. Scott, 1980a, Biological changes of mangrove and sand beach communities at the Peck Slip oil spill site, eastern Puerto Rico: RPI Rept. to NOAA/OMPA (Boulder, Colo.), RPI, Columbia, S.C., 63 pp.

7) Baker, J. M., 1981, The investigation of oil industry influences on tropical marine ecosystems: Mar. Pollut. Bull., Vol. 12, pp. 6-10.

8) Birkeland, C. A., A. Reimer, and J. R. Young, 1976, Survey of marine communities in Panama and experiments with oil: Ecological Research Series, EPA-600/3-76-028, 175 pp.

9) Diaz-Piferrer, M., 1962, The effects of an oil spill on the shore of Guanica, Puerto Rico: in Proc. 4th Mtg., Assoc. Isl. Mar. Labs., Curacao, Univ. Puerto Rico, Mayaquez, pp. 12-13.

10) Getter, C. D., S. C. Snedaker, and M. S. Brown, 1980b, Assessment of biological damages at the Howard Star oil spill site, Hillsborough Bay and Tampa Bay, Florida: RPI Rept. to Dept. of Nat. Res. (Tallahassee, Flor.), RPI, Columbia, S.C., 65 pp.

11) Gundlach, E. R., J. Michel, G. I. Scott, M. O. Hayes, C. D. Getter, and W. P. Davis, 1979, Ecological assessment of the Peck Slip (19 December 1978) oil spill in eastern Puerto Rico: in Proc. Ecol. Damage Assess. Conf., Soc. Petrol. Indus. Biol., pp. 303-318.

12) Lair, M. D., R. G. Rogers, and M. R. Weldon, 1971, Environmental effects of petrochemical waste discharges on Tallaboa and Guayanilla Bays, Puerto Rico: Rept. to Environ. Protec. Agency, Reg. IV, TS 03-71-208-02, 98 pp.

13) Lewis, R. R., 1979, Large-scale mangrove restoration on St. Croix, U.S. Virgin Islands: in Proc. 6th Ann. Conf. on Wetlands Restoration and Creation, Tampa, Flor., pp. 231-242.

14) Lopez, J. M., 1978, Ecological consequences of petroleum spillage in Puerto Rico: in Proc. of a Conf. on Impacts of Oil Spills, Amer. Inst. Biol. Sci., pp. 894-908.

15) Lugo, A. E., G. Cintron, and C. Goenega, 1978, Mangrove ecosystems under stress: in G. W. Barrett and R. Rosenberg (eds.), Stress Effects on Natural Ecosystems, 33 pp. + 7 tables + 5 figures.

16) Lugo, A. E. and S. C. Snedaker, 1974, The ecology of mangroves: Ann. Rev. Ecol. and Systematics, Vol. 5, pp. 39-64.

17) Nadeau, R. J. and E. T. Berquist, 1977, Effects of the March 18, 1973 oil spill near Cabo Rojo, Puerto Rico, on tropical marine communities: in Proc. 1977 Oil Spill Conf., API Publ. No. 4284, Amer. Petrol. Inst., Wash., D.C., pp. 535-538.

18) Odum, W. E. and R. E. Johannes, 1975, The response of mangroves to man-induced environmental stress: in E. J. F. Wood and R. E. Johannes (eds.), Tropical Mar. Pollut., Elsevier, New York, pp. 52-62.

19) Rutzler, K. and W. Sterrer, 1970, Oil pollution damage observed in tropical communities along the Atlantic seaboard of Panama: Bioscience, Vol. 20, pp. 222-224.

20) Davis, W. P., G. I. Scott, C. D. Getter, M. O. Hayes, and E. R. Gundlach, 1980, Methodology for environmental assessment of oil and hazardous material spills: in O. Kinne and H. P. Bulnheim (eds.), Helgolander Meeresuntersuchungen (14th European Mar. Biol. Symp., Protection of Life in the Sea, Biologisches Anstalt Helgoland), Hamburg, Vol. 33(1-4), pp. 246-256.

21) Spooner, M., 1970, Oil spill in Tarut Bay, Saudi Arabia: Mar. Pollut. Bull., Vol. 1(11), pp. 166-167.

22) Snedaker, S. C., J. A. Jimenez, and M. S. Brown, 1981, Anomolous, adventitious, pneumatophores in *Avicennia*

germinans (L) in Florida and Costa Rica: Bull. Mar. Sci., Vol. 31(2), pp. 467-470.

23) Gilfillan, E. S., D. S. Page, R. P. Gerber, S. Hansen, J. Cooley, and J. Hotham, 1981, Fate of the Zoe Colocotronis oil spill and its effects on infaunal communities associated with mangroves: in Proc. 1981 Oil Spill Conf., API Publ. No. 4334, Amer. Petrol. Inst., Wash., D.C., pp. 353-360.

EVALUATION OF AMMONIA TOXICITY IN SITE-SPECIFIC ENVIRONMENTAL ASSESSMENTS

J.A. Fava
R.M. Kapp
W.J. Rue, Jr.
  Ecological Analysts, Inc.
  Sparks, Maryland

ABSTRACT

Ammonia is a naturally occurring inorganic chemical which, in low concentrations, is essential to life. At elevated levels, however, ammonia can be toxic to both aquatic and terrestrial organisms. In surface waters, the toxic form of ammonia is un-ionized ammonia, NH_3.

This paper presents a synthesis and overview of the acute and chronic effects of ammonia on fish and aquatic invertebrates. Discussed are chemical and biological factors that influence the toxicity of un-ionized ammonia. These include such physical/chemical factors as temperature, pH, salinity, and dissolved oxygen, and such biological factors as acclimation, recovery, and differences in ammonia sensitivity among species and life stages.

The current information on the toxicity and fate of ammonia discharged in surface waters is discussed with respect to evaluating environmental impacts of fossil fuel energy development in the United States.

INTRODUCTION

Ammonia is a naturally occurring chemical resulting from the breakdown of organic matter. Although it is ubiquitous in surface waters in the United States, at elevated concentrations it has been shown to be toxic to aquatic organisms. Many industrial point-source categories discharge ammonia into surface waters [API 1981]. In 1980, U.S. petroleum refineries processed an estimated 14.7 million barrels of oil daily, which resulted in an estimated daily surface water contribution of 17,000 pounds of total ammonia nitrogen [Rue and Fava 1981]. This was shown to represent less than 0.5 percent of the total point-source contribution of ammonia to surface waters in this country. However, as the fossil fuel-producing industries bring new technologies into the production phase, these estimates may change. The current economic and political goal of achieving U.S. energy independence is encouraging technological advances in the area of petroleum substitutes, which in turn produce additional waste materials that must be managed in an environmentally acceptable manner.

Potentially significant concentrations of ammonia have been reported for a variety of coal conversion process wastewater streams. For example, Synthane gasification of seven different coals yielded ammonia wastewater concentrations (before treatment) ranging from 2,500 mg/liter for Illinois char to 11,000 mg/liter for Pittsburgh seam coal [Singer et al. 1978]. Bostwick et al. [1979] reported average ammonia concentrations in effluent streams from a number of coal conversion technologies before treatment. Average wastewater concentrations for the Koppers-Totzek, CO_2 Acceptor, Synthane, HyGas, Lurgi Dry Ash, SRC, and H-Coal processes averaged 184 mg/liter, 1,312 mg/liter, 8,050 mg/liter, 4,170 mg/liter, 13,000 mg/liter, 7,900 mg/liter, and 14,400 mg/liter ammonia, respectively. These figures indicate that, among other waste products, ammonia will be an important regulated constituent for producers of synthetic fuels.

EPA has indicated that the various environmental effects of advanced fuel processes will be felt most significantly on the local or regional level [U.S. EPA 1977]. Since the potential toxicity of ammonia has been shown to be influenced by a number of ambient conditions that affect the availability of the toxic form of ammonia to aquatic organisms, scientifically valid assessments of the impacts of the ammonia discharged in effluents from advanced fuel processes can best be conducted on a site-specific or waterbody-specific basis.

The objective of this paper is to present an overview of the chemistry, fate, and toxicity of ammonia in surface waters and to discuss the implication of this information to site-specific ammonia assessments. This work is based upon an in-depth, state-of-the-art assessment of the literature on the sources, chemistry, fate, and effects of ammonia in aquatic environments conducted by Ecological Analysts, Inc. for the American Petroleum Institute [API 1981].

CHEMISTRY OF AMMONIA IN SURFACE WATERS

Ammonia in water exists in two predominant forms: un-ionized ammonia (NH_3) and the ammonium ion (NH^+_4). The relative proportion of total ammonia that exists as un-ionized ammonia depends primarily on temperature and pH [Thurston et al. 1977]. The percentage of un-ionized ammonia as a function of pH and temperature in fresh water is shown in Figure 1. Raising the pH by one unit at temperatures between 0° and 30°C causes nearly a tenfold increase in the concentration of un-ionized ammonia for a given total ammonia concentration. Similarly, a 5°C temperature rise increases the un-ionized ammonia concentration by about 40-50 percent.

The sensitivity of this relationship is such that normal diurnal variation in these parameters can produce substantial fluctuations in the un-ionized ammonia concentration. For example, a realistic daily pH variation of 0.5 unit, and a temperature variation of 2°C, can cause a greater than threefold variation in the level of un-ionized ammonia for a constant total ammonia concentration [API 1981].

FATE OF AMMONIA IN SURFACE WATERS

The fate of ammonia discharged into surface waters depends upon a number of physical, chemical, and biological factors. One of the mechanisms affecting the fate of ammonia that has received considerable attention, because of its potential for depressing the concentration of dissolved oxygen, is nitrification. Other mechanisms include assimilation, adsorption, and volatilization.

Several factors have been reported to affect the rate of nitrification. The importance of a physical substrate for the attachment of nitrifying bacteria is a major principle in the concept of zones of nitrification described by Tuffey [1973] and Tuffey et al. [1974]. Kholdebarin and Oertli [1977b] observed that the presence of suspended particulates enhanced nitrification in laboratory

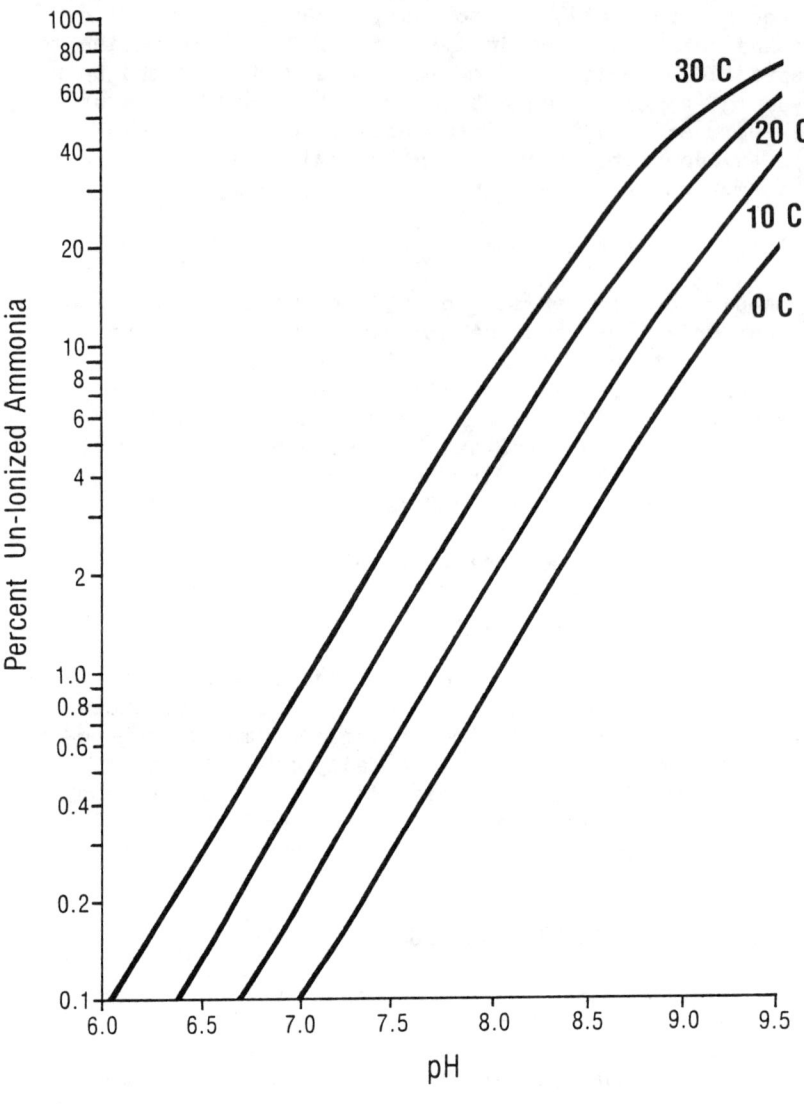

(Drawn from values tabulated by Thurston et al. 1977.)

Figure 1. Percent un-ionized ammonia as a function of pH and temperature in fresh water.

experiments on river water samples. Because nitrification is a biological process, temperature would be expected to have an effect. Knowles et al. [1965] found that the growth constant for Nitrosomonas spp. (converts ammonia to nitrite) increased 9.5 percent per degree increase in the range from 8° to 23°C. The corresponding rate for Nitrobacter spp. (converts nitrite to nitrate) was 5.9 percent per degree. The effect of pH on nitrification has been reported by Alexander [1974], Chen et al. [1972], and Kholdebarin and Oertli [1977a].

In addition to nitrification, living organisms assimilate ammonia as a source of nitrogen. Bell et al. [1979] concluded from laboratory studies of ammonia assimilation rates and from data on phytoplankton abundance that algal assimilation is a significant sink for ammonia in the Wabash River. The authors estimated that this process is capable of removing 34 percent of the ammonia in the water.

Other natural processes have been shown to remove ammonia from aquatic systems. Adsorption of ammonia to particles of suspended matter that subsequently settle is a potential mechanism for removing ammonia from the water column [API 1981]. Loss of ammonia to the atmosphere at elevated pH is another mechanism for ammonia removal [Keeney 1972]. Bell et al. [1979] found that volatilization could account for 67.5 percent of the observed loss of ammonia below an industrial discharge to the Wabash River.

In summary, it is difficult to generalize about the fate of ammonia discharged into surface waters. The implication of the factors described above is that an ammonia discharge that is potentially harmful to a particular waterbody may result in a negligible perturbation in another, depending on a variety of site-specific conditions.

TOXICITY OF AMMONIA TO AQUATIC ORGANISMS

The toxicity of ammonia to aquatic organisms has been studied by numerous researchers. Toxicity data are available for over 50 species of freshwater and marine fish, mollusks, and arthropods. Data have been tabulated and discussed for 58 species in API [1981]. The emphasis of this section is to provide an overview of the factors affecting the toxicity of un-ionized ammonia.

The toxicity of ammonia has been shown to depend principally on the concentrations of the un-ionized fraction, which in turn has been shown to be primarily a function of pH and temperature. In addition to these factors, however, other chemical, physical, and biological factors have been identified in the technical literature. Based upon the review by API [1981] and Thurston and Russo [1979], the following is an overview of the influence of temperature, pH, acclimation, recovery from sublethal effects, and species and life-stage differences on the toxicity of ammonia to freshwater organisms. More detailed discussions can be found in API [1981].

Temperature

The effect of temperature on the toxicity of ammonia to fish has been discussed by Colt and Tchobanoglous [1976], Roseboom and Richey [1977], and Thurston and Russo [1979]. Independent of its well-described effect on the concentrations of the ionized and un-ionized fractions, ambient water temperatures have been shown to affect the resistance of several species.

Roseboom and Richey exposed bluegill sunfish (Lepomis macrochirus) at 22°and 28°C and computed 96-hour LC50 values of 0.49 and 1.3 mg/liter NH_3-N, respectively. Similar results were reported by these researchers for largemouth bass (Micropterus salmoides) and channel catfish (Ictalurus punctatus), indicating that at increasing temperatures, these three species were more tolerant of un-ionized ammonia.

Colt and Tchobanoglous used 50-76 mm fingerling channel catfish in static bioassays at three temperatures. The 96-hour LC50 values for un-ionized ammonia at 22°, 26°, and 30°C were reported to be 2.4, 2.9, and 3.8 mg/liter NH_3-N, respectively, again supporting the results of Roseboom and Richey that the toxicity of un-ionized ammonia decreases at higher temperatures.

Results of acute ammonia toxicity tests on rainbow trout (Salmo gairdneri) and cutthroat trout (Salmo clarki) reported by Thurston [1979] indicate that, within the temperature range 10-20°C, no apparent correlation between temperature and 96-hour LC50 values has been determined. Thurston states that these results agree with those reported for rainbow trout by Herbert [1962] and Lloyd and Orr [1969], but contrast with the (British) Ministry of Technology [1968], which reported that ammonia was much more toxic to both adult and juvenile rainbow trout at 5°C than at 18°C.

It is apparent from this discussion that the effect of temperature on the toxicity of un-ionized ammonia is not a simple direct relationship, but rather is probably species-specific and is probably also related to physiological optimum temperatures of the organisms being tested.

pH

As with temperature, the principal effect of pH is on the dissociation of ammonia into its two primary aquatic fractions: the recognized toxic un-ionized form and the predominant and relatively nontoxic ionized form. At a temperature of 20°C, an increase in pH from 7.0 to 8.0 results in an un-ionized ammonia increase of ~tenfold. However, many researchers have indicated that the effect of pH cannot be explained solely by its effect on the percentage of un-ionized ammonia in solution.

For example, Stevenson [1977] reported that a decrease in water pH from 8.0 to 6.0 would theoretically result in a 97-fold decrease in un-ionized ammonia; however, the observed toxicity reduction to the white perch (Morone americana) was only decreased by a factor of 29 in fresh water.

Tomasso et al. [1980] determined the 24-hour LC50 values for channel catfish at three pH levels. The LC50 at pH 8 (1.82 mg/liter NH_3-N) was found to be significantly higher than at pH 7 or 9 (1.39 and 1.49 mg/liter NH_3-N, respectively).

Experimenting with fingerling rainbow trout, Downing and Merkens [1955] reported that at the lowest concentration of un-ionized ammonia tested (0.86 mg/liter NH_3-N), fish survived longer at pH 7.0 than at pH 8.0. At the higher exposure concentrations (1.38 and 1.96 mg/liter NH_3-N), the reverse was usually true.

Robinson-Wilson and Seim [1975] compared the acute toxicity of ammonia to coho salmon at four pH levels. Their results indicated that although the concentration of (total) ammonia yielding the 96-hour LC50 decreased as pH increased, in terms of un-ionized ammonia the opposite trend was observed--the toxicity decreased as pH increased. This same trend was reported in a study using crustacean larvae (Macrobrachium rosenbergii) by Armstrong et al. [1978]. The acute toxicity of un-ionized ammonia varied inversely with pH while the toxicity of ionized ammonia (and total ammonia) varied directly with pH.

Acclimation to Ammonia

There is limited information available on the potential for aquatic organisms to develop an increased ammonia tolerance following acclimation to low (nonlethal) ambient levels of ammonia.

Thurston and Russo [1979] reviewed the available literature concerning acclimation. Their review notes that some studies have reported that previous exposure of fishes to low ammonia concentrations reduces or does not affect their tolerance to lethal ammonia levels [Steinman 1928; McCay and Vars 1931; Fromm 1970]. However, that review emphasized that there is a larger body of information which indicates that prior exposure of fishes to low concentrations of ammonia increases their resistance to lethal concentrations.

The following information on acclimation was cited by Thurston and Russo [1979]. Vámos [1963] reported that carp (species not given) subjected to high ammonia concentrations, followed by a recovery period, and then resubjected to toxic concentrations, fared better than fish not previously exposed. Malacea [1968] reported that both carp (Rhodeus sericeus amarus) and minnows (Phoxinus phoxinus) acclimated to ammonia at subacute concentrations had a significantly greater survival time when next subjected to acute concentrations than did fish not previously exposed. Schulze-Wiehenbrauck [1976], testing rainbow trout, and Redner and Stickney [1979], testing cichlids (Tilapia aurea), have also reported similar results.

In the study with Tilapia aurea, Redner and Stickney reported a 48-hour LC50 of 2.40 mg/liter NH_3-N for fish not exposed to ammonia prior to acute testing. After another group of fish was exposed to sublethal concentrations of ammonia (0.43-0.53 mg/liter NH_3-N) for 35 days, exposure concentrations as high as 3.4 mg/liter NH_3-N caused no mortalities within 48 hours. These results indicated that a tolerance to ammonia was developed during acclimation.

The effect of a 24-hour acclimation by Atlantic salmon to un-ionized ammonia concentrations just below their reported 24-hour LC50 value was evaluated by Alabaster et al. [1979]. Their results indicate that this short-term acclimation period increased the 24-hour LC50 values by between 38 and 79 percent compared to unacclimated fish. Furthermore, this acclimation phenomenon was reported to be fully effective following exposure periods of 500-1,000 minutes.

Stevenson [1977] reported that, in general, prior exposure to ammonia increased the LC50 for juvenile white perch (Morone americana). He states, however, that under the given test conditions, it was necessary for previous ammonia exposure to exceed 150 mg/liter total ammonia (0.04 mg/liter NH_3-N) before the acclimation effect appeared. At levels 0.04 mg/liter NH_3-N, no acclimation effect was evident.

Hazel et al. [1979] indicated that acclimation was not a factor in the high tolerance of red shiner (Notropis lutrensis) to ammonia. Using two separate red shiner sample populations (one from sewage polluted Mill Creek, the other from clean Cedar Creek), the acute toxicity values for both groups were similar.

Recovery from Sublethal Effects

Evidence of recovery from some sublethal effects of chronic exposure to un-ionized ammonia was reported by Smith and Piper [1975]. Histopathic examination of tissues from rainbow trout exposed as fingerlings to ~0.0186 mg/liter NH_3-N for nine months by the authors revealed severe hyperplasia of gill tissues, hypertrophy of gill epithelium, and some liver cell degeneration. Necrotic lesions were found in only one of the five livers examined. Other changes included some reduction of splenic lymphoid tissue and mild necrosis and sloughing of intestinal mucosa. At 9-1/2 months, a group of fish were returned to control conditions and allowed to recover for 45 days. At the end of that recovery period, tissues in all fish examined were reported to be normal.

Some compensatory mechanisms are evidenced in a study of growth of rainbow trout fry exposed to sublethal levels of un-ionized ammonia. Burkhalter and Kaya [1977] report that mean lengths of fry exposed continuously to 0.05 and 0.10 mg/liter NH_3-N concentrations (as well as higher concentrations) were significantly different from those of controls ($p < 0.05$) 21 days after hatching. However, at 28 days, there was no significant difference in mean lengths between the 0.05 mg/liter NH_3-N treatments and controls, and this condition continued throughout the duration of the 42-day test. At 42 days, mean lengths from the 0.10 mg/liter NH_3-N treatment also approached those of controls (no significant difference). The initial growth response evidenced at 21 days had essentially disappeared for both exposure concentrations.

In the same context, Thurston [1980] reported on a 4-year unpublished study with rainbow trout. He summarized the study as follows: "it was not possible to isolate lasting differences between fish at 0.07 mg/liter NH_3 and control fish, although in shorter time periods retardation in growth of fry has been clearly demonstrated at 0.04 mg/liter NH_3 and gill damage has been demonstrated after six months at 0.02 mg/liter NH_3." Clearly, any effects that may have been evident during shorter time periods could not be distinguished from controls after four years.

Although the topics of acclimation and recovery have not yet been given a great deal of attention, the available evidence does suggest that deleterious effects on fish from short-term sublethal exposures to ammonia are not always lasting. In addition, as Thurston [1980] suggests, there is reasonable evidence that fish with a history of prior acclimation to some sublethal concentration of ammonia are better able to withstand what would normally be considered an acutely lethal concentration.

Species and Life-Stage Differences

The means and ranges of all freshwater 96-hour LC50 values obtained from the literature were grouped by family to demonstrate the effect of apparent genetic differences in the susceptibility of fish to un-ionized ammonia (Figure 2). These data indicate that family groups display differences in sensitivity and react differently, and that, within a family, the range of acute toxicity values may vary dramatically depending not only on experimental exposure conditions (i.e., pH, temperature, DO, and salinity), but also on the particular species being tested [API 1981]. The complexity of the situation is further evidenced by Thurston [1980], who reported that different genetic strains of the same species (<u>Pimephales promelas</u>) may exhibit different toxicity responses to un-ionized ammonia. LC50 values for fathead minnow obtained from different sources, including a commercial hatchery in the Ozarks, the EPA-Duluth Environmental Research Laboratory, and farm ponds and streams in eastern Montana, ranged from 0.66 to 2.87 mg/liter NH_3-N.

Rice and Stokes [1975] observed in a series of static bioassays that the eggs and alevins (before yolk-sac adsorption) of rainbow trout were much more tolerant to un-ionized ammonia than later life stages. No mortality to fertilized eggs, embryos, nor alevins was observed during a 24-hour exposure to 3.58 mg/liter reported as un-ionized ammonia until about the 50th day of development. The authors report that, after the 50th day, susceptibility of alevins increased and continued to do so as the yolk-sac was absorbed. The authors also report 24-hour LC50 values of

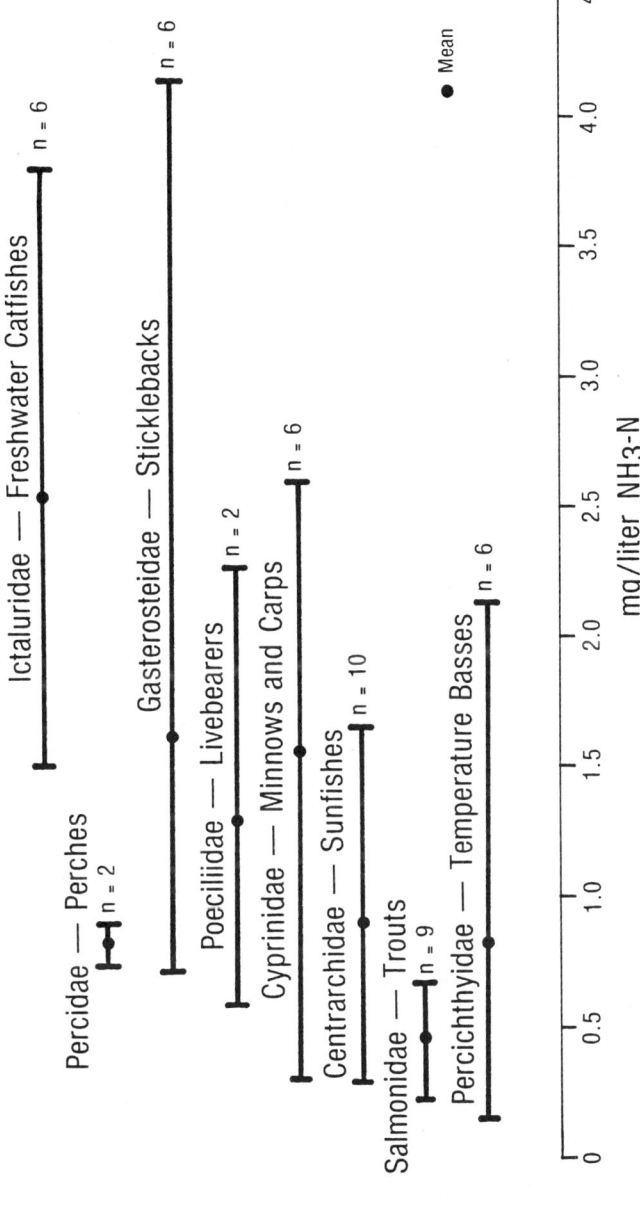

Figure 2. Variability of 96-hour LC50 values for freshwater fishes.

0.068 mg/liter as un-ionized ammonia for 85-day-old fry and 0.097 mg/liter as un-ionized ammonia for adults.

Roseboom and Richey [1977] demonstrated that resistance to un-ionized ammonia among bluegills increased as body weight increased. The 96-hour LC50 values at 22°C ranged from 0.40 to 0.80 mg/liter NH_3-N for bluegill fry, with average weights of 0.072 and 0.646 grams, respectively.

Hemens [1966] demonstrated that adult male mosquitofish were more susceptible to un-ionized ammonia than adult females of similar body length. Males showed statistically significant shorter survival times when tested at a concentration of 2.13 mg/liter NH_3-N, pH of 7.80, and temperature of 21.5°C. Although females of this species are larger and heavier than males, the data showed no indication of a relationship between length and survival time for either sex.

Evaluation of the Relative Importance of Factors Affecting Un-ionized Ammonia Toxicity

It is clear that a number of physical, chemical, and biological factors affect the toxicity of un-ionized ammonia to aquatic organisms. The relative importance of each factor depends on the specific environmental conditions existing at a particular location or waterbody. Each of the factors discussed above (temperature, pH, acclimation, and species and life-stage sensitivities), has been shown to affect the toxicity of un-ionized ammonia, and should be considered in evaluating potential impacts on aquatic life. Depending on the site-specific conditions, all the factors may not need to be examined, but they should be considered in the overall process. The potential role and influence of these factors illustrates the complexity of un-ionized ammonia in terms of its toxicity to aquatic organisms and the large number of factors that must be considered in conducting valid ammonia-toxicity experiments.

APPLICATION OF CHEMISTRY, FATE, AND TOXICITY INFORMATION TO SITE-SPECIFIC ASSESSMENTS

The site-specific approach to assessing the potential impacts of ammonia discharged from petroleum refineries or advanced fossil-fuel facilities requires consideration of a number of general toxicological factors and local biological, hydrological, and chemical conditions. For a detailed discussion on the regulatory framework and policies relevant to the discharge of ammonia into surface waters, see API [1981]. These factors are:

o Natural fluctuations in temperature, pH, and dissolved oxygen

o Species composition and population dynamics

o Life history of local species

o Relative sensitivities of species and life stages

o Nature of the toxic action of ammonia

o Appropriateness of the available laboratory data for application to local conditions.

An integrated assessment of these factors will help to define the relationship between general water quality objectives based on laboratory data and the site-specific water quality that supports the existing uses [Fava et al. 1980]. This relationship, rather than strict numerical standards, should provide the basis for site-specific assessments of the potential receiving water impacts of ammonia discharges.

Historically, regulation of effluents containing ammonia was accomplished primarily by developing a single effluent limitation for ammonia. Two problems are evident with this approach. First, the relationship between effluent limitations for total ammonia and in-stream water quality standards for un-ionized ammonia varies temporally. As discussed above, the pH and temperature ranges associated with a single waterbody will greatly influence this relationship. Second, many discharges are subject to limitations based on an historically low design-stream flow (e.g., Q_{7-10}). However, the actual frequency of occurrence of the design-stream flow is exceptionally low (by definition) based upon the historical record [Fava et al. 1980; API 1981].

Seasonal Considerations: Temperature and pH

In deriving an appropriate ammonia effluent limitation to protect the receiving waterbody, it is essential to recognize that pH and temperature, which determine the proportion of the total ammonia that exists in the toxic un-ionized state, can vary substantially from season to season. It is assumed that water quality-based effluent limits will be for total ammonia and will have been backed up from un-ionized ammonia standards. Because of this variation in temperature and pH, it would be inappropriate to establish a single total ammonia effluent limitation that would have to be met throughout the year. If a discharger were faced with a single effluent limitation based on worst-case conditions, i.e., summer pH and temperature, the cost of meeting this limitation during the cooler months, which also have lower pH values, would not be justified in terms of improvements to water quality. Conversely, an effluent limitation based on average annual receiving water pH and temperature may not afford adequate protection of aquatic life in periods when receiving waters have higher pH and temperature.

If a state is going to adopt an un-ionized ammonia water quality standard, it must also consider establishing seasonal-effluent limitations for total ammonia whereby a discharger would be allowed to release more total ammonia during the winter than in the summer, and intermediate quantities during the spring and fall.

Design-Stream Flow Considerations

By tying effluent limits for ammonia to a single design-stream flow, a certain low probability (e.g., < 1 percent) of exceeding the water quality standard is accepted. However, during periods of the year other than those typically associated with low flow, the same probability of exceeding the standard can be overcompensated for. A wasteload allocation study sponsored by EPA Region VIII [Hsiao et al. 1979] illustrates this point.

Based on monthly projections of low flow (7-day low flow expected once in 10 years during a given month), Hsiao et al. recommended that allocations of ammonia loads to two wastewater treatment facilities on the Colorado River be made on a month-by-month basis. It was also suggested that a generalized seasonal approach could be taken for regulation based on the monthly estimates. These recommendations were based on maintaining an in-stream ammonia standard with the same degree of protection, on a monthly basis, provided by the Q_{7-10} during the September low-flow period.

As such, it can be argued that the degree of protection afforded during low design-stream flows (or other similar dilution scenarios) need not be increased by applying the water quality standard for ammonia to that single flow on an annual basis. At a minimum, the relative degree of protection in proportion to low flow can be the same in proportion to high flow without sacrificing the benefits to be derived by compliance with the standard.

SUMMARY

A basic understanding of the interdependence of the toxicity, chemistry, and fate of ammonia in aquatic environments provides a sound basis for the derivation of acceptable discharges to aquatic systems. Well-defined interactions of temperature and pH on the availability of toxic un-ionized ammonia, and influences of other quantifiable biotic and abiotic factors, can play an important role in evaluating options for the maximum use of scarce water supplies. Modern technologies and water quality standards must be based on realistic targets for protection of aquatic resources. Because predictive methods are available to quantify the seasonal components of the toxic un-ionized ammonia fraction, a greater flexibility in water uses can be obtained. The general approach presented in this paper should help provide a basis for determining the degree of ammonia reduction necessary to maintain the designated use of a particular waterbody.

LITERATURE CITED

Alabaster, J.S., D.G. Shurben, and G. Knowles. 1979. The effect of dissolved oxygen and salinity on the toxicity of ammonia to smolts of salmon, Salmo salar. J. Fish. Biology 15(6):705-712.

Alexander, M. 1974. Microbial formation of environmental pollutants. Adv. Appl. Microbiol. 18:1-73.

American Petroleum Institute. 1981. Ammonia: The Source, Chemistry, Fate, and Effects of Ammonia in Aquatic Environments. Prepared by Ecological Analysts, Inc., Sparks, Maryland.

Armstrong, D.A., D. Chippendale, A.W. Knight, and J.E. Colt. 1978. Interaction of ionized and un-ionized ammonia on short-term survival and growth of prawn larvae, Macrobrachium rosenbergii. Biol. Bull. 154:15-31.

Bell, J.M., D.W. Nelson, and R.F. WuKasch. 1979. Studies on the Fate and Disappearance of Ammonia in the Wabash River. Prepared for Eli Lilly and Co.

Bostwick, L.E., M.R. Smith, D.O. Moore, and D.K. Webber. 1979. Coal Conversion Control Technology. Volume 1. Environmental Regulations; Liquid Effluents. Report No. EPA-600/7-79-228a.

Burkhalter, D.C. and C.M. Kaya. 1977. Effects of prolonged exposure to ammonia on fertilized eggs and sac fry of rainbow trout (Salmo gairdneri). Trans. Am. Fish. Soc. 106:470-475.

Chen, R.L., D.R. Keeney, and J.G. Konrad. 1972. Nitrification in sediments of selected Wisconsin lakes. J. Environ. Qual. 1:151-154.

Colt, J. and G. Tchobanoglous. 1976. Evaluation of the short-term toxicity of nitrogeneous compounds to channel catfish, Ictalurus punctatus. Aquaculture 8;209-224.

Downing, K.M. and J.C. Merkens. 1955. The influence of dissolved oxygen concentration on the toxicity of un-ionized ammonia to rainbow trout (Salmo gairdneri Richardson). Ann. Appl. Biol. 43:243-246.

Fava, J.A., R.M. Kapp, J.J. Gift, C.R. Flynn, and J.A. DelPup. 1980. A Biological Risk Assessment Approach to Establish Water Quality-Based Effluent Limits. American Society for Testing and Materials. Fifth Symposium on Aquatic Toxicology, Philadelphia, Pa. October 7-8, 1980.

Fromm, P.O. 1970. Toxic action of water soluble pollutants on freshwater fish. Water Pollution Control Research Series 18050 DST 12/70, U.S. Environmental Protection Agency, Washington. As cited in Thurston and Russo 1979.

Hazel, R.H., C.E. Burkhead, and D.G. Huggins. 1979. The development of water quality criteria for ammonia and total residual chlorine for the protection of aquatic life in two Johnson County, Kansas streams. Office of Water Research and Technology, Department of the Interior. Washington.

Hemens, J. 1966. The toxicity of ammonia solutions to the mosquitofish (Gambusia affinis Baird and Girard). J. Proc. Inst. Sewage Purif. 65:265-271.

Herbert, D.W.M. 1962. The toxicity to rainbow trout of spent still liquors from the distillation of coal. Ann. Appl. Biol. 50:755-777.

Hsiao, J.S., B. Sheikh-ol-Eslami, and L.H. Botham. 1979. Final Report. Ammonia Investigations in the Colorado River. Grand Junction and Fruita, Colorado. Prepared for U.S. EPA Region VIII by Engineering-Science, Denver, Colorado. EPA 908/5-79-004.

Keeney, D.R. 1972. The Fate of Nitrogen in Aquatic Ecosystems. Literature Review No. 3, Univ. of Wisconsin Water Resources Center, Madison.

Kholdebarin, B. and J.J. Oertli. 1977a. Effect of pH and ammonia on the rate of nitrification of surface water. J. Water Pollut. Control Fed. 49:1,688-1,692.

Kholdebarin, B. and J.J. Oertli. 1977b. Effect of suspended particles and their sizes on nitrification in surface water. J. Water Pollut. Control Fed. 49:1,693-1,697.

Knowles, G., A.L. Downing, and M.J. Barrett. 1965. Determination of kinetic constants for nitrifying bacteria in mixed culture, with the aid of an electronic computer. J. Gen. Microbiol. 38:263-278.

Lloyd, R. and L.D. Orr. 1969. The diuretic response by rainbow trout to sub-lethal concentrations of ammonia. Water Res. 3:335-344.

Malacea, I. 1968. Untersuchungen über die Gewöhnung der Fische an hohe Konzentrationen toxisher Substanzen. (Studies on the acclimation of fish to high concentrations of toxic substances.) Arch. Hydrobiol. 65(1):74-95. [In English translation.] As cited in Thurston and Russo 1979.

McCay, C.M. and H.M. Vars. 1931. Studies upon fish blood and its relation to water pollution, in A Biological Survey of the St. Lawrence Watershed, pp. 230-233. Supplement to 20th annual report, New York Conservation Dept. As cited in Thurston and Russo 1979.

Ministry of Technology. 1968. Water Pollution Research 1967. H.M. Stationery Office, London.

Redner, B.D. and R.R. Stickney. 1979. Acclimation to ammonia by *Tilapia aurea*. Trans. Am. Fish. Soc. 108(4):383-388.

Rice, S.D. and R.M. Stokes. 1975. Acute toxicity of ammonia to several developmental stages of rainbow trout, *Salmo gairdneri*. Fish. Bull. 73:207-211.

Robinson-Wilson, E.F. and W.K. Seim. 1975. The lethal and sublethal effects of a zirconium process effluent on juvenile salmonids. Water Resour. Bull. 11(5):975-986.

Roseboom, D.P. and D.L. Richey. 1977. Acute Toxicity of Residual Chlorine and Ammonia to Some Native Illinois Fishes. ISWS/RE-85/77. Department of Registration and Education Report of Investigation 85. Illinois State Water Survey, Urbana.

Rue, W.J. and J.A. Fava. 1981. The relative contribution of point sources of ammonia to surface waters, in Industrial Wastes--Proceedings of the Thirteenth Mid-Atlantic Industrial Waste Conference (C.P. Huang, ed.), pp. 608-617.

Schulze-Wiehenbrauck, H. 1976. Effects of sublethal ammonia concentrations on metabolism in juvenile rainbow trout (Salmo gairdneri Richardson). Ber. dt. wiss. Komm. Meeresforsch. 24:234-250.

Singer, P.C., F.K. Pfaender, J. Chinchilli, A.F. Maciorowski, J.C. Lamb, and R. Goodman. 1978. Assessment of coal conversion wastewaters: Characterization and preliminary biotreatability. Report No. EPA-600/7-78-181.

Smith, C.E. and R.G. Piper. 1975. Lesions associated with chronic exposure to ammonia, in The Pathology of Fishes (W.E. Ribelin and G. Migaki, eds.), pp. 497-614. Univ. Wisconsin Press, Madison.

Stevenson, T.J. 1977. The effect of ammonia, pH and salinity on the white perch Morone americana. Ph.D. dissertation. University of Rhode Island.

Steinmann, P. 1928. Toxikologie der Fische. Handbuch der Binnenfischerei Mitteleuropas 6:289-342. (Cited in Chipman 1934). As cited in Thurston and Russo 1979.

Thurston, R.V. 1980. Statement of Robert V. Thurston. Prepared for the Minnesota Pollution Control Agency Proposed Ammonia Standards.

Thurston, R.V., R.C. Russo, and K. Emerson. 1977. Aqueous Ammonia Equilibrium Calculations. Technical Report No. 74-1, Fisheries Bioassay Laboratory, Montana State Univ., Bozeman.

Thurston, R.V. and R.C. Russo. 1979. Some Factors Affecting the Toxicity of Ammonia--Implications For Water Quality Standards. Presented at the American Water Resources Association Symposium on The Uses of Scientific Information in Planning for Environmental Quality Objectives, Las Vegas.

Tomasso, J.R., C.A. Goudie, B.A. Simco, and K.B. Davis. 1980. Effects of environmental pH and calcium on ammonia toxicity in channel catfish. Trans. Am. Fish. Soc. 109:229-234.

Tuffey, T.J. 1973. The Detection and Study of Nitrification in Streams and Estuaries. Ph.D. thesis. Rutgers Univ., New Brunswick.

Tuffey, T.J., J.V. Hunter and V.A. Matulewich. 1974. Zones of nitrification. Water Resour. Bull. 10:555-563.

U.S. EPA. 1977. Advanced Fossil Fuels and the Environment. U.S. EPA, Washington. Report No. EPA-600/9-77-013.

Vamos, R. 1963. Ammonia poisoning in carp. Acta Biol. Szeged 9(1-4):291-297. As cited in Thurston and Russo 1979.

VEGETATION RESPONSE TO BRINE
SPILL RECLAMATION MEASURES:
BOREAL FOREST, ALBERTA,
CANADA

L. R. Hettinger
    Dames & Moore
    Golden, Colorado

INTRODUCTION

Problems and methodology associated with reclamation of lands affected by high salt concentrations have received considerable attention. The primary focus of studies has been the reclamation of agricultural lands due to their economic importance. Little information, however, is available regarding reclamation of forested lands contaminated with brine. Since field conditions and flora in forested areas differ from agricultural areas, reclamation procedures for salt affected areas may also differ.

Production of brine from primary and secondary recovery of crude oil in Western Canada continues to increase with time as does the quantity of brine spilled in association with crude oil production. Since most of these spills occur in the forested areas of Alberta, the Canadian Petroleum Association requested that studies be implemented to develop methods for their reclamation [1].

Most ions contained in the brine solution, such as Ca^{++}, Mg^{++}, Na^+, K^+, Cl^-, SO_4^{--} can be toxic if concentrations are above levels which plants can tolerate [2], even though most elements which form salts are essential to plant growth. Treshow [3] defines salinity as occurring when a soil saturation extract has an electrical conductivity (EC) value of 4 mmhos/cm (mS/cm) or greater. Effects on even sensitive species are reported to be negligible below 2 mmhos/cm (mS/cm) and only very few salt tolerant species remain productive above 16 mmhos/cm (mS/cm). Salts of sodium, magnesium and calcium are the most common in saline soils. High sodium content soils commonly cause problems for agricultural species. In addition, chloride, often found in association with sodium or calcium, can also be toxic. Much of the damage ascribed to salts may actually be due to the chloride ion which causes leaf burn or tip burn [4].

Holmes [5] observed that chloride concentrations of 0.2 percent in leaf tissue were toxic to a number of tree species.

Excessive salt affects plants primarily by cell plasmolysis due to high osmotic pressure imbalance between the root and soil and within the root from outer to inner tissue between the epidermis and xylem [2]. Sodium also causes an imbalance in other ions such as calcium in plant metabolism [6]. However, the mechanism for nutrient imbalance as it causes reductions in plant growth, is not well established in relation to salt toxicity [2].

The objectives of this study were to:
(a) assess the effects of a simulated brine spill on native vegetation, and
(b) assess the effectiveness of several reclamation-mitigative measures which will assist in returning soil conditions to a favorable environment for plant growth.

The results reported here are part of larger investigations reported in Innes and Webster [1].

STUDY SITE DESCRIPTIONS

Two study sites typical of the vegetation of the Boreal Forest Region of Alberta [8] were selected in 1977. Distinctive tree species of this region include lodgepole pine *(Pinus contorta)*, aspen *(Populus tremuliodes)* and balsam poplar *(Populus balsamifera)*. Relatively mature vegetation is indicated by the prominence of large white spruce *(Picea glauca)* trees with black spruce *(P. mariana)* as a lesser constituent [7]. White birch *(Betula papyrifera)* and tamarack *(Larix laricina)* are important locally in older stands, with the latter occurring in wet depressional areas. Both balsam fir *(Abies balsamea)* and alpine fir *(A. lasiocarpa)* occur in this Section on higher plateaus containing fairly old (200 years or greater) stands [7].

The vegetation of the sites and in the general vicinity of the sites was described by selecting visually homogeneous stands for data collection representing the range of vegetation types (communities) of each test site.

Test Site 1 was located in a lodgepole pine dominated forest, although white spruce and balsam fir are present. It appears that the site is similar to the *Picea glauca/Ptilium crista-castrensis* community type described by Achuff and LaRoi [7].

Test Site 2 was located in a slightly wetter site and contains black spruce in addition to lodgepole pine. This forest type is typical of much of the vegetation in the Swan Hills area having a heath-moss dominated understory. Therefore the results from this test site are used here to assess the various treatments.

METHODS

A series of 4 treatment plots measuring 15 m by 30 m were established in 1977 at two Test Sites (Figure 1). Species percent cover and a number of soil parameters (conductivity, pH, texture, sodium, potassium, calcium, magnesium, chloride, sulphate, bicarbonate and carbonate) were measured in-situ and via samples prior to treatments in each plot. These results are presented in detail by Innes and Webster [1].

Treatments at both sites were implemented in the spring of 1978 (as indicated in Figure 1) and included:
  Treatment 1: Control;
  Treatment 2: Application of brine, flush with two applications of fresh water within one week;
  Treatment 3: Treatment 2, followed by gypsum;
  Treatment 4: Application of brine.

A number of subtreatments (see Figure 1) were superimposed on the four main treatments including:
  Subtreatment A: Control on upper third of plot;
  Subtreatment B: Seeded to a grass-legume mixture plus fertilizer application on center portion of plot;
  Subtreatment C: Fertilizer on lower third of plot.

The volume of brine applied to the sites ranged from 24-25 m^3 (150 to 160 barrels) which is equivalent to a coverage depth of 17-18 cm. The electrical conductivity of the brine ranged from 56 to 60 mmhos/cm [1].

The subtreatment seed mixture contained, on a weight percentage basis, Orbital tall wheatgrass (50), Frontier reed canary grass (20), Boreal creeping red fescue (20), Yukon yellow sweet clover (6), and Aurora alsike clover (4), and was applied at the bulk rate of 56 kg/ha (50 lb./acre).

Fertilizer was applied in 1978 and 1979 and consisted of calcium nitrate (34, 67, 101 kg/ha; 30, 60, and 90 lb./acre of N_2), superphosphate and potassium (22 kg/ha; 20 lb./acre of P_2O_5, K_2O) (see Figure 1).

Permanent plots were established in the summer of 1977 to monitor plant responses to the various treatments and subtreatments (Figure 2). An area of 1 X 3 m in size was used to randomly locate four 0.5 X 1.0 m quadrats.

Absolute cover values (% ground cover) of living and damaged or dead tissue of each species were used to assess treatment effects. Quadrat vegetation was observed prior to treatment (July, 1977) and a number of times after treatment to ascertain recovery and ameliorating effects of reclamation measures (July and September, 1978, August, 1979; August, 1981). Four ubiquitous species were designated to monitor treatment effects over the study time period, although only

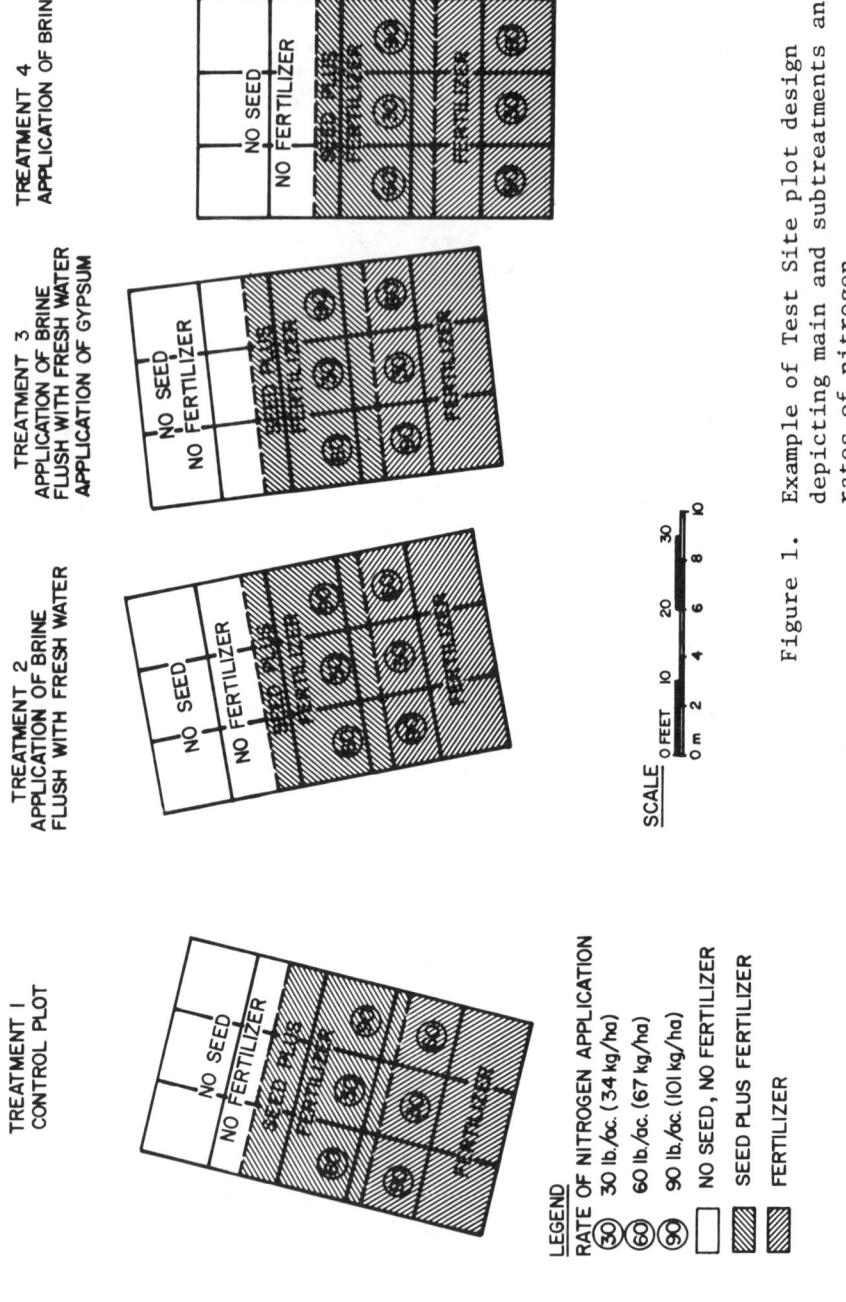

Figure 1. Example of Test Site plot design depicting main and subtreatments and rates of nitrogen.

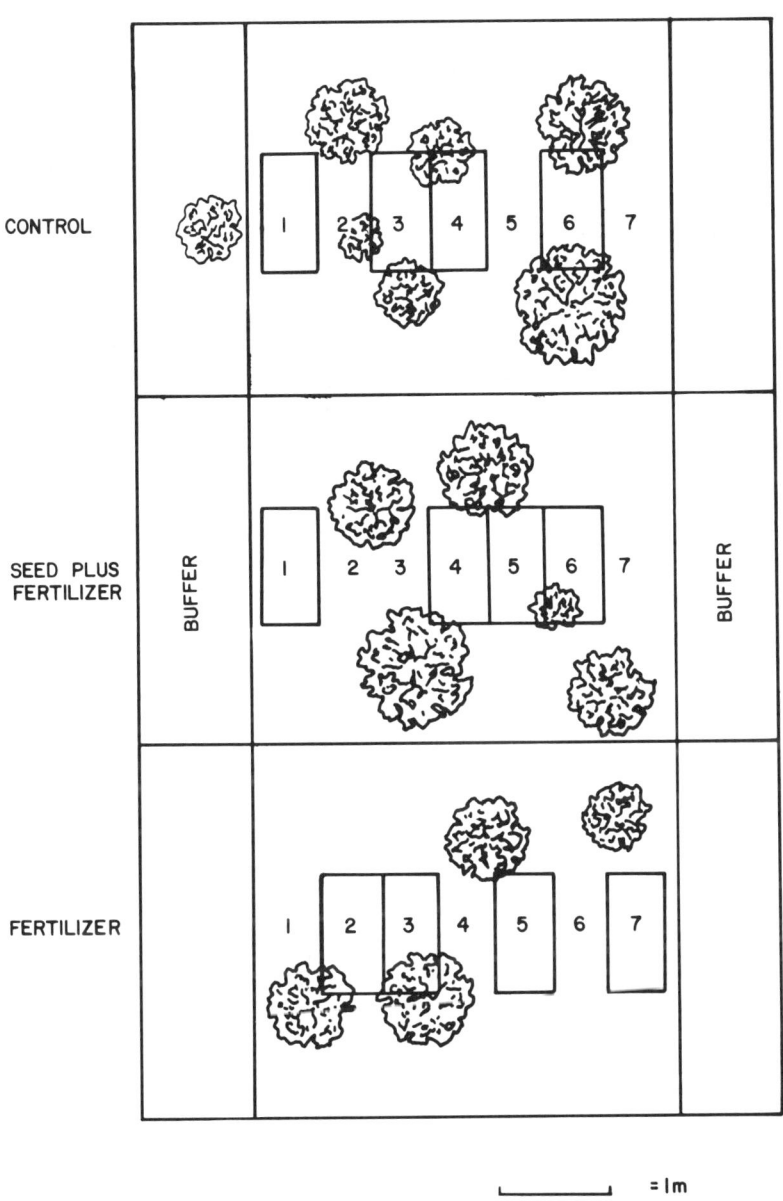

Figure 2. Schematic diagram of four permanent quadrats (0.5 X 1.0 m) for assessing treatment effects on vegetation.

those at Test Site 2 (Labrador tea, *Ledum groenlandicum;* blueberry, *Vaccinium myr tilloides;* feathermosses, *Hylocomium splendens* and *Pleurozium schreberi)* are reported here.

RESULTS

Necrotic tissue became evident soon after the application of brine on tree as well as most understory species. Necrotic tissue generally increased in abundance through the 1978 growing season for most species and, several salt tolerant species had invaded the plots by late summer that were not treated with fresh water and gypsum, salt tolerant species were evident by late summer *(Psoralea lanceolata, Geranium bicknellii, Rubus strigosus, Hordeum jubatum).*

As anticipated, vegetation damage is closely correlated with salt concentration of the soil as expressed by conductivity and concentration of sodium and chloride ions (Figures 3, 4, 5, 6). The reduction of salt concentration by flushing with fresh water, and especially by application of water and gypsum decreased either the amount of initial damage to vegetation or the length of time required for recovery (Figures 5, 6). Conductivities were measured at about 5 mmhos/cm in the top 30 cm of soil after application of brine. This was reduced to about 1.5 mmhos/cm after application with fresh water (Figure 3). Sodium levels also decreased in response to water application, however the most marked decrease occurred in plots also treated with gypsum, indicating an active Ca/Na ion exchange (Figure 4).

Both Labador tea and blueberry had a high percentage dead of the total cover in relation to the control plots after one year (Figure 5). The response of *Pleurozium schreberi* was more variable but the most severe tissue damage[a] was associated with the brine only plots. *Hylocomium splendens* showed a more consistent injury pattern to the brine application (Figure 6).

Cover estimates indicate that the vascular species responded to secondary treatments of flushing with water and gypsum application. Gypsum appears to have reduced the amount of damage the first year, and in Labrador tea, appears to have aided recovery to near original cover values by the end of the second year (Figure 5). The reduction of dead material among the treatments was less striking in blueberry the first year. However, the proportion of damage by the second year in the plot treated with water and gypsum had decreased to near that of the control plot (Figure 5).

Pleurozium schreberi exhibited a marked recovery by the end of the second growing season (Figure 6). Although this moss seems more resilient than *Hylocomium splendens*, mosses in general are difficult to use as monitoring tools. Mosses

---

[a]"Severely damaged" is used instead of "dead" to describe effects to the mosses as it is difficult to determine when a moss will not recover from a dormant state.

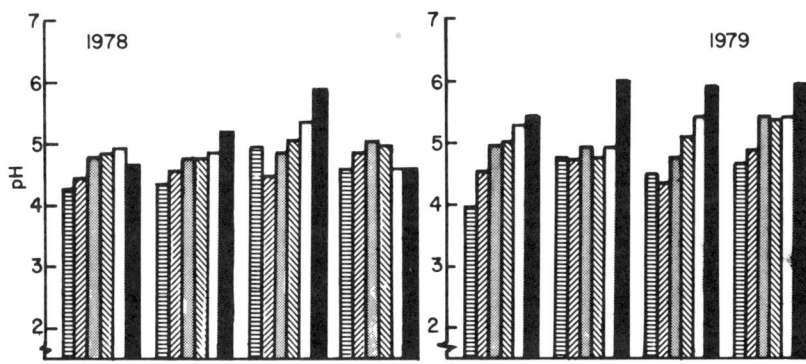
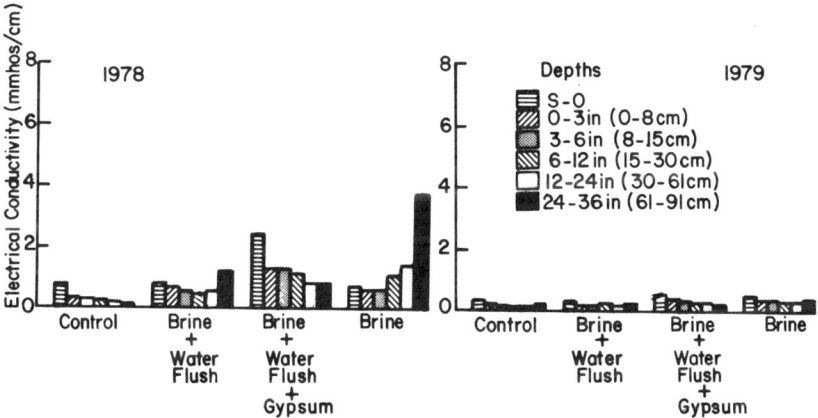

Figure 3. Soil pH and electrical conductivity of saturation extracts for soils sampled in the fall of 1978 and 1979 at Site 2.

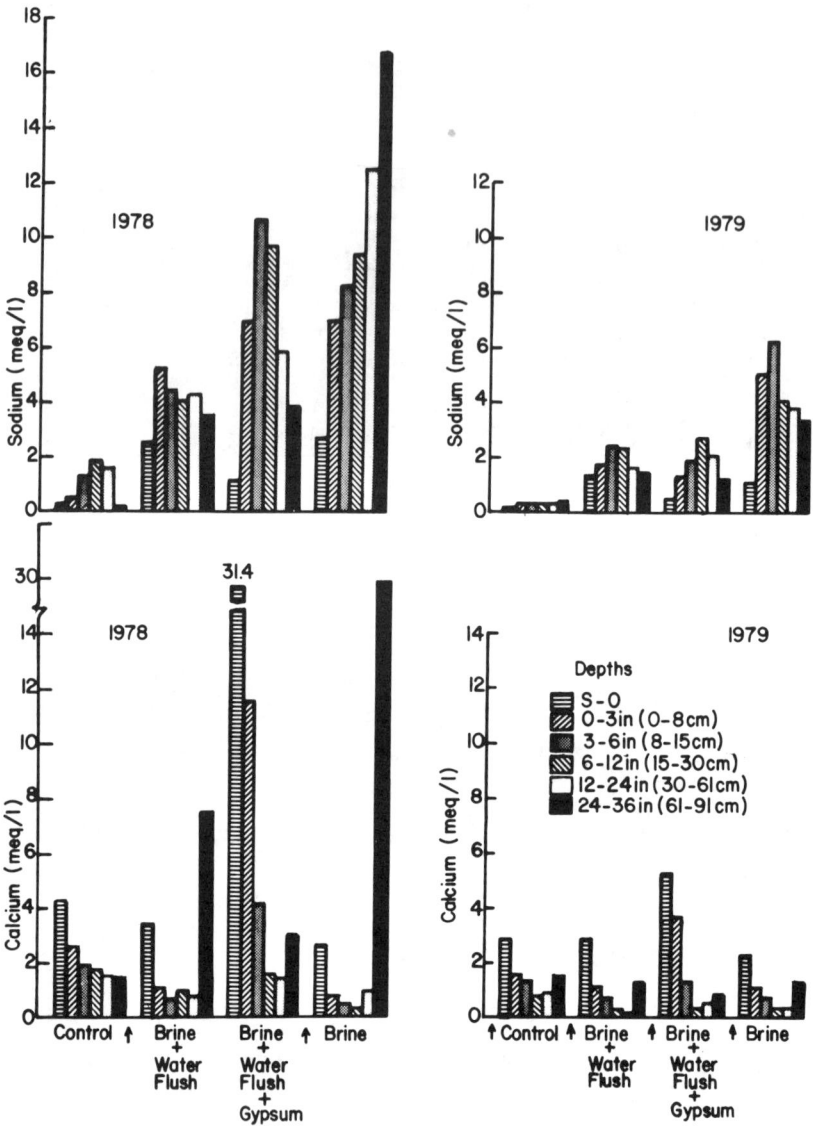

Figure 4. Sodium and calcium concentrations of saturation extracts for soils sampled in the fall of 1978 and 1979 at Site 2.

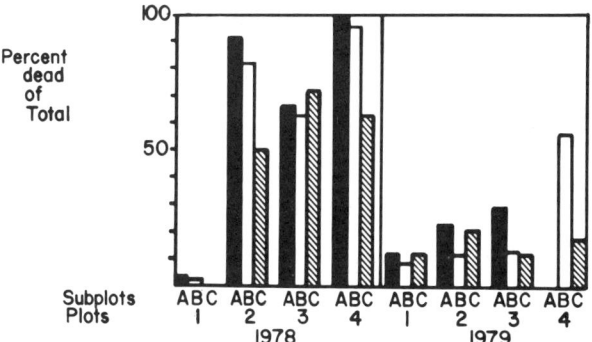

A. Vaccinium myrtilloides - Site 2

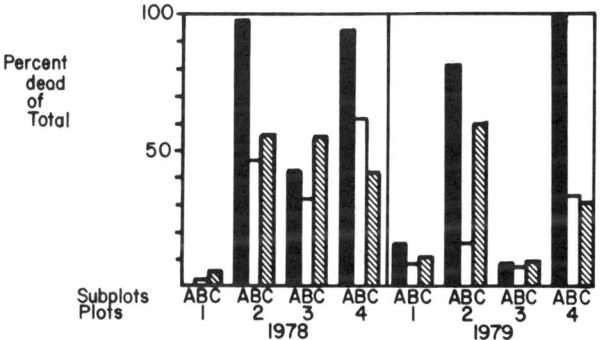

B. Ledum groenlandicum - Site 2

Plots 1. control
2. brine & flush
3. brine & flush & gypsum
4. brine

Subplots A. no seed, no fertilizer
B. seed & fertilizer
C. fertilizer

Figure 5. Percent dead of the total cover for: A. Vaccinium myrtilloides (blueberry) and Ledum groenlandicum (Labrador tea) for Test Site 2 in July 1978 and August 1979 (Total cover determined prior to brine application)

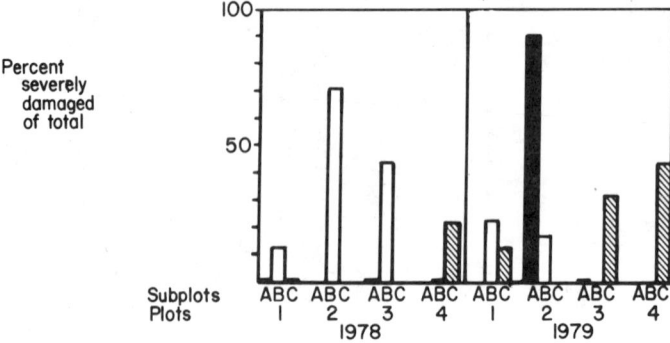

A. Hylocomium splendens - Site 2

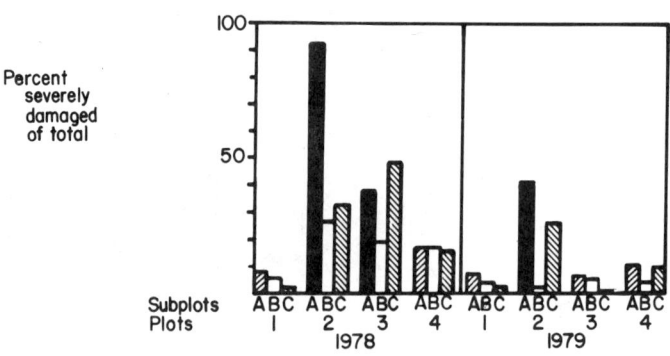

B. Pleurozium schreberi - Site 2

Plots 1. control
       2. brine & flush
       3. brine & flush & gypsum
       4. brine

Subplots A. no seed, no fertilizer
          B. seed & fertilizer
          C. fertilizer

Figure 6. Percent severely damaged of the total cover for: A. Hylocomium splendes and B. Pleurozium schreberi (feather mosses) for Test Site 2 in July 1978 and August 1979 (Total cover determined prior to brine application).

and lichens naturally survive adverse periods such as droughts through dormancy, "greening-up" as conditions improve. However, most upland mosses have an advantage over vascular species to survive brine spills by not having a rooting system which is affected by prolonged salt concentration in the soil. Species such as *Sphagnum*, however, which occur in wet, depressional areas are more prone to damage than some of the other moss species.

The fertilizer and seed application subtreatments appear to have aided recovery of the native species in some instances. Both Labrador tea and blueberry have a high proportion of dead to live tissue in the unfertilized subplots (A) except where gypsum was also applied (Figure 5). The moss species do not consistently show the same trend, although some of the untreated subplots have the highest component of damaged material (Figure 6). The different rates of nitrogen application were reflected by an increase in luxuriance of native species and especially seeded grass cover. No difference could be ascertained between the 67 and 101 kg/ha (60 and 90 lb./ac) rates. However, cover was less at the 34 kg/ha (30 lb./ac), but all rates of fertilizer had a positive effect on plant cover.

DISCUSSION

The levels at which salt ions become toxic to native species has not been well documented. However, many agronomic species, due to economic considerations, have been studied in detail. Richards [6] indicates that most agronomic species are sensitive to salt concentrations above electrical conductivities (EC) of 10 mmhos/cm (mS/cm). More salt tolerant grasses such as alkali sacaton, Bermudagrass, Canada wild rye and western wheatgrass had reduced growth at EC of 18 mmhos/cm (mS/cm) [6].

A number of factors, in addition to individual species' tolerance, are operative in causing salt toxicity. Two of these are soil texture and acidity [3]. Boyko [9] found that high clay content in soils correspondingly increased the effect that a particular saline solution had on plants. This factor is important in relation to the Swan Hills area since many soils, including those at the Test Sites have a high clay content [1]. Any factor that increases the time plant roots are exposed to a saline condition will be important in determining toxicity. Thus, the return of a favorable environment should begin at the soil surface as the salt is leached downward. Rate of species recovery could therefore also be dependent upon rooting depth. Theoretically, the exposure period should be shortest for most bryophytes and lichens, followed by shallow rooted vascular species. This assumption is supported by the electrical conductivity (EC) data (see Figure 3). Plots treated with gypsum and/or water did not exhibit this salt gradient to the same degree. Balsam fir and Labrador tea, with a comparatively deep root system, continued to exhibit a high proportion of dead tissue into the second year in the brine only plots. However, also recovered relatively slowly at both test sites (Figure 6); probably indicating a greater initial sensitivity to salinity.

The degree of toxicity may also be related to a high ratio of sodium to calcium in the soil [3]. The addition of calcium as gypsum helped reduce the high sodium to calcium ratio imparted by the brine. This factor is important in correlating reduction in plant damage the first year and, in general, promoting a more rapid recovery the second year in plots and subplots where calcium was applied (Figures 5, 6).

CONCLUSIONS

Both application of fresh water and calcium will aid returning the soil to a more favorable environment for plant growth. Since it does not appear that many native species in the Swan Hills area are highly salt tolerant (exceptions include raspberry, hairy wild rye, scurf pea, foxtail barley) seeding with salt tolerant agronomic species would help in providing an interim vegetative cover until a suitable soil environment allows for regrowth or invasion of the native species. Mitigative measures such as flushing with water, application of gypsum and the use of a calcium base fertilizer appear to decrease the time required for reduction of salinity below toxic levels. A combination of fresh water, gypsum and fertilizer applications appears to be quite effective.

Observations in August, 1981 indicated that fertilizer must be applied judiciously when agronomic species are used. Dense swards of grass have apparently retarded the reestablishment of native species which are becoming quite evident in non fertilized plots. Both Labrador tea (10-40% cover) and blueberry (10-25% cover) have reestablished in the more open plots. One application of fertilizer in conjunction with fresh water and gypsum applications may be sufficient to minimize saline concentrations in the soil and promote reclamation of the site with a return of native species by the third growing season.

SUMMARY

Results of the vegetation investigation, and those of Innes and Webster [1] of which this study was part, indicate that a combination of procedures can be used to reduce the impact of brine spills on boreal forest soils and vegetation. These include:
1) Brine should be contained to as small an area as possible using small catchment ditches and a tank-pump system;
2) The area should be flushed with fresh water (again using the ditches to control the spread of saline

material) until conductivity of seepage is at or below 2 mmhos/cm;
3) The use of gypsum (4.5 tonnes/ha) will curtail long-term damage to vegetation and promote soil and vegetation recovery. In addition, gypsum allows for a more rapid movement of brine out of the rooting zone. Several annual applications of gypsum may be required to replace Na^+ to below levels toxic to plant growth.
4) Where initial brine contact causes severe plant damage, eg. predominant moss cover, the use of agronomic, salt tolerant species and a calcium base fertilizer may be necessary to provide an erosion curtailing plant cover. However, fertilizer applications should be discontinued after the young plants are well established to facilitate reestablishment of native species.

LITERATURE CITED

1. Innes, R.P. and G.R. Webster (eds.) 1980. Reclamation of brine spills in upland boreal forest areas. Part 1: Research. Canadian Petroleum Association, Calgary, Alberta, Canada.

2. Meyer, B.S., D.B. Anderson and R.G. Bohning. 1960. Introduction to plant physiology. D. Van Nostrand Co., Inc., New York, N.Y.

3. Treshow, M. 1970. Environment and plant response. McGraw-Hill Book Co., New York.

4. Bernstein, L. and H.E. Hayward. 1958. Physiology of salt tolerance. Annual Rev. Pl. Phys. 9:25-46.

5. Holmes, F.W. 1966. Salt injury to trees. II. Sodium and chloride in roadside sugar maples in Massachusetts. Phytopathol. 56:6, 633-636.

6. Richards, L.A. (ed.) 1954. Diagnosis and improvement of saline and alkali soils. Agric. Handbook No. 60. U.S.D.A., Washington, D.C.

7. Achuff, P.L. and G.H. La Roi. 1977. Picea-Abies forests in the highlands of northern Alberta. Vegetatio 33:127-146.

8. Rowe, J.S. 1972. Forest regions of Canada. Can. Dept. Env., Can. For. Serv. Publ. No. 1300. Ottawa, Ontario, Canada.

9. Boyko, H. 1967. Salt-water agriculture. Sci. Amer. 216:89096.

LAND IMPACT
OF DRILLING FLUIDS RESERVE
PITS IN SOUTH WEST FRANCE

B. Tramier
Elf Aquitaine Pau FRANCE

ABSTRACT

Most of the wells drilled in south west of France are deep ones where a large volume of waste effluents is recovered. Several treatments which will be very shortly described are used to lower the impact of these effluents. However, there is always a residue that cannot be eliminated. This residue can then be :
- left alone in the pit
- treated by mechanical thickening, sludge fixation or soil stabilization.

The objective of the study was to look on the impact of 10 reserve pits which were to have been abandoned without treatment or where different treatments have been performed. The study specially includes :
- the evolution of the water quality
- the description of flora and fauna which appeared naturally in and in the vicinity of the pits
- the artificial seeding of some pits.

The results show clearly that most of the time there is no (or only a minor) impact of the pits on the environment.

INTRODUCTION

South West France is the main oil and gas area of the country. A large gas field (LACQ) and several small oil fields have been developed. The drilling activity has been and is still very intense. Hundreds of wells have been drilled in an area which is well known for its agricultural resources. The wells in this region are usually deep ones, most of them are deeper than 4000 metres, a few of them are deeper than 6000 metres. As a consequence, the amount of waste produced by these wells is high and several measures have been taken in order to lower the impact of the drilling activity on the environment. These measures which are briefly described below, have reduced the amount of waste produced. However, when the well is completed, there is still a residue stored in a reserve pit, which can only be :

- left in the reserve pit
- treated by a sophisticated and expensive treatment such as mechanical thickening, sludge fixation, soil stabilisation.

This residue is mainly made of the waste drilling mud and the cuttings. It is usually not toxic by itself but it could be very hazardous to anyone entering into the pit, because it behaves like quick sand. This is the reason why reclamation treatment must be carried out. However, as several reserve pits were still abandoned, research has been made into the environmental impact of these pits.

At the same time, it was decided to see if seeding and/or planting trees in the residue itself was practicable. This presentation summarizes most of the results currently available.

PREVENTIVE MEASURES

Several processes have been developed to lower the volume of residues produced during the drilling phase or to eliminate the residues when the well is completed, if the site must be restored. It is not the objective of this paper to enter into the details of these processes, we shall only mention the following which are currently used in FRANCE :

- the mechanical treatment of drilling muds is fundamental. An efficient treatment, performed with a high speed centrifuge, may save up to 50 per cent of the drilling mud consumption.

- a treatment of the waste effluents can be conducted in two or three steps according to the quality of raw effluent. It may include :
  . Gravity separation
  . Flocculation
  . Biological treatment

The treated effluent can then be recycled to the rig for industrial water or discharged to a river when the quality meets French regulations, the main requirements of which are summarized below in Table I.

Table I. Main regulations in FRANCE

pH	5.5 to 8.5
Temperature	30° C
Suspended Solids	30 mg/l
BOD	40 mg/l
COD	120 mg/l
Oil (gravimetry)	5 mg/l
Nitrogen	10 mg/l

- a treatment of residual viscous wastes (called residue in this paper) by one of these processes :
  . sludge fixation
  . mechanical thickening
  . soil stabilization

Using these measures, it is possible to restore a field to its owner in its original state. However, such a procedure is highly expensive and is not always justified by the local considerations. Several drilling places have been abandoned without any treatment and it is interesting to look at the evolution of some of them.

EVOLUTION OF SIX RESERVE PITS

Six old drilling locations have been visited (A to F), several years after the drilling activity has been finished. These locations have been selected according to their age, to the depth of the well (correlated of course to the duration of the activity on the site), to the types of muds used during the drilling activity. On two of these locations a reclamation treatment has been performed by soil stabilization (on a part only) on location B and by sludge fixation on location C.

Table II. Evolution of

	A	B	C
Field Data			
Age (years)	12	6	5
Depth (metres)	5664	3169	5048
Duration (days)	261	66	150
Drilling Mud	Bentonite Lignosulfonate Barite	Bentonite	Bentonite Lignosulfonate Gypsum Lime
Reclamation Treatment	No	Soil stabilization (partial)	Sludge Fixation
Water Quality			
pH	7.2	7.5	
COD (mg/l)	24	40	
BOD (mg/l)	1	5	
Oil (mg/l)	< 5	< 5	
Flora and Fauna	*Keratella* *Brachionus* *Cyclops*	*Keratella* *Brachionus* *Cyclops*	
	Rushes Reeds Trees	Rushes Reeds	Bushes Trees
	Invertebrates Frogs Ducks	Invertebrates Frogs Ducks	

six reserve pits

	D	E*	F*
Field Data			
Age (years)	4	3	2
Depth (metres)	4856	3920	3204
Duration (days)	172	94	56
Drilling Mud	Bentonite Lignosulfonate Barite Oil based	Bentonite Lignosulfonate	Bentonite Biopolymers
Reclamation treatment	No	No	No
Water Quality			
pH	7.2	7.5 (7.8)	7.5 (7.9)
COD (mg/l)	40	27 (4080)	48 (372)
BOD (mg/l)	3	3 (35)	5 (47)
Oil (mg/l)	< 5	< 5 (8)	< 5 (8)
Flora and Fauna		*keratella* *Brachionus* *Cyclops*	
	Rushes	Rushes Reeds Willows	Rushes Reeds Willows
	Invertebrates Frogs	Invertebrates Frogs Fish	Invertebrates Frogs

* Figures in brackets are the initial values when the pit was abandoned.

Only natural processes are involved in the evolution of these locations i.e. no fertilizers have been added, no seeding nor planting has been carried out.

From each location, samples of water still stored in the reserve pit were taken for chemical and plankton analysis. A brief inventory of the flora and fauna living in and in the immediate vicinity of the pit was also made. The results of all these investigations are in Table II from which the following conclusions can be drawn.

1. Waters in the pits are of good quality and meet the requirements of the French regulations for discharge into a river for instance. When the location has been abandoned, the waters were still polluted, as it can be seen for locations E and F. Due to natural processes, the water quality improves to an acceptable level on each location.
   Because of this good water quality, many species are now living in and around the pits :

   - Zooplankton species are abundant, including *Keratella, Brachionus, Cyclops*

   - Invertebrates are also very well represented by many insects (larvae and adults) and worms

   - Superior animals such as frogs, ducks and even fish.

2. Vegetation grows everywhere. Where it does not, it is usually because of the presence of an oil layer on the water surface. When this oil is removed, vegetation starts immediately. The first species to appear are rushes and reeds. These can cover the whole surface of the reserve pits. Trees (poplars, willows, acacias) appear rapidly initially on the boundary of the pits. Vegetation has been found living on the cutting piles, on oily soils. On the oldest location (A), vegetation has so covered the area that it becomes difficult now to imagine that a well had been drilled there.

3. No significant differences have been noticed in relation to the duration of the drilling phase or to the type of drilling muds used. The only important parameter is the age illustrated by the fact that on old locations there is more vegetation than on recent ones. Locations where a reclamation treatment has been performed, do not appear to have a better evolution than the others. Of course the vegetation and the fauna are different. Terrestrial vegetation and fauna appear everywhere but water vegetation and fauna appear only on untreated locations.

SEEDING AND PLANTATIONS IN RESERVE PITS

Vegetation appears and grows easily in most of reserve pits once floating oil has been skimmed. This was a confirmation of the fact that usually pits do not contain elements which can prevent the development of life. A second step has been conducted in order to look at the possibility of encouraging the revegetation of the reserve pits.

Artificial seeding has been studied on a laboratory scale. Plantation experiments have been carried out on a field scale.

Artificial Seeding

Laboratory experiments have been conducted in glass beakers in which were successively introduced
- a layer of residue collected from a pit
- eventually a layer of sawdust or of corn waste in order to stabilize the soil
- a layer of soil.

Several species have been seeded, among them corn which is a very popular agricultural plant in south west France and peas. Blank beakers containing only soil have also been seeded.

After seeding, beakers were regularly watered and the growth of the different plants has been monitored for several weeks.

The results have been very encouraging, no significant differences have been noticeable among all the beakers used during the tests :
- no delayed growth of the plants was noticeable
- all the plants of a same species were roughly the same size.

Of course all the seeds started their development in the soil layer, but no stress appeared when the roots penetrated into the residue layer. Rootlets developed normally in the residue without any visible damage to the plant.

This study should be completed by chemical analysis of the developed plants to see if some components of the residue are accumulated by the plants.

Such a study has not been performed because the direct mixture of soil is not practicable and the reclamation of a pit by soil stabilization with corn wastes for instance appeared to have only a small operational interest.

Plantation experiments

Planting trees directly in a reserve pit could be a very interesting option in order to restore the pit. Tree plantations have a least two main advantages.

1. Most of the trees are water consuming plants, therefore they could utilize the water contained in the residue and facilitate the soil stabilization.

2. Tree roots could also enhance the soil stabilization.

A pit has been selected to conduct an experiment. On this location a part of the residue had been solidified by a sludge fixation process and had been used to restore a part of the location. Plantations have been made in the pit itself and on the reclaimed area. Various species of trees have been used for the experiment.

70 trees have been planted in the pit and 20 on the reclaimed area. Six months after planting the trees, they have been checked and classified into three classes :

- living : trees which have shown good growth during the six months, they have grown in height or leaves have appeared.

- dead

- doubtful : trees which are not dead but which have shown little or no growth i.e. they have lost their leaves.

In the pit area over the 70 planted trees, including mainly poplars, willows, birches, pines, 23 (33 %) were living, 17 (24 %) were dead and 30 (43 %) were doubtful (see table III). The most resistant trees seem to be : oaks (marsh),poplars, willows (Marsault),and hornbeams. The least resistant ones are : willows (weeping), pyracanth, cypress and plum trees.

On the reclaimed area over 20 planted trees including mainly pines, piceas, poplars, cypress, eucalyptus, 10 (50 %) were living, 5 (25 %) were dead and 5 (25 %) were doubtful (see table IV). The most resistant trees are pines and eucalyptus, the least resistant ones are cypress.

Looking at the pattern of living and dead and doubtful trees, it is not possible to select areas where trees are living, dead or doubtful. May be when the doubtful trees have evolved in one way or another it will be possible to draw more precise conclusions. But up to now the resistance seems to depend only on the nature of the tree.

Some trees seem to acclimatise very well to the soil conditions found in the reserve pits. Further evolution must of course be monitored, but we believe that trees which have started to develop in the pit, will probably continue and tree plantation could be an interesting process to reclaim a drilling area including reserve pits.

Table III. Pit plantations

Species	Planted	Living	Dead	Doubtful
Poplar	13	4	2	7
Willow	6	2	3	1
Pine	3	1	0	2
Picea	3	0	0	3
Plum	3	0	2	1
Hornbeam	2	2	0	0
Oak (Marsh)	2	2	0	0
Alder	2	1	0	1
Plane	3	1	1	1
Birch	4	1	1	2
Pyracanth	3	0	2	1
Ash	3	1	0	2
Tulip	2	1	0	1
Robinia	2	1	0	1
Cypress	3	1	2	0
Maple	1	0	0	1
Rowan	2	0	0	2
Other shrubs	13	5	4	4
	70	23	17	30

Table IV. Reclaimed area plantation

Species	Planted	Living	Dead	Doubtful
Pine	6	4	0	2
Picea	5	2	1	2
Poplar	3	2	1	0
Cypress	3	0	3	0
Eucalyptus	3	2	0	1
	20	10	5	5

CONCLUSION

The cost of reserve pit reclamation in France is very expensive, in the range of 20 USD/m^3 of residue. At such a cost, treatment of every pit will obviously reduce the number of wells drilled every year.

The study discussed here shows that most often, old reserve pits do not have a severe impact on the local environment and according to the future use of the location where a well has been drilled, a reclamation treatment is not always justified. If the pit is well skimmed for oil, vegetation appears, rapidly followed by various species of fauna. Of course, the consistency of the residue stored in the pit may create a hazard ; this must be taken into consideration and may justify by itself a reclamation treatment. When this is necessary, the planting of trees, especially poplars, oaks, hornbeam and some willows, may be an interesting way of overcoming the problem.

THE EFFECT OF HYDROCARBON
DEVELOPMENT ON ELK AND
OTHER WILDLIFE IN
NORTHERN LOWER MICHIGAN

J. P. Bennington
    Standard Oil
       Company (Indiana)

R. L. Dressler
    Commonwealth Associates
       Incorporated

J. M. Bridges
    Commonwealth Associates
       Incorporated

ABSTRACT

    An environmental study was conducted to determine the impacts of hydrocarbon development activities on elk, bear, and bobcat in the north-central portion of Lower Michigan. Field data were collected during the period of August, 1979, to December, 1980, in a regional study area northwest of Atlanta, Michigan. Intensive surveys were conducted in four subregional areas and at 26 study sites within the regional area.
    Standard biological techniques, including collection of track data from designated routes in each subregion, were employed. The track data were collected twice monthly for elk, bear, and bobcat and were used as an index of animal use over time. The data were subjected to statistical analysis. Other methods of data collection, more qualitative in nature, indicated animal presence and abundance.
    Levels of elk activity were recorded throughout the regional area. Hardwood upland areas in the Vienna and Montmorency Township study areas contained the greatest elk concentrations, especially in the winter. These areas were also areas of ongoing hydrocarbon development activities. Only limited elk activity was recorded in the southern portion of the Pigeon River Country State Forest, an area where hydrocarbon activity had taken place prior to a

court order banning further development. Elk activity, however, has been historically low in this area.
Bear and bobcat were generally widespread throughout the regional study area. A preference for lowland habitats was noted for both species.

Analyses of the collected data indicated that levels of hydrocarbon development activity occurring during the study period had only short-term, localized impacts on the species studied. Elk, bear, and bobcat activity decreased within one-quarter mile of active drilling operations but returned to predrilling levels within two to four weeks after the site was abandoned, i.e., drilling was complete. Based on the numbers of wells completed in one study area (Montmorency Township), one well per 2.0 square miles appears to have had little adverse effect on the species studied.

INTRODUCTION

Hydrocarbon development activities were initiated in the early 1970's in the Pigeon River Country State Forest (PRCSF), located in the northern part of Michigan's Lower Peninsula. Despite efforts by interested oil companies and the Michigan Department of Natural Resources (MDNR) to design an environmentally sound approach to development, drilling in the PRCSF was halted by court order in 1979. One of the main reasons for halting development activities centered over the concern for the potential impact on the elk herd located in the region.

The elk *(Cervus elaphas)* were introduced to the area in 1918 from Wyoming following extirpation of the indigenous elk in the late 1800's(1). The elk herd expanded from less than a dozen introduced elk to between 1,200 and 1,500 in 1961. The primary range of the herd includes 300 to 400 square miles in Otsego, Cheboygan, Presque Isle, and Montmorency Counties.

To reduce the elk herd and eliminate some of the problems associated with a large herd, e.g., overbrowsing, competition with deer, and crop depredations, two hunts were authorized by the MDNR in 1964 and 1965. The hunts resulted in a total kill of 472 elk. The herd declined further and in March of 1975, 159 animals were counted in an MDNR census. The most recent MDNR count (December, 1980) was 437 elk.

Until the late 1970's, conclusions regarding the impact of hydrocarbon development activities on elk were based on personal observation and professional opinion rather than solid scientific study. Knight (2) addressed the question of impacts on elk in a limited study conducted in a portion of Montmorency County (to the east of the PRCSF) where development activities were underway. Results of Knight's study were not released until 1980. In the meantime, Amoco Production Company, recognizing the need for additional studies with broader scope, contracted Commonwealth Associates, Incorporated (CAI) biologists to complete such studies. Though the study was funded by Amoco, CAI was responsible for its design and implementation.

The field studies were initiated in August, 1979, and continued until December, 1980. The effort included analyses of the effects of hydrocarbon development activity on elk, bear, and bobcat in portions of Montmorency, Cheboygan, and Otsego Counties.

SAMPLING METHODOLOGY

A sampling program was designed to obtain data for analysis at three levels--regional, subregional, and site-specific. The types of information collected included: track counts, animal sightings, and various other indicators of wildlife usage such as droppings, elk rubs and bugling, and bear tree markings (scratches).

Regional-level analyses employed general observations and existing data. The data were used to determine the size and distribution of the elk population. Track and observational data from four subregions, designated on the basis of their hydrocarbon development activity and the presence of subject species, provided a comparison of wildlife activity at various stages of petroleum development. Intensive site observations around drill sites were used to determine, more specifically, the effects of development operations.

Study Area

The study area included portions of Montmorency, Otsego, and Cheboygan Counties (approximately 200 square miles) in the north-central part of Michigan's Lower Peninsula (Figure 1). The area encompasses much of the traditional elk range (1). Data at the regional level were obtained from available literature, aerial censuses, and general observations pertaining to the entire study area. At the subregional level, field data were collected in four areas (Figure 2): Montmorency Township (almost entirely private lands); Vienna and Vienna/East Township (mainly state owned); Pigeon River Country State Forest/

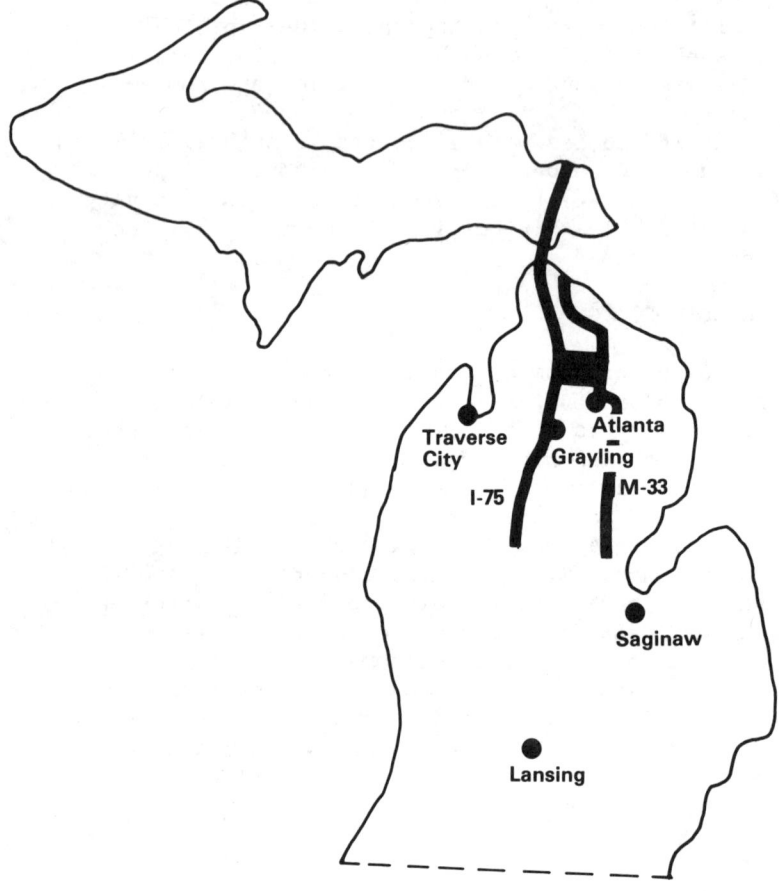

Figure 1. Location of regional study area in north-central lower Michigan.

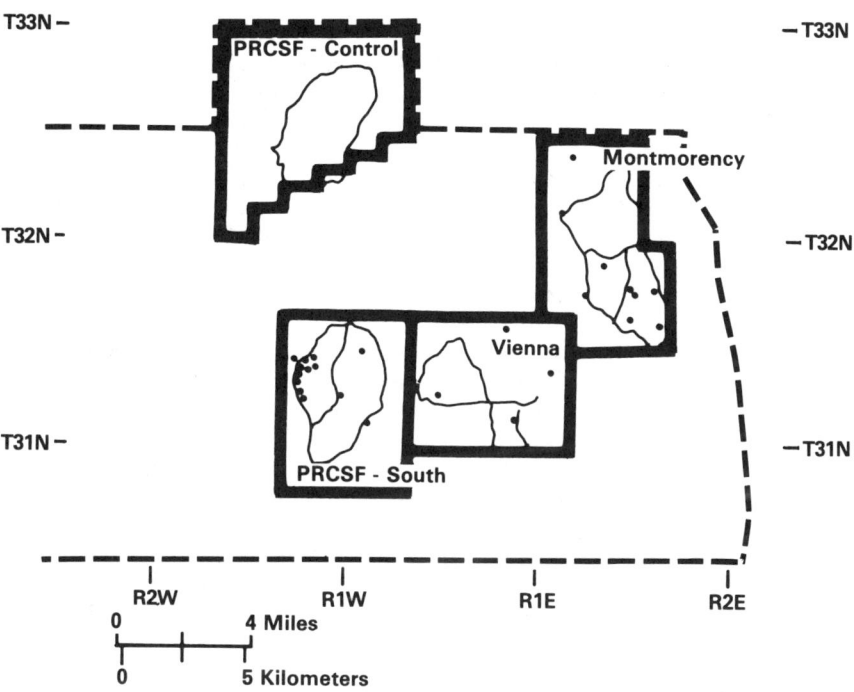

Figure 2. Subregional study areas showing track plot routes and study sites (---Regional study area boundary).

South (PRCSF/South); and Pigeon River Country State Forest/Control (PRCSF/Control). Three of the subregions were selected to include areas of existing or proposed hydrocarbon development and the fourth served as a control area. Site-specific observations were made at 26 proposed or existing hydrocarbon sites (Figure 2), nine of which were selected for intensive data collection.

Aerial Observations

Regional and subregional aerial observations were made in March of 1980 using methods adopted from Bergerud (3), Cahalane (4), and Montfort (5). A Bell Helicopter (Model 47 G2A1), capable of carrying two observers and the pilot was used to fly transects at various altitudes depending on the sampling format. One-fourth-mile-wide transects, located at one-mile intervals, were flown at an altitude of 300 to 500 feet above ground level over the eastern portion of the regional study area to provide an indication of elk distribution.

A more intensive effort in which overlapping transect strips were flown at 100 to 300 feet provided an estimate of the total elk in the subregional study areas. All elk observed along the transects were counted and their location noted. A ground crew preceded the aerial crew to identify high-use areas based on track observations. Hand-held, two-way radios were used to coordinate air and ground searches and to relay information on elk activity from one crew to verify the elk count made by the other.

In addition to the helicopter, a single-engine, high-wing plane was used at various times throughout the winter of 1979 to 1980 to survey the regional study area. Early-morning or late-afternoon flights were made in January, February, and April, 1980, to determine the regional winter range of the elk population. Locations of winter concentrations of elk were also noted.

Track Plot Routes

Subregional study areas were quantitatively sampled using methods suggested by Overton (6) and described by Knight (2). Track plot routes were designated using existing dirt roads near individual study sites in each of the subregional study areas (Figure 2). The routes were selected to provide a representative portion of each of the four subregional areas. The lengths of each route were:

Montmorency Township	17.5 miles
Vienna Township	
Vienna Township	12.7 miles
Vienna Township/East	7.8 miles
PRCSF/South	15.6 miles
PRCSF/Control	11.9 miles

The routes were driven approximately every two weeks. The number of sets of bear, bobcat, and elk tracks were recorded for each quarter-mile segment. The data were used to determine seasonal activity for each subject species. The examination of data from specific quarter-mile segments helped to define normal activity levels and determine the effects of hydrocarbon development activities.

In addition to the track data collected over the term of the study, several other data were analyzed. The vegetative composition of the area along each track plot route was determined including dominant plant community, relative age of the dominant forest community, presence of grassy openings, and the diversity of vegetative types. Topographic data were developed for each quarter-mile segment of the track plot routes. Information gathered included the direction of slope and the general nature of the area (lowland, upland, or transitional). Weather conditions were recorded during the study periods. Air temperature, cloud cover, wind speed and direction, and presence of precipitation were noted. In addition, elk track plot data collected by Knight (2) for Montmorency Township during three May-to-August periods in 1976 to 1978 were considered.

Track counts and vegetation data collected along track plot routes were coded for computer analysis. The Statistical Analysis System (SAS) developed by the SAS Institute, Incorporated (7) was used to analyze the data.

The track data were correlated with hydrocarbon development activities and vegetation conditions. Elk track activity in a given area during periods when petroleum development activities were underway were compared, using a T-test procedure, with those data from times when or areas where no development activity was occurring. The statistical procedures were used to test the hypothesis that the difference in numbers of tracks for the two comparable conditions was equal to zero, i.e., that elk distribution is not affected by hydrocarbon sites present within their range.

Site-Specific Observations

Systematic observations were made at each of the study sites prior to, during, and after hydrocarbon development activities. Records of elk, bear, and bobcat tracks, sightings, droppings, and territorial signs such as bull elk rubs and bear scratchings on trees were kept. Such qualitative data were used to determine species activity in relation to specific hydrocarbon development operations.

Prior to site clearing, a vegetation map of the proposed site was developed. Data indicating subject species' activity within one-quarter mile of the site were collected during site visits conducted several times a month until drilling commenced.

During site clearing, drill pad preparation, and subsequent drillings, regular visits (at least once daily during the week-long surveys conducted twice a month) were made. Evidence of elk, bear, and bobcat was recorded similar to the predevelopment visits.

Following conclusion of drilling operations, less frequent site visits were made, e.g., several times a month. Species' activity, including return to areas previously used, as well as increased use of new areas, was noted.

In addition to the monitoring of drilling activities, a production simulation was established at one of the sites. A production pump was set in place and a natural-gas-powered engine was run daily during a three-month period in late 1980. Daily visits were made to refuel the engine and reset a timing mechanism. During the regular surveys of the study area, daily visits were made to the simulation site prior to morning start-up, while the engine was in operation, and after evening shut-down to record elk, bear, and bobcat activity.

RESULTS AND DISCUSSION

Elk

Aerial Surveys

Regional Level. The aerial sample conducted on March 6, 1980, covered about 22.5 square miles (six transects, one-quarter-mile wide by 15 miles long) of the approximately 200-square-mile study area. Fifty-seven elk were observed, most of which were found in upland hardwood vegetation. This compares with MDNR total count data for the same general area of 75 elk in March, 1975, 87 elk in December, 1977, and 95 elk in December, 1980. Total elk counted by the MDNR over the primary elk range (about 600 square miles) was 159 in 1975, 255 in 1977, and 437 in 1980.

General aerial observations made during the present study (winter of 1979 to 1980) showed that the elk moved from isolated harems scattered throughout the regional study area in early winter and concentrated in large herds of cows and calves with smaller herds of bulls by mid-winter. These herds were usually found in upland hardwood areas located near large openings. As spring approached, the elk dispersed throughout the study area.

Several trends are revealed in analyzing the aerial survey data. The elk herd has increased steadily over the last five years despite continued human encroachment into the "traditional" elk herd range. The encroachment has been in the form of increases in the number of homes and cabins, increased recreational activity, increased logging activity, and increased hydrocarbon development activity. Management activities contributed to the increased herd size. These activities have included the creation of stabilized grassy openings, careful location of logging cuts, and anti-poaching measures.

The importance of grassy openings and recently cut areas to elk is evidenced in a shift in herd distribution to areas such as the Vienna uplands. Scattered well sites, many of which are abandoned, have had little effect on the overall herd distribution compared to the changes resulting from managed habitat alterations. However, abandoned drill sites, when adequately revegetated, are frequently used by elk and other grazing species (Figure 3).

Subregional Level. Subregional analysis of aerial survey results included data obtained during this study (March, 1980) and from MDNR surveys in 1975, 1977, and 1980. The MDNR data are summarized in Table I. Though the census procedures used by the MDNR and CAI biologists were basically the same, the intensity and coverage varied somewhat.

Table I. Elk Population Trends

	Numbers of Animals		
Area	1975 (DNR)	1977 (DNR)	1980 (DNR)
Montmorency Township	72	60	60
Vienna Township	17	28	138
PRCSF/South	1	2	8
PRCSF/Control	22	14	78

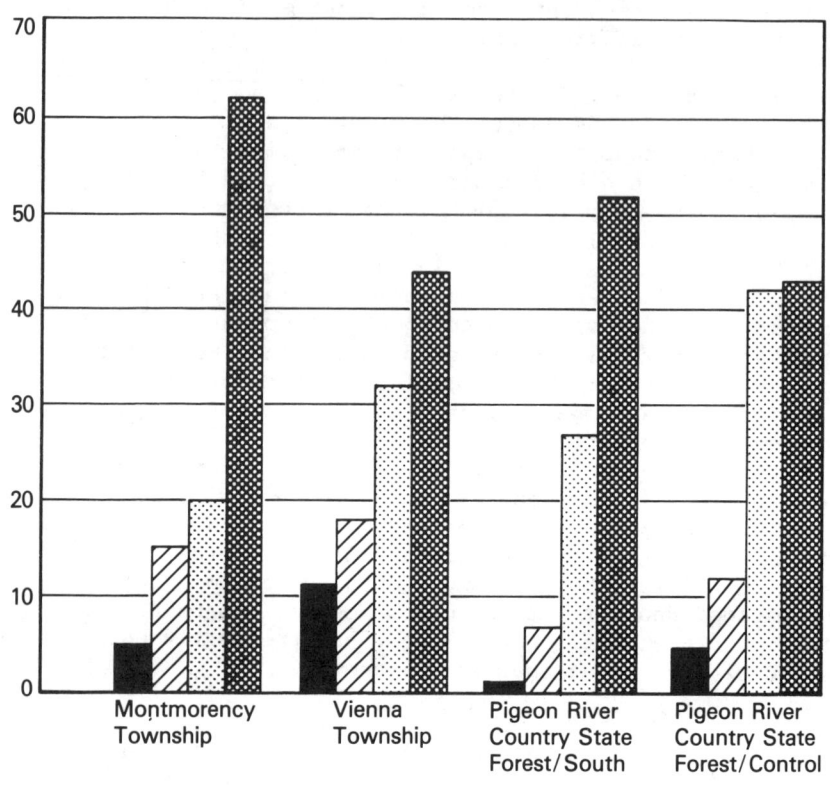

Number Of Elk Per Acre (X10⁻³)
Percentage Of Openings
Percentage Of Upland Hardwoods
Percentage Of Conifers

Note: Number of Elk Based on MDNR Census Dec. 1980

Figure 3. Vegetation composition vs elk per acre.

The aerial census information shows a somewhat progressive increase in the elk population in the subregional areas over the past five years. This is in line with the overall increase in herd size from 159 in 1975 to 437 in 1980 based on MDNR censuses. Populations in Montmorency Township are presently holding steady after experiencing a slight decline between 1975 and 1977. The PRCSF/South area continues to be little used. This area has historically been a low use area (1). Populations in the northern portion of the Forest (PRCSF/Control) have shown fluctuations in use with a general increase between 1975 and 1980.

As is evident from Table I, there has been a general shift in the population to the east, towards Vienna Township. Within the Montmorency Township study area, the elk counted in 1975 represented 71 per cent of all the elk recorded in Montmorency County. In 1977, the 60 elk counted represented 40 per cent, and in 1980, 25 per cent. Comparable figures for the Vienna Township study area in Montmorency County were 17 per cent for 1975, 19 per cent for 1977, and 58 per cent for 1980. This apparent shift from the Montmorency to the Vienna Township study area may be attributed to the intensive management of the latter study area. Several large openings have been created in the Vienna area to draw elk off nearby private lands and onto state lands. In addition, the trend in population shift began before the series of well drillings (Fall 1979 to Spring 1980).

Track Plot Surveys

The track plot data provide an index of elk use and habitat preference. The same data can be used to determine alterations in elk activity resulting from petroleum development operations.

Activity Versus Vegetation Type. Analysis of vegetation data collected along the track plot routes resulted in the designation of 13 dominant vegetation types. The frequency at which these types were distributed varied from one subregional area to the next.

The most common types found along the Montmorency Township route were aspen, hardwoods, and upland conifers. In addition, about 66 per cent of the quarter-mile segments had some type of opening or clearing associated with them.

Elk activity during the winter was generally low along the Montmorency Township route. The elk were generally concentrated off the route in the winter in the south and west portions of the subregion. Most of the activity occurred in the upland hardwood areas and in the openings.

Elk were frequently noted on the sunny, southwestern slopes in the winter.

Other habitat types were used more frequently in the spring and summer. Upland conifer areas experienced high use apparently due to the abundance of openings associated with them along the route. Aspen areas were less used due to their advanced stage of growth.

Aspen and aspen-conifer were the two most common types of vegetation along the Vienna Township route. About two-thirds of the quarter-mile segments designated aspen or aspen-conifer were middle-aged stands. Aspen areas were regularly used throughout the year, but especially in spring and fall. Hardwoods were used heavily in the fall due to their production of mast, and hardwood areas near openings were used more often then pure stands. More widespread use of all habitat types was evident during the spring to fall due to the dispersal of elk from their wintering areas. Elk activity was consistently high in those route segments associated with openings.

In the Vienna Township/East area, upland conifer areas comprise approximately 66 per cent of the route. Elk use is generally low along the route except for late spring (calving) through the rutting season. The upland conifer areas were used throughout the year at varying levels, apparently due to the presence of numerous small openings associated with the cover afforded by conifers. Aspen-conifer, cherry, and shrub areas were used especially when leaves were present. The low-growing nature of such areas provides browse in the predominantly coniferous area.

The PRCSF/South area is comprised mainly of aspen-conifer, upland conifer/hardwood, and hardwood vegetation types. Openings were located near aspen-conifer areas.

Elk activity is generally low in this subregion, the greatest activity associated with the spring and summer months and with aspen-conifer and aspen areas. Because there was no drilling activity in the State Forest during the course of the study, hydrocarbon development activity levels were constant. In addition, no baseline data are available for pre-hydrocarbon development periods. Track data, therefore, could not be analyzed statistically.

Elk activity along the track plot route in the control area was relatively low except for the period spring through rutting season. The aspen-conifer, aspen, lowland hardwood/conifer, and upland conifer hardwood areas were most frequently used. The openings were little used due to their proximity to more heavily-traveled roads.

Activity Versus Hydrocarbon Development Activity. Track plot data were also analyzed to determine what effect hydrocarbon development operations might have on elk movement.

Operations were monitored along routes in Montmorency Township and the two routes in Vienna Township.

In Montmorency Township, activity around a site drilled in spring of 1980 was monitored. Track data from those quarter-mile segments near the drill site were tested against segments with similar vegetation characteristics but which were located outside the influence of development activity. Results of a T-test analysis indicate that elk track activity in segments near the drill site was significantly higher than that in nonaffected segments during nondrilling periods, i.e., before and after drilling (T value = 1.82, PR>{T} = 0.0001). During drilling, the T value was equal to 0.0, indicating no difference in elk activity between similarly vegetated segments within or outside the influence of drilling activities. It appears that elk activity was normally higher in the drill site area for most of the study period than it was away from the site. During the actual drilling, the activity levels diminished to those normally found in similarly vegetated areas along the route. Several weeks after drilling and site abandonment, elk activity returned to levels similar to those found before drilling. A short-term drilling-activity-related impact is thus indicated.

Knight's track data (2) were included in an analysis to determine long-term trends in elk use in Montmorency Township from 1976 to 1980. The data were limited to those collected between May and August (calving and summer seasons). Knight's analysis indicated a drop in elk population estimates for the area from a high of 92 in the fall of 1976 to a low of 59 in the winter of 1976. The spring 1977 population was also the lowest spring population reported by Knight. Knight attributes the decrease to elk movements out of the area.

Since then, elk activity, as indicated by track data and aerial surveys, for the study area in general has definitely increased (Table I). Interestingly, this increase has occurred during a period when five additional drilling sites were completed in the area.

Analysis of elk track data along the Vienna track plot route during a drilling operation provided results similar to those seen in Montmorency Township. While none of the T-tests indicated significant differences between similarly-vegetated route segments near or far from drilling sites, general observations suggested a lower elk use around a drill site during drilling compared to nondrilling periods. The effect was short-term, with elk activity returning to predrilling levels around the abandoned sites in a few weeks.

The Vienna/East area is generally a low-use elk area with the exception of late spring or calving season through the rutting season. Widespread elk use during this period coincides with the dispersal of elk from nearby concentrations. Aspen-conifer areas near the single drill site during nondrilling periods were used more than similar areas elsewhere. The difference, however, was not statistically significant (t = -0.05 PR>$\{T\}$ = 0.5615). During the drilling period, there was no difference in elk activity for similarly-vegetated track plot segments in nondrilling areas.

Track data were not analyzed statistically in the PRCSF/South subregion because hydrocarbon activity had ceased following a court order. In the Control subregion, track data revealed no trends which could be attributable to hydrocarbon development activities in other areas.

Site-Specific Monitoring

Studies conducted on specific sites further documented the effects of hydrocarbon development on elk behavior. Since the investigation was limited to the areas around selected sites, the data collection was more intensive. The results of the monitoring efforts are summarized in Table II.

Activity Versus Drilling and Related Operations. Data collected during the site-specific surveys in the four subregional areas indicate that elk activity is not permanently affected by current levels of hydrocarbon development. The initial tendency appears for elk to move away from human activity. The relocation, however, is short-term. Although elk activity was noted in the vicinity of only two active site clearing and drilling operations, activity was noted, after varying lengths of time, at nearly every site following completion of drilling and reclamation activities. Decreased activity was apparent during development and up to two to four weeks thereafter, but then returned to pre-drilling levels. The activity did not significantly differ between pre-development and post-development levels. In fact, tracks were found crossing one site in the Vienna Township study area while clearing operations were underway. Elk tracks were also found within 100 yards of an operating drill rig, and elk tracks were found on several producing wells studied.

The short-term effect of drilling activities is similar to that noted around intensive logging operations which occurred in one section of the PRCSF/South area during the study. Elk avoided the immediate vicinity of logging operations conducted during the early fall of 1979. Following logging activities, elk utilization of the area increased over previous levels.

Table II. Drill-Site, Vegetation, and Elk Activity Data

Subregional Area	Site	Predominant Vegetation Around Site	Elk Activity Within 1/4 Mile of Drill Site*		
			Pre-Drill	During	Post-Drill
Montmorency	4	aspen, upland hardwood	−	−	+
Montmorency	5	hardwood-conifer	+	NA	+
Montmorency	6	hardwood, aspen-hardwood	+	+	+
Montmorency	7	hardwood-aspen	+	NA	+
Montmorency	10	open-conifer	0	0	+
Montmorency	11	hardwood, aspen-hardwood	0	0	+
Montmorency	12	hardwood, aspen-hardwood	0	0	+
Vienna	1	aspen, aspen-pine	+	−	+
Vienna	3	open-hardwood	+	+	+
Vienna	13	conifer	0	0	+
PRCSF/South	18	aspen	0	0	+
PRCSF/South	P**	conifer-hardwood	0	0	−

* + = sightings, tracks, and/or droppings; − = no signs; 0 = occurred before study; NA = no access to site
** Site of several wells plus production facilities.

Site clearing and drill pad preparation does result in the loss of some habitat. Though new habitat is created, the new habitat (revegetated grassy clearing) may not compensate for the old in all respects. For example, for calving purposes, existing cover may be more important than a newly created opening. In some ways, the new habitats resulting from development operations, e.g., the two-acre openings, can have a positive effect on elk. Elk use such openings for foraging. Elk were found to use all the abandoned and producing drill sites in the Montmorency and Vienna Township study areas at sometime during the year. Such openings are similar in principal to those maintained by the MDNR.

Based on the total number of drill sites now located in the Montmorency Township study area, a density of one site per two square miles appears to have little effect on elk activity. This is basically the same conclusion reported by Knight (2) earlier for the same area when site density was one site per 2.4 square miles.

Activity Versus Production Operations. Limited production operations in the overall study area precluded the determination of the effect such activity might have on elk behavior. A "horse-head" pumping unit was, therefore, moved to an abandoned drill site in the Vienna Township study area to simulate production activities. A small engine placed on a timer simulated some of the noise associated with such operations during the day. Visits to the site each morning to restart the engine and reset the timer simulated the human activity found at producing sites.

Track plot data for that subregion indicated that track activity during the simulation period (October through December, 1980) was higher than during the drilling operations conducted at that site (October, 1979). Though the levels were less than those associated with nonhydrocarbon areas, T-tests applied to the data indicated that the differences were not statistically significant.

Site-specific monitoring also indicated the localized extent of production activity impacts. Six elk were observed and photographed on the site during the period when the pumping equipment was being installed. During simulated operations, elk activity was noted in the area primarily during the nonoperational periods. However, these are normally the highest activity periods for elk use in the absence of hydrocarbon development activity. Elk tracks were observed on or adjacent to the site during each of the two survey periods every month the simulation was in operation, except deer hunting season and the late

December survey. Thirty-four elk were counted approximately one-half mile northeast of the simulation site in December during the MDNR elk census.

Production facilities at current levels do not appear to adversely affect the elk. This is based not only on visits during the simulation, but on observations made at producing wells in the PRCSF/South and Montmorency Township study areas as well.

Bear

Information on the effects of hydrocarbon development activity on bear is sparse due to their habitat preferences (8) and behavior (9). Only one sighting was made during the course of the study. A female and three yearling cubs were observed near a production facility in the PRCSF/South study area in April of 1980. However, tracks and/or droppings were identified in all the subregional study areas.

Track Plot Surveys

Bear tracks along the Montmorency Township route were recorded only during the period June to October, 1980. All tracks were observed in vegetation types having some coniferous cover. Along the Vienna Township route, bear tracks were found in scattered locations between May and November of 1980. All tracks were near lowlands or upland coniferous cover. No bear tracks were recorded along the Vienna Township/East route. Bear activity along the PRCSF/South route appeared to be associated with nearby lowland conifer areas. Tracks were noted throughout the study period except January to March, 1980. Only two observations of bear tracks were made in the control area.

Site-Specific Monitoring

Site-specific data were recorded for the various subregions. Bear markings were found on trees located within one-quarter mile of a drill site in the Montmorency Township study area. The scratches indicate a bear's territorial boundary. Two of the trees had fresh scratches on them in April, 1980, while drilling was underway. Other sites in the subregion contained bear tracks and droppings prior to development.

No tracks were observed on abandoned sites located in the Vienna Township study area, but bear activity was noted around many of the sites.

Moderate bear utilization on and around study sites in the PRCSF/South area was noted. Bear tracks were seen on several occasions on one site. Most bear activity occurred around sites where extensive lowland conifer areas were present.

Hydrocarbon development activities at current levels have no long-term effect on bear based on observations at abandoned drill sites in the Montmorency Township and PRCSF/South study areas. Some loss of cover and food sources occurs.

Regional analysis of bear population shifts in relation to human encroachment is difficult, since populations of such a wide-ranging species are difficult to measure.

The track plot data, though indicative of widespread use of the study areas by bear, were not sufficient to permit statistical analysis. Although little bear activity was noted near active drilling operations, the limited drilling which has taken place to date in the regional study area appears to have had little effect on this species.

Bobcat

Bobcats in Michigan have been studied by Erickson (10) and others. The effect of hydrocarbon development activity on bobcats has not been addressed in any great detail. Within the study region, bobcat tracks and/or droppings were noted on every track plot route and during general observations in all the subregions.

Track Plot Surveys

Bobcat activity was widespread along the Montmorency Township route. Tracks were observed throughout the study period. More tracks were observed during the May to September, 1980, period than at other times of the year. Though no consistent use of any one vegetation type was apparent, consistent observations along portions of the route indicate regular use of the area by resident bobcats.

Widespread activity was also noted along the route in the Vienna Township study area. Activity was scattered throughout most seasons. Highest usage was during the June through October period in 1980. Similarly, activity was noted throughout the study period along the Vienna/East route with most tracks seen between May and November, 1980. Vegetation types utilized by bobcats varied widely and no apparent preference was indicated by the track data.

Along the route in the PRCSF/South study area, track activity was scattered. Most activity occurred between May and October, 1980. No vegetation type preference was indicated. Track data did indicate that the home ranges of a number of bobcats overlap the route.

Bobcat activity was noted throughout the study period along the Control route. Most occurred in the northern section of the area.

Site-Specific Monitoring

Bobcat activity was not noted on any of the study sites in the Montmorency Township study area. The absence of evidence of activity on the sites compared to the abundance of it along the track plot route probably reflects the bobcat's predilection for existing trails. Some tracks were noted near abandoned sites in the Vienna Township and PRCSF/South study areas. In the forest, activity was particularly evident along trails and roadways near lowland conifer stands.

Analysis of bobcat populations and potential impacts to them is difficult. As with the bear activity data, bobcat data were insufficient to permit statistical analysis. No bobcat sign was found on any abandoned or producing study site. Again, this lack of evidence is probably more a result of the animal's preference for open linear features than a negative impact from human activities. This is supported by the frequency of tracks seen along trails and roads, even those near abandoned and producing sites. The loss of habitat from drill site clearing is only temporary.

CONCLUSIONS

Analysis of the data collected during the aerial surveys, track plot route surveys, and site-specific monitoring efforts has led to several conclusions regarding the effects of hydrocarbon development activities on elk, bear, and bobcat in northern Lower Michigan. In addition, elk behavior in relation to other forms of human activity has been observed.

Levels of hydrocarbon development currently found in areas such as Montmorency and Vienna Townships have generally had only short-term, localized impacts on the subject species. Elk, bear, and bobcat activity decreased within one-quarter mile of drill sites during development, but returned to predrilling levels within two to four weeks following site abandonment. The relatively high levels associated with some areas prior to drilling were again noted after drilling was complete.

In his studies, Knight (2) reported that a well-site density of one per 2.4 square miles did not seem to affect elk movements or distribution. Data collected during the present study indicate that a density of one site per 2.0 square miles in the same area similarly appears to have little impact. Both conclusions are based on the densities of well sites found in the Montmorency Township area.

Reduction in elk activity was also noted when production equipment was present at a site. This reduction, however, was restricted to the immediate area and apparent, for the most part, only during periods of operation. At the site of production simulation in the Vienna Township study area, elk were observed within one-quarter mile of the site after drilling and abandonment. Elk were observed less often during the simulation experiment, and especially while the engine was operating. This reduction in activity, however, appeared to be much less than that seen during the actual drilling of the well.

Habitat loss resulting from development of a drill site is greatly mitigated through the regrading and revegetation of the site. In fact, a properly regraded and revegetated site can be beneficial to wildlife species such as elk. The small openings provide grazing areas for both elk and deer. They also provide "edge" habitat where wildlife species move easily while remaining near safe cover. A correlation does exist between elk activity and the presence of such openings (Figure 3). The MDNR has been creating large openings for elk in areas such as the Vienna Township, and forest cuttings contribute to the creation of additional openings used by elk.

In addition to hydrocarbon development activities, other forms of human activity were found to influence elk behavior. Habitat seemed to have more effect on elk activity than the presence of gravel or dirt roadways associated with hydrocarbon development activities. For example, roads infrequently used by humans were frequently crossed by elk. More heavily-traveled roadways were also crossed by elk on a regular basis. In fact, elk were generally tolerant of vehicular traffic, including snowmobiles, and were reluctant to change their direction of movement unless approached directly by a moving vehicle. This was the case for all seasons except during deer hunting season when human activity in the study area is at its highest.

Elk moved away when approached by a human on foot at any time of the year. Some individuals or groups of elk were tolerant though of human approach. Such behavior is in contrast to that exhibited by hunted elk in the western states.

The limited extent and short-term nature of impacts on elk, bear, and bobcat from hydrocarbon development activity levels currently found in the north-central portion of Lower Michigan clearly suggests that our search for and production of petroleum reserves can be compatible with the wildlife species studied in that area. An industry cognizant of the presence and importance of our wildlife resources can greatly reduce its impacts on the environment in which it operates. Continued improvement of revegetation and reclamation techniques will promote the timely return of drill sites to useful habitat. Avoidance of areas frequented by elk, especially during winter, calving, and rutting periods, will also reduce the effect of development activities on their behavior. Lowland coniferous areas frequently used by bear and bobcats for seclusion and protection should also be avoided. This study, in addition to identifying the habitats and periods most critical to the well-being of the subject species, has also helped to characterize the short-term nature and limited extent of hydrocarbon development impacts measured during the study period.

LITERATURE CITED

1. Moran, R. J. 1973. The Rocky Mountain elk in Michigan. Michigan Department of Natural Resources, Wildlife Division Research and Development Report, No. 267. Lansing, Michigan, 93 pp.

2. Knight, J. E. 1980. Effect of hydrocarbon development on elk movements and distribution in northern Michigan. University of Michigan Ph.D. dissertation.

3. Bergerud, A. T. 1963. Aerial winter census of caribou. J. Wildl. Manage., 27(3): 438-449.

4. Cahalane, V. H. 1938. The annual northern Yellowstone elk herd count. Trans. North America Wildl. Conf., 3: 388-389.

5. Montfort, A. 1975. Les techniques de denombrement adaptees a l'etude quantitative des populations d'ongules savages. Terre Vie, 29: 3-19.

6. Overton, W. S. 1971. Estimating the numbers of animals in wildlife populations. *In* R. G. Giles (ed.)

Wildlife Management Techniques. Wildlife Society, Washington, D.C.

7. Statistical Analysis System Institute. 1979. SAS user's guide, 1979 edition. Cary, North Carolina.

8. Burt, W. H. 1977. Mammals of the Great Lakes Region. University of Michigan Press. Ann Arbor, Michigan.

9. Burt, W. H., and R. P. Grossenheider. 1976. A field guide to the mammals. Houghton Mifflin. Boston, Massachusetts.

10. Erickson, A. W. 1955. An ecological study of the bobcat in Michigan. Michigan State University M.S. thesis.

PIPELINE CROSSINGS OF STREAMS:
BENTHOS RECOVERY AND HABITAT
ENHANCEMENT

Donald K. Gartman
   Columbia Gas System,
   Wilmington, Delaware 19807

ABSTRACT

   Four buried natural gas pipeline stream crossings were investigated in Pennsylvania, New York and Maryland relative to density and diversity of the macroinvertebrate community at the crossing. Benthos populations were sampled from the streams at the crossings and at points up- and downstream. Time intervals since instream construction activities and sampling dates varied from several decades to 26 days. The latter stream crossing was sampled before construction and at approximately monthly intervals for one year.
   Results indicate that benthos recovery is relatively rapid and the pipeline crossing itself often results in favorable aquatic habitat, especially where excavation and backfilling creates a more diverse substrate available for colonization by macroinvertebrates. No long-term damage to the benthic community was noted.
   Pipeline construction techniques can be modified on a site-specific basis as indicated by stream substrate and bank soil conditions to minimize sedimentation impacts to downstream reaches.

INTRODUCTION

   Regulatory agencies responsible for protection of fish and wildlife resources are quite sensitive to requests to construct a buried pipeline crossing of a stream, especially a stream with valuable sportfishing potential. Although little empirical evidence exists, it would seem intuitive that instream excavation and pipeline burial would cause increased levels of suspended solids, direct loss of the benthos community within the right-of-way work area of the crossing, and downstream siltation of fish spawning areas and benthos habitat.

There are numerous studies on the impacts of dredging on aquatic habitat with laboratory studies dealing with direct mortality effects of suspended sediments on various fish species, but these are not directly relevant to pipeline crossings. Pipeline crossings typically require one to five days for streams less than about 30 m in width and less than about 1 m average depth. In streams with rocky or solid substrate, siltation is of concern only when the stream banks are being cut and exposed soil comes into contact with water through equipment movement or water flow changes.

Streams that are susceptible to siltation impacts because of substrate characteristics and stream bank contour can be crossed by means of "fluming" of the stream flow through pipes temporarily placed in the stream. These "flume pipes" transport the water away from the work area while the pipeline trench is being excavated from the banks with a backhoe. Obviously, where blasting is required this procedure is severely limited. The environmental aspects of this technique in stream crossings have been evaluated by Crabtree et. al. (1) and Baddaloo (2).

During construction, factors such as stream flow, channel width and depth affect the construction procedures required to install the pipeline across a stream.

Other concerns of pipeline stream crossings are caving and recession of stream banks. If this kind of erosion is extensive, stream "scour" could expose the pipeline in the vicinity of the crossing and result in damage to the pipeline. Thus, sound engineering and environmental design compliment each other for a successful pipeline stream crossing.

PRESENT STUDY

In 1980 it was decided to evaluate the aquatic macroinvertebrate communities of the streams in the vicinity of several of Columbia's natural gas pipeline crossings.

In addition, a more intensive study of a crossing of a Pennsylvania trout stream (Bushkill Creek) was undertaken to monitor the reaction of the benthos on a pre- and post construction basis.

The construction of this pipeline across this productive trout stream provided the opportunity to investigate the changes to the benthic community resulting from the instream construction activity on a short-term basis.

Three of the streams experienced pipeline crossings in the past at various times. The benthos communities of these streams were investigated at the crossings and at similar reaches at up-and-downstream stations with collections from one day's sampling. It is recognized that this limited sampling gives only a "snapshot" of the benthic community, but that any major changes to the substrate habitat should be reflected by the density and diversity of the macroinvertebrates.

STUDY AREAS

Chautauqua Creek

Chautauqua Creek in Chautauqua County, New York is located within the Lake Erie drainage near Mayville, New York. The pipeline crossing at this location was completed in September of 1979. Construction of the buried 14-inch (35.6 cm) natural gas pipeline required blasting through the slate substrate which dominates the stream bottom. The stream at this point is entrenched within a gorge referred to locally as "The Gulf," (Figure 1). The crossing is located approximately 10 miles (16 km) from the mouth of the stream at Lake Erie. The stream receives runs of lake-run rainbow trout (Salmo gairdneri) and is stocked with trout by the New York Department of Environmental Conservation.

Zekiah Swamp Creek

Zekiah Swamp Creek is found in Charles County, Maryland and is located at the headwaters of the Wicomico River which discharges into the Potomac River and eventually the Chesapeake Bay (Figure 2). The 36-inch (91 cm) pipeline crossing was constructed in August 1976. Zekiah Swamp is a

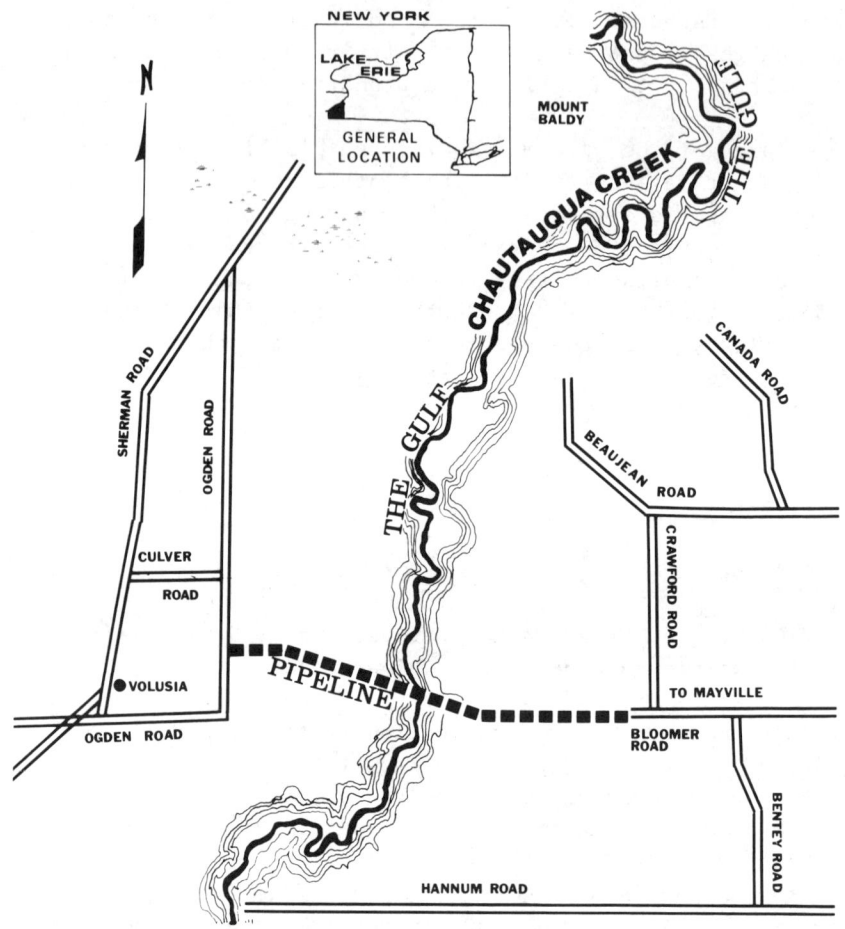

Figure 1. Map of Chautauqua Creek pipeline crossing location.

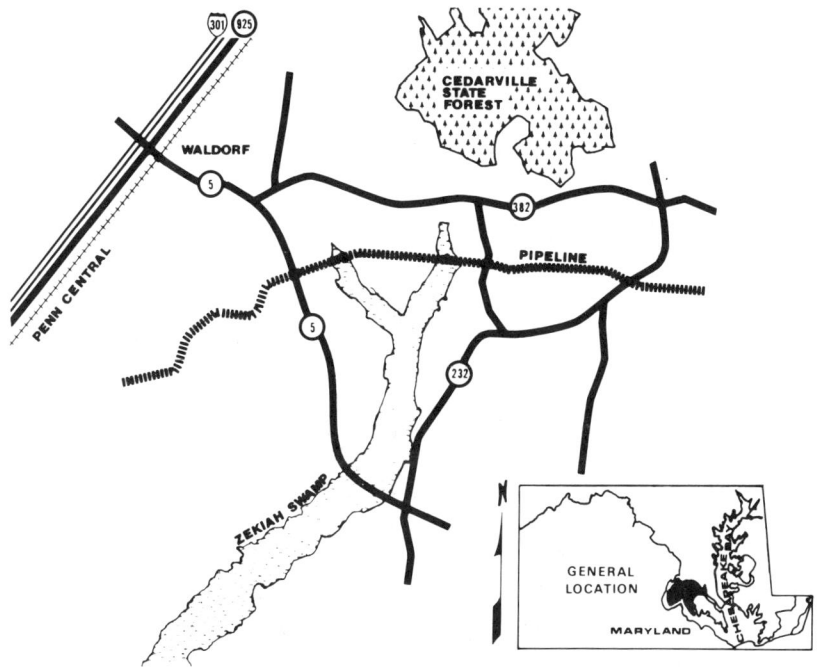

Figure 2. Map of Zekiah Swamp pipeline crossing location.

5400 ha. hardwood swamp which has been the subject of several environmental controversies in the recent past. The area is rich in sand and gravel as well as timber resources. This stream is dominated by gravel with a smaller amount of sand and silt. Most of the tributary streams of this extensive swamp are shaded with a closed canopy of water-tolerant tree species. Benthic substrate diversity is greatly limited by the shifting sand and gravel substrate.

Broad Run

Broad Run in Chester County, Pennsylvania is a small, second-order stream which drains into the Brandywine River (Figure 3). This 20-inch (51 cm) pipeline was constructed in 1932 and is scheduled for replacement in the near future. This stream has a diverse benthos community and is generally less than 15 cm deep in the area of the crossing. An

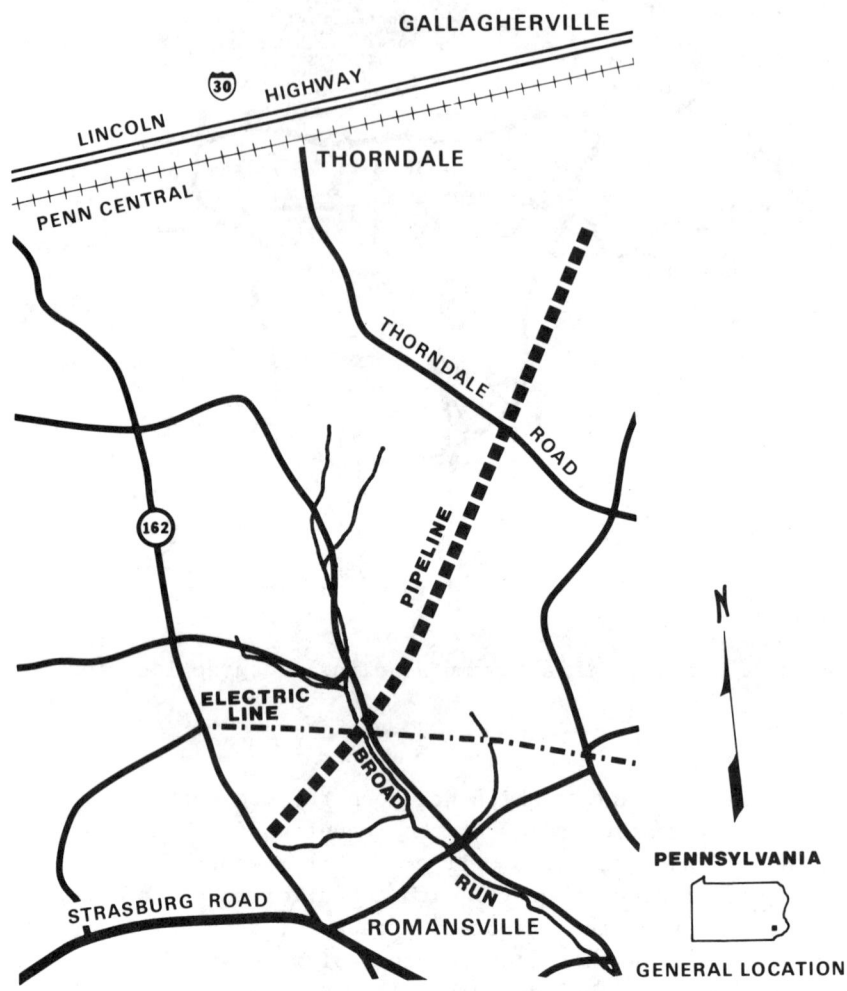

Figure 3. Map of Broad Run pipeline crossing location.

electric transmission line also crosses the stream at this point and land use in the area is pasture for livestock.

Bushkill Creek

Bushkill Creek, Northampton County, Pennsylvania has a natural population of brown trout, Salmo trutta, and discharges into the Delaware River near Easton (Figure 4).

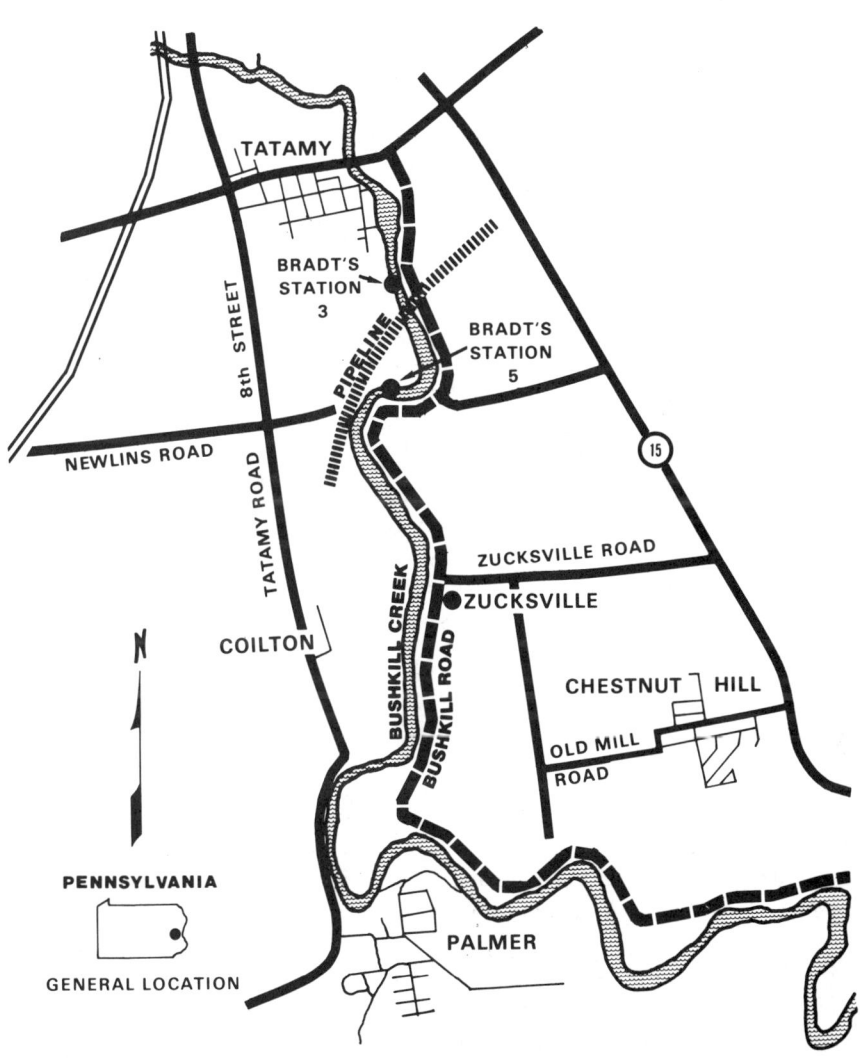

Figure 4. Map of Bushkill Creek pipeline crossing location.

In August 1980, Columbia constructed a 20-inch (31 cm) pipeline across the stream within the Pennsylvania Fish Commission's "fly-fishing only" section. At the crossing the Bushkill is approximately 20 m wide with an average depth of 0.3 m. Blasting was required to open the trench through the dolomite substrate. Precautions were taken to direct water away from the construction area although "fluming" was not feasible. Bushkill Creek has been studied in the past by P. Bradt (3, 4) in the vicinity of the crossing which made comparison of benthos populations from previous years possible.

METHODS

Surber square foot (0.09 m^2) samples were taken at the pipeline crossings and at representative points above and below the crossings for Chautauqua Creek, Zekiah Swamp and Broad Run in the summer of 1980. Sample numbers varied from 6 to 9 depending on width of the streams.

Bushkill Creek was sampled before pipeline construction and at approximately monthly intervals after construction. Also, Bradt's upstream and downstream sampling stations (1978) were compared with the present 1980-81 data from the pipeline crossing.

The samples were preserved in 95% ethanol and returned to the laboratory for identification and enumeration. The macroinvertebrates were identified to genus where possible.

RESULTS AND DISCUSSION

Table 1 presents the macroinvertebrate densities and Shannon-Weaver diversity estimates for the pipeline crossings and up-downstream sampling locations for Chautauqua Creek, Zekiah Swamp Creek and Broad Run.

Since it is generally true that relatively unstressed environments support communities with numerous species represented, the Shannon-Weaver index was chosen for comparison as recommended in the EPA Biological Field and Laboratory Methods (5). Generally, $\bar{d}$ ranges between 3 and 4 with clean, favorable conditions, while $\bar{d}$ is usually less than 1 in

Table 1. Mean numbers of aquatic macroinvertebrates (N/m^2) and diversity ($\bar{d}$) estimates for stream benthos communities of study streams, 1980

Stream	Upstream N/m^2 ($\bar{d}$)	Pipeline N/m^2 ($\bar{d}$)	Downstream N/m^2 ($\bar{d}$)
Chautauqua Creek Chautauqua Co., NY crossed in 1979	527 (3.13)	732 (3.56)	269 (2.59)
Zekiah Swamp Charles Co., MD crossed in 1976	32 (1.37)	430 (2.38)	2,023 (2.77)
Broad Run Chester Co., PA crossed in 1932	882 (3.65)	3,970 (3.19)	1,786 (3.84)

polluted or stressed conditions (6). The index is subject, however, to "false low values" if one species has an extraordinarily large number of individuals even in samples with many species.

Chautauqua Creek

Chautauqua Creek samples included 24 taxa from the pipeline area with 14 and 17 taxa at the downstream and upstream sites respectively. The dominant organisms at all locations were chironomids. The pipeline crossing had eight ephemeropteran genera while the upstream samples contained nine and the downstream site recorded six. Trichopteran taxa at the crossing were more numerous (six) compared with upstream (four) and downstream (two) samples.

The blasting required for pipeline burial resulted in more diverse habitat with a correspondingly diverse benthos population (Figure 5).

Figure 5. Chautauqua Creek pipeline crossing, Chautauqua County, New York.

Zekiah Swamp Creek

    Zekiah Swamp Creek samples demonstrated a depauperate benthos community above the pipeline crossing where the substrate was predominately sand. The pipeline crossing resulted in an opening of the swamp canopy with growths of sedges and grasses along the banks. Substrate at the crossing was an equal mixture of sand and pea-gravel. Chironomids were the most common organism found although a total of 12 taxa were collected including one aquatic lepidopteran, <u>Parapoynx</u>. The downstream site contained

riffle areas of pea-gravel with growths of <u>Potomogeton</u>.
Benthos samples indicated 15 taxa with chironomids and
hydropsychid trichopterans dominating.

The major change to the swamp ecosystem resulting
from the pipeline construction was reflected by the thick
growths of water lil ies, sedges, and other wetland plants
which invaded the right-of-way opening, (Figure 6). This
open area has been found to be used heavily by waterfowl.

Figure 6. Zekiah Swamp pipeline crossing, Charles County,
Maryland.

Broad Run

Broad Run had the longest "in-place" natural gas pipeline crossing of the 4 streams investigated. The small riffle at the crossing had a more numerous macroinvertebrate community than either the up-or-downstream sites although diversity (d) was slightly lower. Large populations of trichoptera were found at the crossing which may be a result of more available light at this opening in the stream border vegetation, (Figure 7).

Figure 7. Broad Run pipeline crossing, Chester County, Pennsylvania.

Bushkill Creek

Pre- and post-construction data for Bushkill Creek from August 1980 through April 1981 are presented in Figure 8. Bushkill Creek samples noted 25 taxa with a mean density of $872/m^2$ in August before the pipeline crossing. The diversity index (d) for these samples was 2.97. Bradt's (1978) sampling station 3 which is located approximately 80 meters upstream from the pipeline crossing recorded a summer density of $4,842/m^2$ with a standard deviation of ± 3,999, indicating the typically high variation found in benthos samples. Diversity (d) was estimated at 2.9.

The area of stream crossed by the pipeline initially had a less diverse "riffle habitat" than Bradt's upstream transect. After construction, however, this changed.

Blasting was required to open the stream bottom so a 7-foot (2.1 m) deep trench could be excavated to receive the pipeline. Also, large rock segments from the east bank were removed to facilitate boring under the highway which parallels the stream. These large "boulders" were later placed in the stream crossing reach with the advice of the Pennsylvania Fish Commission. These boulders created areas of quiet "pocket water" behind them which could be used by the resident brown trout population (Figure 9).

Water velocities were greatly altered in this reach which had previously been a smoother "run" with less diverse water flow patterns.

Post construction samples were taken 26 days after pipeline construction. A total of 27 taxa and a mean density of $5,391/m^2$ were recorded with d = 2.5. Diversity was somewhat suppressed by an extraordinary increase in Hydropsyche which colonized the newly exposed dolomite substrate made available by blasting and pipeline placement procedures. Hydropsyche numbered $2,129/m^2$ while the ephemeropteran Baetis was recorded at $1,261/m^2$.

Bradt's downstream station (approximately 500 m below the crossing) noted a December 1981 density of $2303/m^2$ with d = 3.4. Her 1978 study noted a winter mean density of $3260/m^2$ at this station with d = 3.3.

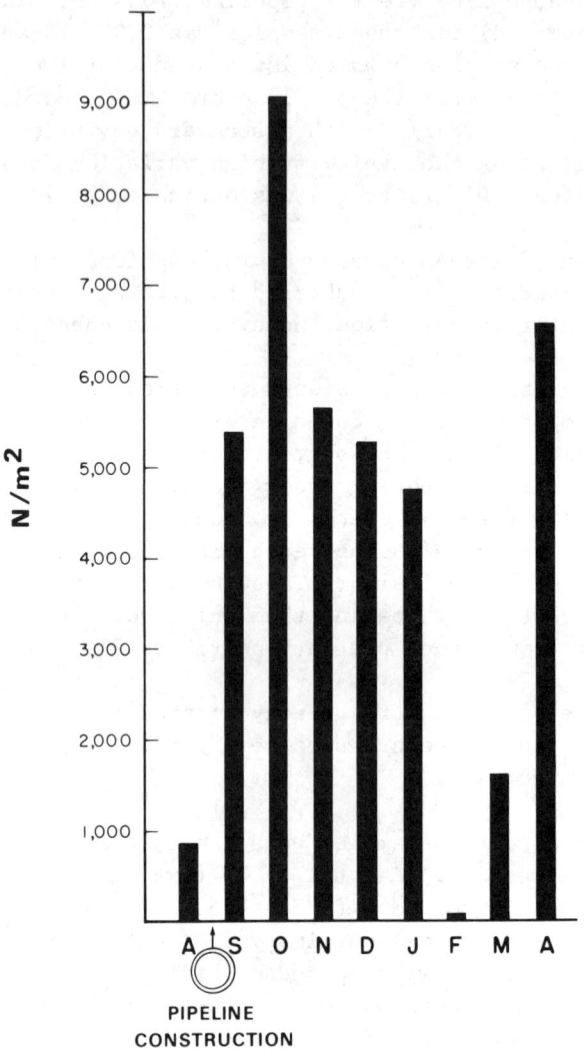

Figure 8. Mean numbers of benthic macroinvertebrates per square meter from riffle area over pipeline, Bushkill Creek, Northampton County, Pennsylvania, August 1980 - April 1981.

Figure 9. Bushkill Creek pipeline crossing, Northampton County, Pennsylvania.

Mean numbers of benthic macroinvertebrates over the pipeline crossing gradually decreased throughout the fall and winter months with a low of $165/m^2$ recorded under "ice-scour" conditions in February. Densities were up 10-fold in March and the April sampling indicated $6531/m^2$ with 23 taxa.

The main point indicated by the stream crossing studies is that no long-term damage to the benthos communities was noted. Macroinvertebrate populations were at least comparable, or in fact, more numerous and diverse at pipeline stream crossings when compared to similar up-and-downstream reaches. The Bushkill study noted that benthos recovery at the crossing is rapid.

Areas of stream bottoms disturbed by natural or man-induced activities are readily recolonized by aquatic macroinvertebrates as long as suitable habitat is available.

Waters (7) noted that each day large portions of the benthic community were carried downstream as "drift" (the downstream movement of organisms transported by water currents). The phenomenon of "drift" allows disturbed stream areas to be reinvaded rapidly. Müller (8) studied a 150 m reach of Skravelbäcken in Sweden which had been cleared with a tractor. Within 11 days the benthos population was up to 9,240 organisms/m^2. Allen (9) in his study of Cement Creek, Colorado noted that pans set in the stream bottom with various substrate sizes were colonized within 24 hours - with a total of 31 species observed in one pan.

Pipeline construction across streams need not be deleterious to the aquatic community - in fact, with careful planning by biologists and engineers, it can be stated that a more productive "reach" of stream is possible with minimal short-term impacts.

LITERATURE CITED

1. Crabtree, A. F., Bassett, C. E., and L. E. Fisher. 1978. The impacts of pipeline construction on stream and wetland environments. Michigan Public Service Commission. Lansing, Michigan. 154 pp.

2. Baddaloo, E. G. 1978. An assessment of effects of pipeline activity in streams in the Durham and Northumberland counties of Ontario. In, Energy/Environment '78, a symposium on energy development impacts. International Society of Petroleum Industry Biologists. ARCO, Co. Los Angeles, CA, 321 pp.

3. Bradt, P. T. 1974. The Ecology of the benthic macroinvertebrate fauna of the Bushkill Creek, Northampton County, Pennsylvania. Lehigh University, Ph.D. Thesis 180 pp.

4. Bradt, P. T. and Wieland, G. E. 1978. The impact of stream reconstruction and a gabion installation on the biology and chemistry of a trout stream. Completion Report for Grant No. 14-34-0001-6225, USDOI, Office of Water Research and Technology 61 pp.

5. Biological field and laboratory methods. EPA 670/4-73-0011. 1973. Edited by C. I. Weber. Natl. Enviro. Research Center. USEPA, Cincinnati, Ohio 45268.

6. Wilhm, J. L. 1970. Range of diversity index in benthic macroinvertebrate populations. Jour. Water Poll. Cont. Fed., 42(5): R 221-R 224.

7. Waters, T. F. 1965. Interpretation of invertebrate drift in streams. Ecology 46. 327-334.

8. Müller, K. 1954. Die Drift in fliessenden Gewässern. Arch. Hydrobiology. 49. 539-545.

9. Allen, D. J. 1975. The distributional ecology and diversity of benthic insects in Cement Creek, Colorado. Ecology '56. 1040-1053.

SOME EFFECTS OF OIL AND GAS
EXPLORATION ACTIVITIES ON
TUNDRA VEGETATION IN NORTHERN ALASKA

P. C. Reynolds
Bureau of Land Management
Fairbanks, Alaska

ABSTRACT

Bureau of Land Management (BLM) staff monitored winter cross-country travel associated with oil and gas exploration activities on the North Slope of Alaska in the National Petroleum Reserve-Alaska between 1978 and 1981. Different effects on four vegetation types were documented. Tractors pulling sled-mounted trailers did more damage than did low-ground-pressure seismic vehicles. A dry upland meadow was moderately affected by tractor trains, but recovered within 16 months. Sedge tussocks and riparian willows recovered in 16 months from the effects of low-ground-pressure seismic vehicles, but significant effects were still present along a tractor trail after 28 months. Trails created by tractor trains will be visible through riparian willows for several years. A major effect of winter cross-country travel was to mar aerial scenic values of a wilderness area. Impacts had no apparent effects on wildlife as the amount of plants killed or altered were small compared to the total habitat available.

OBJECTIVES AND METHODS

Between 1978 and 1981, the U.S. Geological Survey (USGS), through its operator, Husky Oil, conducted an oil and gas exploration program in the National Petroleum Reserve-Alaska (NPR-A) located on the North Slope of Alaska. A total of 21 exploratory wells were drilled and about 5000 miles of seismic line were shot. The Bureau of Land Management (BLM), which is responsible for surface management of the NPR-A, closely monitored this program.

The 23-million-acre Petroleum Reserve contains no major road system and all equipment and supplies must be transported by air or carried cross-country. As the land is underlain by continuous permafrost, most exploration activities were conducted during the winter. When the tundra surface layer is frozen and snow covered, vehicles moving equipment to well sites and conducting seismic surveys travel cross-country more easily and disturb the tundra less than when the tundra is unfrozen. "Winter" conditions of snow cover and frozen ground usually last from October until May in the NPR-A. The use of low-ground-pressure vehicles with wide tires or tracks also helps to minimize impacts to the tundra. Observations made during the BLM monitoring effort of the exploration program suggested, however, that winter cross-country travel could cause some surface disturbance.

The objective of this study was to examine winter trails and seismic lines created during cross-country travel associated with the exploration program, and to document the effects that different types of travel had on different arctic ecosystems.

Several activities associated with the exploration program required cross-country travel. During construction of well pads, winter trails were created between borrow sites, water sources, ice airstrips and well pad construction sites. Ice roads, built on winter trails when large amounts of traffic were anticipated, were made by spraying water along trail routes until 12 to 18 inches of ice were built up on the surface. When equipment was being moved from one site to another, winter trails between sites were traversed only once or twice each season by vehicles with wide tracks or tires and by tractors pulling sled-mounted trailers ("tractor trains"). Trails were also created during seismic operations. Low-ground-pressure vehicles with wide tracks were used for surveying, drilling seismic holes, laying geophones, and recording seismic detonations along "seismic lines." Seismic crew support trains ("tractor trains"), consisting of sleeping and eating facilities in sled-mounted trailers pulled by tractors, traveled adjacent to seismic lines and created "tractor trails". Fuel and supplies were flown into small airstrips bladed on frozen lakes and were transported to the seismic trains by tractor-pulled sleds. The emphasis of this study was to look at seismic lines and associated tractors trails in the foothills of NPR-A.

Surface disturbance resulting from cross-country travel in arctic areas underlain by permafrost has been discussed by many authors. Hok (1) described effects of vehicle operation on arctic tundra. Brown and Grave (2) summarize current literature in their discussion of physical and thermal disturbance of permafrost. Some of these studies describe summer travel on tundra (3, 4, 5) or document disturbances which occurred several years ago (6). Most of these studies described impacts to the arctic coastal plain.

For the purposes of this study, I defined "impact" as a significant difference in the amount of live plants present on a trail or seismic line compared with an adjacent undisturbed area (control site). I defined "recovery" to a trail or seismic line which had received "impact" as no significant difference in the amount of live plants present on the trail or line compared with the adjacent control site.

During the winter and spring of 1977-78, I made aerial observations of some seismic lines and tractor train trails during routine monitoring of the seismic program and traveled with a seismic train to observe and photograph portions of lines and trails from the ground. I revisited these areas during late June 1978 to examine effects on plants. During the winter and spring of 1978-79, I made aerial and ground observations of the same seismic lines, and seismic tractor trails, and looked at cross-country winter trails associated with an exploratory well (Lisburne #1) being drilled in the foothills of NPR-A.

In late June, 1979 I returned to these areas to quantify the effects I had observed the previous summer. I selected study sites along seismic lines and tractor trails in four vegetation types which are associated with major arctic ecological systems (7). These are described as follows:

1. Wet sedge meadow: The sedge Carex aquatilis was the dominant plant species present. Equivalent to "3.C.1.a. wet sedge meadow tundra" (8), and "upland wet sedge meadow" (7,9), wet sedge meadows are important habitat for nesting waterfowl and shore birds. Brown lemmings are major herbivores (7).

2. Dry upland meadow: Dryas integrifolia, mosses, lichens and lupine (Lupinus arcticus) were the dominant plant species present. Patches of vegetation were interspersed with patches of bare ground caused by natural frost heaves. Equivalent to "2.C.1.c. Dryas-herb mat and cushion tundra" (8), "dry upland meadow" (9) and "alpine tundra" (7), dry upland meadows are utilized by most large mammal species (7).

3. Sedge tussocks: Eriophorum vaginatum, a tussock-forming sedge was the most common plant species present. Dwarf birch (Betula nana), willow (Salix sp.), mosses, labrador tea (Ledum palustre) and lingonberry (Vaccinium vitisidaea) were also

major components. Equivalent to "3.C.2.d. sedge tussock mixed shrub tundra" (8), "upland tussock tundra" (6), and "cotton grass tussocks" (9), sedge tussocks are the most abundant vegetation type on the NPR-A (7). This is important habitat for large mammals, particularly caribou who use the new tussock growth in early summer during calving.

4. Riparian willows: Willows, primarily Salix alaxensis, growing in dense stands along river gravel bars, were the only overstory species present. Equivalent to "2.A.1.a. closed tall shrub willow" (8), riparian willows are important habitat for moose (Alces alces) and willow ptarmigan (Lagopus lagopus) (7).

I conducted most of the study in the southwestern foothills of the NPR-A near the Utokok River where seismic surveys had created seismic lines and tractor trails during the previous two winters. In late June 1979 I selected eight study sites where wet sedge meadow, dry upland meadow, sedge tussock and riparian willow habitat had been bisected by a seismic line and by a seismic tractor in March 1979. I also selected two study sites where a seismic tractor trail had crossed sedge tussocks and riparian willows in March 1978. The sites represented typical effects from seismic vehicles and tractor trains.

I selected two study sites near the Lisburne No. 1 exploratory well in the southeastern foothills of NPR-A. One was on a winter trail which had been used daily by vehicles hauling water to the well site between March and April 1979. The other was on a winter trail which had been used to bring equipment and supplies into the well site in February 1979 and had been used by seismic crews in late April 1979 after the snow cover had melted. This study site represented some of the worst surface disturbance observed during the study. Both of these study sites were located in sedge-tussocks.

The type of cross-country travel which occurred at each study site is summarized as follows:

Study Site	Vegetation Type	Date of Occurrence	Description of Activity
1	Wet sedge meadow	3/79	1979 seismic line:
2	Dry upland meadow		a single pass of 10 to
3	Sedge tussocks		15 low-ground-
4	Riparian willows		pressure wide-tracked vehicles

5	Wet sedge meadow	3/79	1979 seismic tractor trail: single pass of one tractor pulling four to six sled-mounted trailers
6	Dry upland meadow		
7	Sedge tussocks		
8	Riparian willows		
9	Sedge tussocks	3/78	1978 seismic tractor trail: single pass of 3 tractors each pulling four to six sled-mounted trailers
10	Riparian willows		
11	Sedge tussocks	3/79-4/79	1979 winter trail to water source: three to six passes daily of vehicle with low-ground-pressure tires hauling water to well site
12	Sedge tussocks	4/79	1979 winter trail: single pass of 10 to 15 low-ground-pressure wide-tracked vehicles and three tractors each pulling four to six sled-mounted trailers, across snow-free area

At each of the 12 study sites I ran a pair of transects: one on the trail or line created by the cross-country traffic where plants were likely to be impacted and one parallel to the trail in an undisturbed area of the same type of vegetation.

Along each 30.5-meter (100-foot) transect, I recorded amounts of live plants, dead plants and bare ground present in 10 quadrants, each one square meter in area. Percent plant coverage by different species was also recorded.

In June 1980, I reran pairs of transects at study sites 3, 6, 7, 8, 9, 10 and 12, and visually examined and photographed the other study sites.

In June 1981, I photographed seismic lines and trails from the air.

Statistically significant differences were determined by "T" tests which compared the mean percentage of live plants along transects in potentially disturbed areas with the mean percentages of live plants found along transects in adjacent undisturbed sites.

RESULTS

Preliminary observations made during the summer of 1978 suggested that effects of winter cross-country travel varied with the type of travel and the type of vegetation traversed. Low-ground-pressure vehicles traveling along seismic lines appeared to cause less impact than did seismic tractor trains. The wet sedge meadow and dry upland meadow appeared to be the least impacted by cross-country traffic. Faint "green trails", visible from the air, occurred where tractor trains crossed wet sedge meadows. Trails across dry upland meadows were often difficult to detect. Effects on sedge tussocks and riparian willows were more obvious. Tussocks were crushed and broken, resulting in brown trails, which were conspicuous from the air. Willows were broken and uprooted by moving tractor trains.

Data from transects run during the summer of 1979 verified these initial observations. Figure 1 compares the effects of four different kinds of winter vehicle traffic traversing sedge tussocks during the first growing season (two to four months) after the travel had occurred. Although significant impacts ($P=.05$) occurred at all sedge tussock sites, I observed the least impact along the seismic line (Site 3, Figure 1) which was created by single passes of about 15 low-ground-pressure seismic vehicles. At this site four months after the line was made, 60% of the plant cover was alive compared with 80% live plant cover in the adjacent non-disturbed site. Portions of sedge tussocks were brown, possibly killed by freezing when vehicles compressed the insulating snow layer covering the plants. A few tussocks had been crushed and broken by the low-ground-pressure vehicles, resulting in a faint brown track visible on the ground and from an airplane.

Impacts on the seismic tractor trail (Site 7, Figure 1) were the result of a single pass of one tractor pulling four to six sled-mounted trailers. In spite of a snow cover of at least six inches, the heavy sleds crushed and broke the tussocks, leaving a brown trail, which was conspicuous from the air after the snow had melted. Four months after the impact occurred 35% of the plant cover was alive compared with a 83% live plant cover in the adjacent non-disturbed site. I observed the highest percentage of dead plant (63%) at this site.

The trail between the Lisburne exploratory well and a water supply lake (Site 11, Figure 1) was traversed from three to six times each day by low-ground-pressure trucks with large tires hauling water. The amount of live plants was 33% along the trail and 69% in adjacent non-disturbed sedge tussocks. The amount of bare ground (31% along the

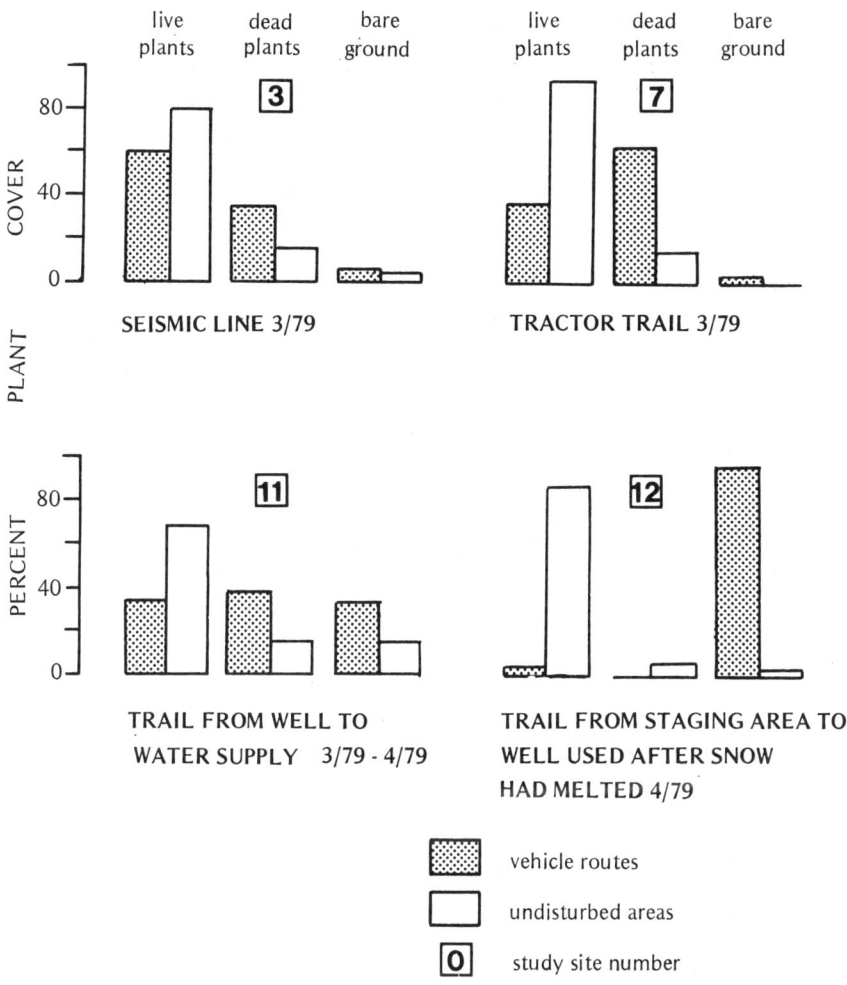

Figure 1. Effects of four types of winter vehicle traffic on sedge tussocks, in late June 1979, two to four months after the travel had occurred.

water supply trail compared with 15% in the non-disturbed area) was relatively higher than that observed along the seismic line and the seismic tractor trail. The repeated impact of daily travel along this trail removed more plant cover than did single passes of wide-tracked seismic vehicles and tractor trains.

Parts of the trail between the Betty Lake supply point and the Lisburne exploratory well showed substantial impact to sedge tussocks. This trail was initially created by the passage of numerous low-ground-pressure seismic vehicles and tractors pulling sleds during an equipment move into the well site in February 1979. But most of the impact occurred in late April when seismic vehicles and tractor trains moved along this trail, after the snow had melted. In the area of the worst impact (Site 12, Figure 1) tussocks were obliterated, leaving an organic mat of soil and debris for about 200 yards. Two months after the impact, only a few sedge shoots were present; 97% of the trail was bare of plants compared to 4% bare ground in adjacent non-disturbed sedge tussocks.

Figure 2 compares the effects of two types of winter vehicle traffic (wide-tracked seismic vehicles and tractor trains) in four types of vegetation four months after the travel had occurred.

In both the wet sedge meadow (Site 1, Figure 2A) and the dry upland meadow (Site 2, Figure 2A), low-ground-pressure seismic vehicles traveling along a seismic line caused no significant impacts. The amount of live plants along the seismic line was not significantly different (P=.05) from the amount observed in adjacent non-disturbed areas.

In sedge tussock and riparian willows, I observed significant impacts on the seismic line four months after the line had been created (Sites 3 and 4, Figure 2A). As discussed in the previous section, some of the tussocks had been broken and crushed resulting in a faint brown track through this type of vegetation. A brown track was also visible through riparian willows; the tops of the willows were bent, broken and leafless, but plants had not been uprooted. Only 29% of the canopy cover consisted of live plants in riparian willows on the seismic line compared to 58% live canopy cover in adjacent undisturbed willows.

By comparison, the trail created by seismic tractor trains caused a significant impact in three of the four types of vegetation studied. Impact to the wet sedge meadow (Site 5, Figure 2B) again was not significant (P=.05) as the amount of live plants observed along the trail (60%) was not statistically different from the amount observed in an adjacent non-disturbed area (52%).

The dry upland meadow (Site 6, Figure 2B) was moderately impacted. Although the amount of live plants present was significantly less (53%) compared with an adjacent undisturbed area (79%), the trail was difficult to see on the ground and from the air. The presence of large amounts of bare ground caused by naturally occurring frost perturbations made the trail less obvious than if the area contained a more continuous plant cover.

Both the sedge tussocks and riparian willows sustained significant impact where they were traversed by tractor trains (Sites 7 and 8, Figure 2B). As discussed in the previous section, about 50% of the sedge tussocks were

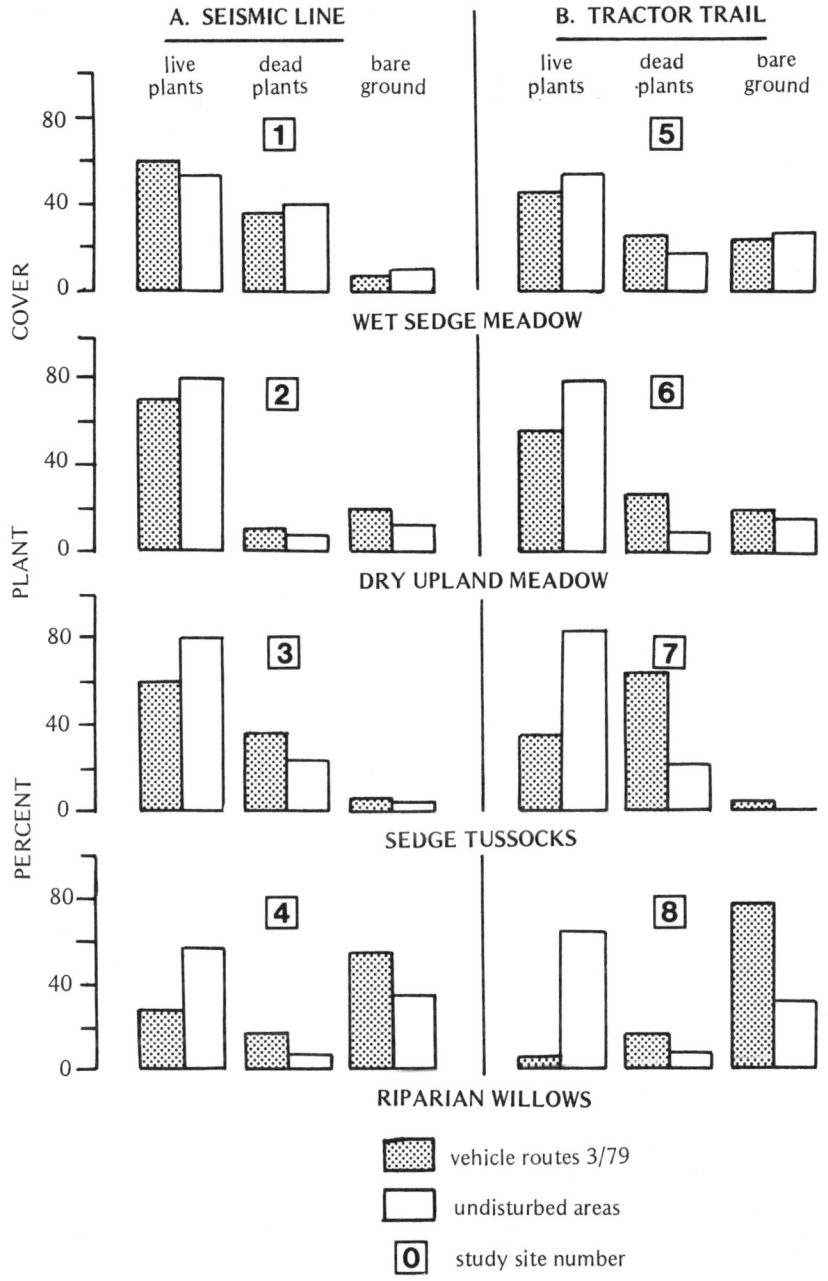

Figure 2. A comparison of the effects of winter vehicle traffic along a seismic line and a seismic tractor trail on four types of vegetation, measured in late June 1979, four months after the travel had occurred.

damaged or killed by crushing. This left a distinct brown track which was visible on the ground and conspicuous when observed from an airplane. In the riparian willows, almost 50% of the willows were uprooted and removed by tractor trains. Most of the remaining shrubs had broken branches. Destruction of the willow canopy cover left open areas which were invaded by grasses, sedges, lupine (<u>Lupinus arcticus</u>) and bearberry (<u>Arctostaphylos rubra</u>.) The resulting open area was very visible both on the ground and from an airplane.

Revisits to study sites in June of 1980 showed that recovery depended upon the type of vegetation impacted and the type of vehicle traffic that caused the impact. Along the seismic line, the two type of vegetation--sedge tussocks and riparian willows--that showed significant impact four months after the seismic line was shot, had recovered after 16 months. There was no significant difference ($P=.05$) in the amount of live plants in sedge tussocks (85%) and in the adjacent undisturbed area (84%). A few crushed tussocks were still visible, but new sedge shoots were appearing and the line was very difficult to find. Riparian willows which had been brown and bent in the summer of 1979 were not distinguishable from adjacent willows in 1980. The line was not detectable from the ground or visable from the air.

Recovery along the seismic tractor trail varied with the type of vegetation impacted. The dry upland meadow site recovered relatively quickly. I observed no significant difference ($P=.05$) between the amount of live plants present along the tractor trail (78%) and in an adjacent undisturbed area (83%) 16 months after the tractor trains had traveled across this area.

Recovery of sedge tussocks was slower. Figure 3 shows a comparison of three different sedge tussock areas subjected to different amounts of impact. A seismic tractor trail created in sedge tussocks in March 1978 (Site 9, Figure 3) was only about 75% recovered 28 months after this impact. A seismic tractor trail created in March 1979 (Site 7, Figure 3) showed a faster recovery; 16 months after the impact this area was about 85% recovered. Apparently the impact along the 1978 tractor trail was greater than that which occurred along the 1979 tractor trail. In 1978 the area had been traversed by passes of three tractor trains. In 1979 the impact resulted from the passage of only one train. The snow cover in 1979 was deeper than in 1978 and helped protect plants. The "worst case" impact (Site 12, Figure 3) seen at the trail into the exploratory well site was almost 50% recovered 16 months after vehicles had traversed the area with no snow cover, and obliterated most of the plant cover. Because of the apparent disturbance, the area had been treated with one application of 10-20-20

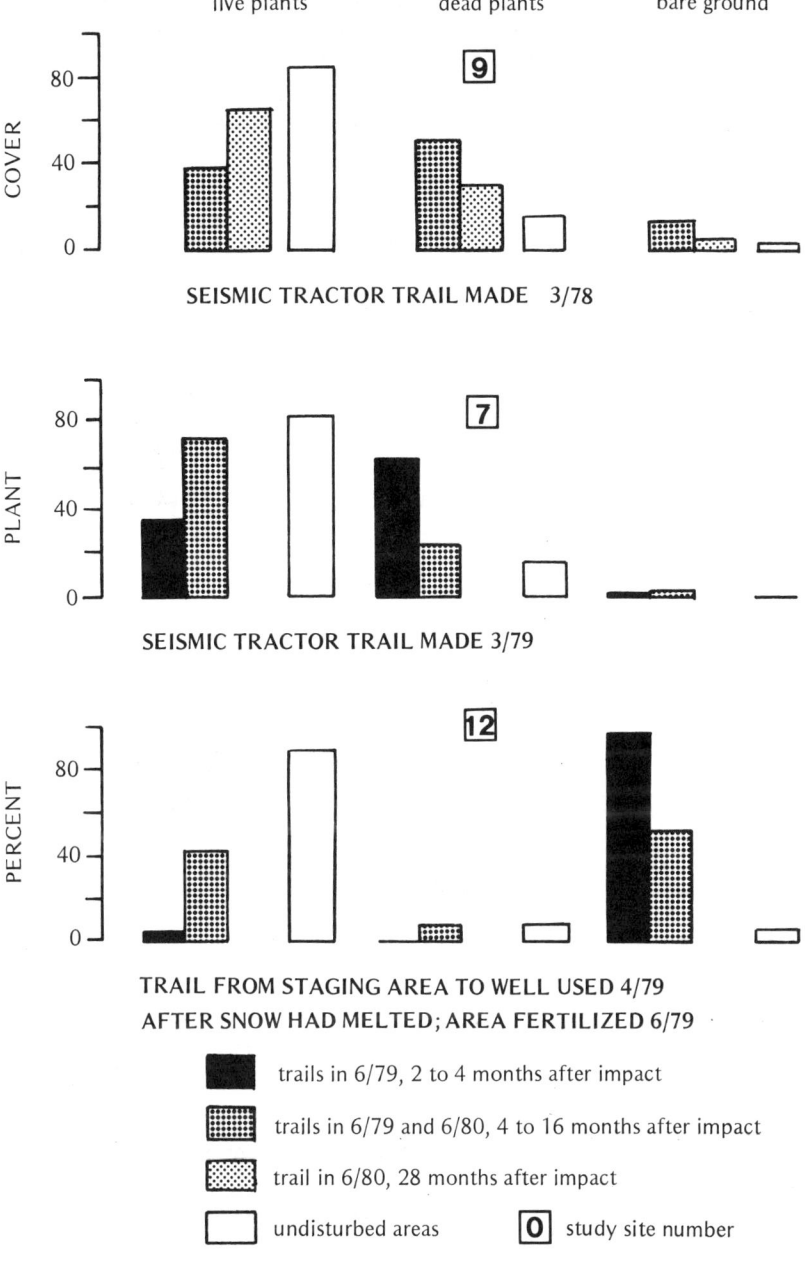

Figure 3. Partial recovery of sedge tussocks four, sixteen, and twenty-eight months after impact by seismic tractor trains traveling in winter and traveling across a snow-free area.

413

fertilizer, applied by aircraft at a rate of 600 pounds per acre about two months after the damage had occurred. Within three months, a heavy cover of cloudberry Rubus chamaemorus was observed growing in fertilized areas and after 16 months tussock-forming sedges were recolonizing the damaged area.

Figure 4 shows partial recovery of riparian willows

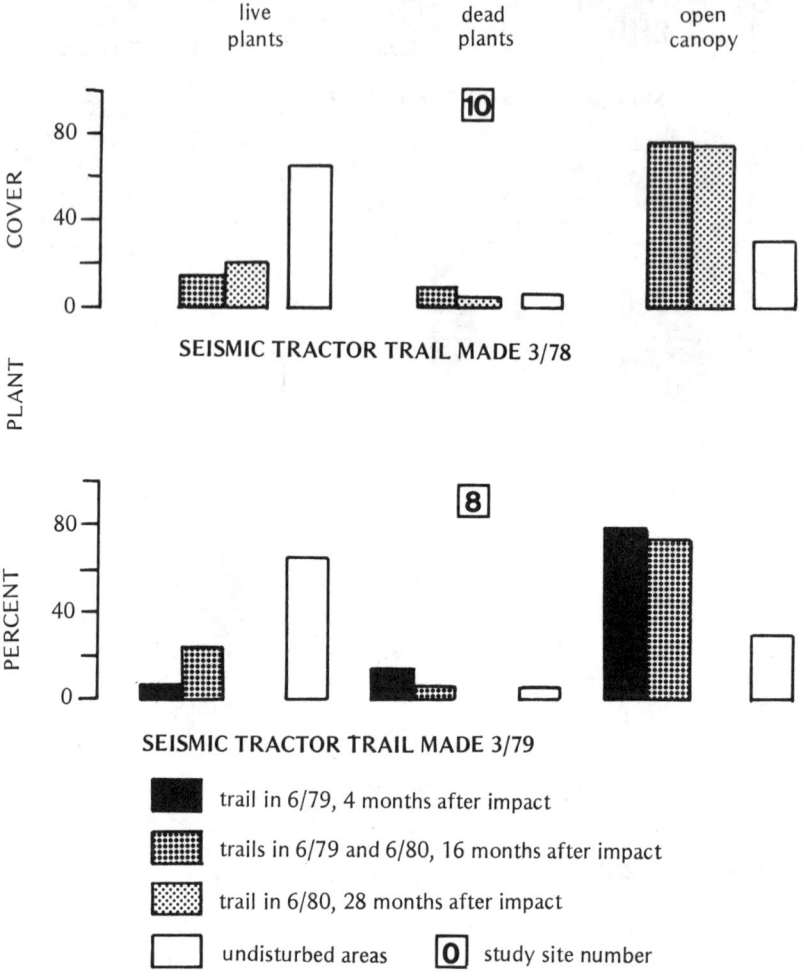

Figure 4. Partial recovery of riparian willows, four, sixteen, and twenty-eight months after impact by seismic tractor trains traveling in winter.

damaged by seismic tractor trains in March 1978 and in March 1979. Because a large percentage of the canopy cover was uprooted and removed, recovery to this type of habitat was slow. New willow shoots were observed sprouting from remaining willows that were broken and crushed. New growth on broken branches averaged only 26 cm in height after 16 months and 42 cm in height after 28 months, compared with a mean height of 98 cm for willows in the adjacent non-disturbed areas. The percentage of open canopy was still high and the trail was very visible even after 28 months. It will be several years before the regrowth reaches the height of the adjacent willows and the area traversed by the tractor trains is no longer an obvious trail very visible from the ground and from an airplane.

As was observed in the sedge tussocks, the impact created by the 1978 tractor trains was greater than that resulting from the 1979 trains, and recovery was slower. A deeper snow cover in 1979, compared with 1978, apparently provided more protection for plants being traversed by vehicles.

With the exception of changes in understory vegetation which occurred when the riparian willow canopy cover was removed, I observed no apparent change in species composition at sites where significant amounts of live plants had been killed or removed. Recovery involved recolonization of the same species present prior to the impact.

DISCUSSION & CONCLUSIONS

Impacts to different types of vegetation resulting from winter cross-country travel associated with seismic and exploratory well drilling activities varied with the type of vehicles and amount of traffic. Single passes with low-ground-pressure vehicles with wide flexible tracks or large tires did little damage even to the most sensitive types of vegetation if the ground was frozen and snow covered. Trains of tractors pulling sled-mounted trailers that make one or a few passes across sedge tussocks or riparian willows can damage or kill a significant percentage of plants along the route traversed, even when several inches of snow cover are present. But the depth of snow cover apparently does affect the degree of impact and rate of recovery. Along trails created in the winter of 1978 when snow cover was light in the NPR-A foothills, plants recovered more slowly than did plants impacted during the winter of 1979 when a heavier snow cover was present. The worst impacts observed occurred when vehicles traversed a snow-free area.

The impacts described in the study occurred in limited areas of the NPR-A. During four years of this exploration program, an estimated maximum of 214 square miles was subjected to winter cross-country travel, about .6% of the total 37,000 square miles within the NPR-A. Much of this travel was across wet sedge meadows on the coastal plain. A major effect of winter cross-country travel was to mar the aerial scenic values of this large area detracting from its wilderness quality. Trails, particularly tractor trails crossing stands of willows, are often highly visible from the air and will be for several years in the future.

The difference in impacts observed in the four types of vegetation studied were due in part to the growth forms of plants comprising these types. In the wet sedge meadow, the dead blades of grasses and sedges from previous summer's growth were bent and flattened beneath the weight of the vehicles. When new growth emerged, it appeared to be greener than plans in undisturbed areas where dead plants were still standing amid the new growth. When this standing dead is felled, it provides nutrients in the disturbed area, allowing for increased growth (10). The result is the characteristic "green trails", which are commonly seen in wet sedge meadows. In the dry upland meadow, many plants are low, prostrate species which are protected by wind-packed snow. Naturally occurring frost heaves cause patches of bare ground that help mask the visibility of surface disturbances. In sedge tussocks, the relatively high dessicated hummocks are vulnerable to surface impacts. Even beneath several inches of snow, the top-heavy tussocks are crushed by the weight of sleds pulled by tractors. Low-ground-pressure vehicles do less damage than do tractor trains. Tall riparian willows are also likely to be damaged by tractors pulling heavy sleds. The upper branches of willows are not protected by snow and the lower parts are covered with deep loosely-packed snow which cannot support the weight of tractor trains. The trains break off and uproot willows as they bog down in the deep snow.

Impact by winter cross-country travel associated with the current petroleum exploration program has not significantly altered wildlife habitat on the NPR-A. Although sedge tussock habitat and riparian willow habitat are important for caribou and moose, as well as other species, the quantities of plants which have been killed or damaged is small compared to the total habitat available. Sedge tussocks are particularly abundant in the NPR-A foothills. An expanded exploration program requiring a concentration of winter vehicle activity along the willow-covered gravel bars of major river systems could eventually affect winter moose habitat.

REFERENCES CITED

1. Hok, J. R. 1971. Some effects of vehicle operation on Alaskan arctic tundra: Univ. of Alaska, M.S. Thesis, 86 p.
2. Brown, J. and N. A. Grave, 1979. Physical and thermal disturbance and protection of permafrost. Special Report 79-5. U.S. Army Corps of Engineers, CRREL, Hanover, New Hamshire. 42 p.
3. Radforth, J. R. 1973. Long term effects of summer traffic by tracked vehicles on tundra. Environmental Social Committee Northern Pipelines, Task Force on Northern Oil Development Report No. 73-22. Muskeg Research Institute, Univ. New Brunswick. 60 p.
4. Walker, D. A., P. J. Webber, K. R. Everett, J. Brown. 1977. The effect of low-pressure wheeled vehicles on plant communities and soils at Prudhoe Bay, Alaska. Special Report 77-17. U. S. Army Corps of Engineers CRREL, Hanover, New Hampshire. 42 p.
5. Abele, G., D. A. Walker, J. Brown, M. C. Brewer, and D. M. Atwood. 1976. Effects of low-ground-pressure vehicle traffic on tundra at Lonely, Alaska. Special Report 76-16. U.S. Army Corps of Engineers CRREL, Hanover, New Hampshire. 63 p.
6. Lawson, D. E., J. Brown, K. R. Everett, A. W. Johnson, V. Komarkova, B. M. Murray, D. F. Murray and P.J. Webber. 1978. Tundra disturbances and recovery following the 1949 exploratory drilling, Fish Creek, Northern Alaska. Report 78-28. U.S. Army Corps of Engineers CRREL. Hanover, New Hampshire. 81 p.
7. NPR-A Task Force. 1978. NPR-A Ecological profile study report 4. U.S.D.I. NPR-A 105(c) Land Use Study. Anchorage, Alaska. 118 p.
8. Viereck, L.A., C. T. Dyrness, and A. R. Batten. In preparation. Revision of Viereck, L. and C. Dyrness. 1980. A preliminary classification system for vegetation in Alaska. USDA For. Service Gen. Tech. Report PNW-106. Institute of Northern Forestry. Fairbanks, Alaska. 38 p.
9. Selkregg, L. L. 1975. Alaska regional profiles - Arctic region. Univ. of Alaska Arctic Environmental Information and Data Center. 218 p.
10. Tieszen, L. ed. 1978. Vegetation and production ecology of an Alaskan arctic tundra. Ecological Studies 29. Springer-verlag. New York. 686 p.

A HYDROLOGICAL INVESTIGATION
OF A MOSAIC ECOSYSTEM

Steve Cox, Billy Hall,
Bert Langley, Allison Winter,
Tom Simpson, and Dave Hawkins
  Dames & Moore
  Atlanta, Georgia

INTRODUCTION

This study was required by the U.S. Army Corps of Engineers (COE) following the review of permit applications for a petroleum industry refinery expansion. This COE requirement was based on comments during agency review, some agencies determining that part of the area to be utilized by the industry had importance as a wetlands, and as a source of nutrient export to an adjacent estuary. The COE determined that a decision on the applications for this area (designated as Area 15) be delayed until additional data were gathered on its importance, both in situ and to an adjacent estuary. This type of study has not been conducted previously in habitats in this area and the data may also be used by agencies in subsequent reviews of other industry applications.

AREA DESCRIPTION

The 81-hectare study area is characterized as a slash pine (Pinus elliottii) forest containing a mosaic of numerous intermittent, shallow surface depressions which often remain waterlogged long after precipitation events and commonly support an abundance of hydrophytic plants. These surface depresssions form large, poorly defined swales which provide a means of drainage to the adjacent tidal marsh. Black rush (Juncus roemerianus) comprises almost 100 percent of the vegetative cover in the tidal marsh although a narrow ecotone of salt meadow cordgrass (Spartina patens) forms a band 10 to 20 meters in width between the pine forest and tidal marsh.

MATERIALS AND METHODS

The principal focus of the study is the estimating of sources and quantities of nutrients being exported from Area 15 (Figure 1) to the adjacent estuary and the ecological importance of these energetics. The results of the study

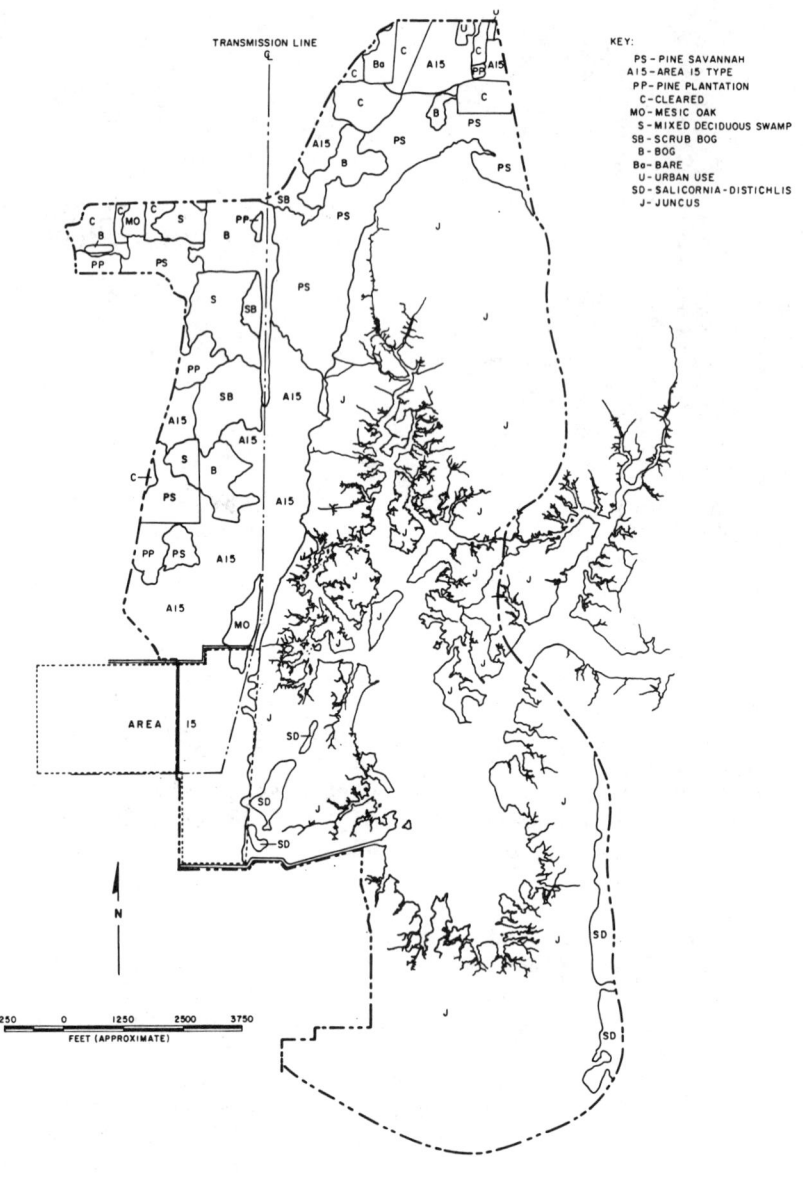

Figure 1. Vegetation Map of Drainage Basin.

will then be superimposed upon an ecosystem model framework.

Ecosystem Model

An ecosystem model will portray those ecological functions considered most important to processes occurring within the pine forest and to interactions occurring between the pine forest and tidal marsh. A preliminary ecosystem model is shown in Figure 2 using symbols developed by Odum and Odum (1976). Three types of symbols are used to represent compartments, or storages, of energy and materials: one for plants, another for animals, and a third for nonliving materials. Lines connecting these compartments signify the flow of energy and materials. Boxes with an "x" represent a pumping or positive interaction; boxes with a "-" represent a negative interaction. Inspection of the diagram in Figure 2 shows a simple, but logical movement of material and energy within and through the system. This ecosystem model also provided the basis for selection of the study methodologies, or tools, that were selected to investigate the various functional processes.

Water Resources

Quantifying materials exported and identifying the ecological importance of matter and energy transfer require both a quantitative and qualitative understanding of the hydrologic cycle. These investigations seek to define nutrient transport vectors of surface and ground water regimes and relate these vectors to habitats in the study area.

The specific hydrologic functions being investigated are:
o ground water gradients relative to tidal elevations and rainfall;
o ground water depths and fluctuations relative to the various vegetative communities;
o occurrence of saline water intrusion;
o relationship between rainfall and the material transport vectors (surface runoff, interflow, and ground water recharge); and
o the range of surface, base, and ground water flow volumes and rates, and the direction of these flows.

The hydrologic budget is quantified through investigation of: precipitation, surface runoff, interflow, base flow, ground water recharge, and evapotranspiration.

The term interflow refers to any nonvertical subsurface flow in the absence of boundary conditions. Baseflow refers to nonvertical flow associated with boundary conditions such as a water table or an impervious soil layer above the water table. Analysis of the interflow component will provide an assessment of lateral flow in unsaturated soil produced by

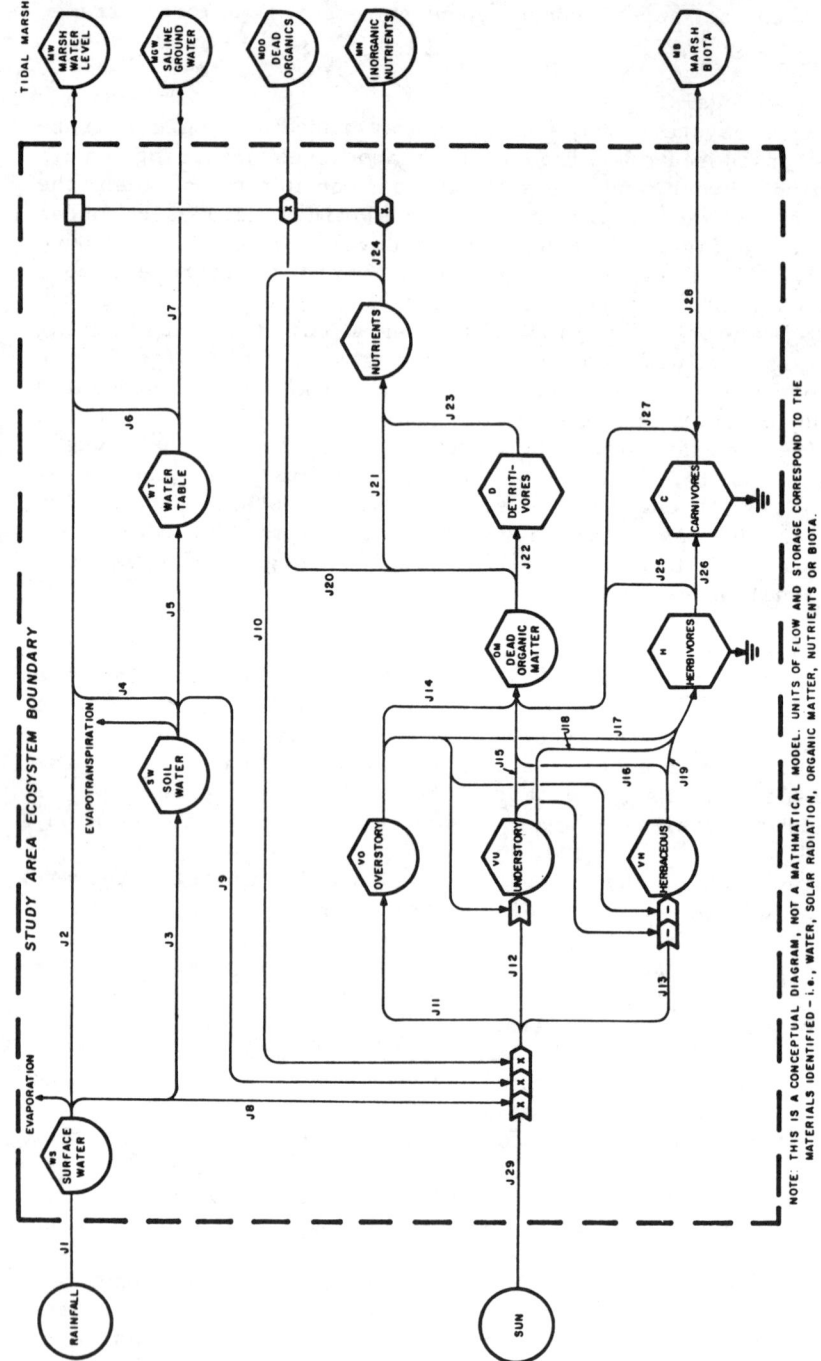

Figure 2. Ecological Systems Diagram of Study Area.

anistropic soil conditions. The overall hydrologic cycle is being investigated through analysis of key parameters on a sample catchment area. Extrapolation of these data are then made to the rest of the study area. The monitoring system consists of four ground water wells, a runoff monitoring flume, a pipe weir, a rain gage, and a tide gage.

All surface and ground water samples are analyzed for total organic carbon (TOC), dissolved organic carbon (DOC), total suspended solids (TSS), dissolved solids (DS), nitrate-nitrite nitrogen, total Kjeldahl nitrogen, ammonia nitrogen, total phosphate, and orthophosphate. In addition, pH, temperature, dissolved oxygen, salinity, and specific conductance are obtained during each sample collection.

Rainfall, tidal elevation, and surface runoff are monitored on a continuous basis, while ground water elevations are monitored both manually, during discrete rainfall events and dry periods, and continuously.

Permeability tests and infiltration tests are performed in the field, supplemented with grain size distribution analyses and density tests on representative soil samples.

Hydrologic field and laboratory data are analyzed in conjunction with the biological findings to quantify the range of seasonal transport of nutrients to the tidal marsh and to assess the relationships between vegetation types and the surface ground water regimes.

The hydrologic information is provided to the biological study team as a data base for quantifying the nutrient transport into the tidal marsh estuary. By combining the surface, base, and ground water flow magnitudes with the measured organic and inorganic nutrient concentrations, a series of values are derived for seasonal organic and inorganic nutrient flux into the tidal marsh per unit area and per unit rainfall.

Biological Resources
Vegetation Mapping

Using topographic data and color infra-red photography, complemented by ground-truthing, a vegetation map was prepared delineating the major vegetation associations of the study area and of the tidal marsh drainage basin. The size of these associations is measured by polar planimetry and preliminary estimates of their productivity made from existing literature and field measurements of peak biomass. The estimate of productivity for the marsh ecosystem also allows an estimate to be made of the comparative productivity value of the adjacent pine forest to the marsh.

Vegetative Species Composition Survey

Sampling for species composition was conducted in late spring of 1981 to coincide with the reproductive stages of important plant species. Seven transects were sampled in the study area, comprising up to 30 quadrats each. Transect lines were perpendicular to drainage patterns in the pine forest (parallel to the tidal marsh) to assure sampling within all habitat types in the Study Area.

Quadrat sampling of the understory and herbaceous layers was conducted using methods similar to those described by Daubenmire (1968). Concomitant sampling of the overstory was conducted using the plotless, point centered quarter (PCQ) method of Cottam and Curtis (1956). Calculations from this sampling effort included density, frequency, basal area or dominance, relative frequency, relative dominance, and relative density (Mueller-Dombois and Ellenberg, 1974; Phillips, 1959; Cox, 1976). An Importance Value, derived by summing the relative frequency, relative density, and relative dominance (Curtis, 1959), was used to rate overall dominance by each species for each vegetative layer.

Peak Biomass

Although peak live biomass underestimates true net primary production (Gosselink, et al.,1977), the method requires only one field study, is relatively easy to make, and provides quantitative information from which comparisons with other studies of similar habitats can be made. Peak live biomass is estimated from the harvest of 50 to 60, randomly selected 0.25 meter square quadrats during a time when biomass is expected to be at a maximum. Standing dead litter and root biomass are also collected from the same quadrats.

Estimates of peak litter are based on several autumn collections when leaf and needle litter fall are expected to be at a maximum. Litter fall is collected in 20, one meter square, elevated screen baskets positioned randomly in the study area. The litter is removed from the baskets every month, oven-dried, and weighed.

Additional sampling was conducted in the tidal marsh, and in portions of the study area that were not part of the tidal marsh drainage, to provide comparative data. By sampling the most mobile components for potential materials exported, an estimate of the area's capacity for detrital and nutrient export can be developed. The values can then be compared with both literature values and the values obtained from the field monitoring studies.

Wildlife Inventory

This phase of the project determines the wildlife usage of Area 15 in comparison with the adjacent salt marsh. The wildlife populations which are dependent upon the study area are inventoried over a one year period. A variety of census techniques are employed to provide quantitative field estimates of the use of these habitats by mammals and birds. Snap-trapping grids and scent posts are employed to estimate use of the area by mammals, and birds are censused visually by a series of transects during one season, augmented by a roadside census every other week. These data document the use of the site as a temporary refuge for migrating birds. Surveys of reptiles and amphibians are based on collections obtained using 30-meter drift fences.

Soil Surveys

Soil samples were collected with soil corers to a depth of 20 cm, along with the sampling of root material for biomass determination. Collections were made along the linear transects used for the vegetative species composition survey. Samples were analyzed for salinity, soil moisture, pH, total organic carbon, total Kjeldahl nitrogen, nitrate-nitrite nitrogen, ammonia nitrogen, orthophosphate, and total phosphate. These determinations will serve to characterize the physical and chemical conditions and degree of aerobic/anaerobic decomposition in the soil. These data will aid in describing wetland functions attributed to the study area.

RESULTS

Although the monitoring program is still in progress, considerable data have been collected and evaluations of these findings are presented below. An unusually dry spring and summer have resulted in insufficient rainfall data to fully define hydrological parameters. However, monitoring of these events will continue through fall and winter to characterize and quantify hydrological boundaries.

The results of the vegetative survey are illustrated in Figure 1, and indicate the dominant species occurring throughout the drainage basin. The vegetation type identified as 'Area 15' is a community typical of the primary study site and is dominated by slash pine, poison ivy, and honeysuckle, with a mosaic of drainage swales along which various grasses (e.g. Spartina patens), ferns, and cypress occur as dominants. As this figure indicates, the open waters of the estuary are completely surrounded by juncus marsh with various pine habitats forming the dominant community on terrain slightly higher in elevation and peripheral to the marsh.

The analyses for soil nutrients occurring in the study area are summarized in Figure 3. These data are arranged along an apparent drainage gradient beginning at a road along the western edge of the study area and ending in the juncus marsh to the east. The concentrations of nutrients illustrated in Figure 3 are normalized to a common scale by expressing as the percent of maximum value for each parameter.

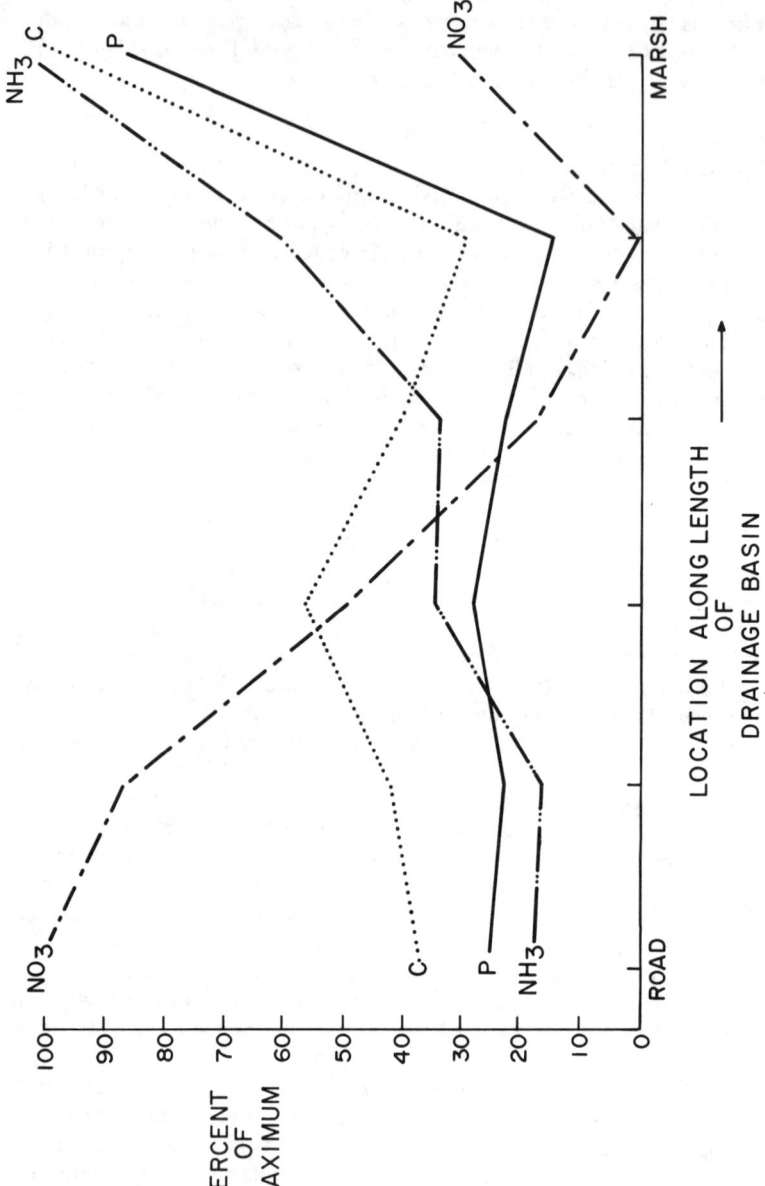

Figure 3. Soil Nutrients.

As these data illustrate, most nutrient concentrations are relatively similar, for each component measured, throughout much of the study area. The only exception is nitrate-nitrogen (NO_3) which decreases along the drainage gradient. However, at the edge of the primary study area, indicated by the juncus marsh (Figure 1), all nutrients increase in concentration, with ammonia-nitrogen (NH_3), total organic carbon (C), and total phosphate (P) attaining highest levels. The decrease in NO_3 (and increase in NH_3) may be related to the increase in soil moisture conditions along the gradient, with concomitant anaerobiosis resulting in a reducing environment.

The water quality analyses for surface waters transported through the primary study area (Area 15) and the estuarine system, are summarized in Figure 4. These data are normalized to a common scale by plotting as percent of maximum value for each parameter.

The surface water nutrient concentrations vary from one sampling location to another, but extreme ranges are evident for each major nutrient, to include:
- o A peak in NO_3 and NH_3 concentration at the flume (east edge of drainage area) and a peak in total Kjeldahl nitrogen (TKN) in the juncus marsh,
- o A peak in TOC at both the marsh and the flume,
- o Two peaks in P concentration - one within the drainage system of Area 15 (pipe weir) and another in the marsh canal draining the primary study area,
- o A general decrease in the concentration of all nutrients in the tidally influenced marshes of the study system.

The movement of nutrients from the study area through ground water appears to be an important route of transport. Although more data will be collected to verify the importance of ground water in carrying nutrients, Table I provides a comparison for two rainfall events, and for two locations - Station 1, on the western, upgrade edge of the study area, and Station 2, on the eastern, downgrade edge.

As these data indicate, measurements for surface water and ground water (during same rainfall event and location) resulted in concentrations for most parameters that were as high or higher in ground water as compared to surface waters.

DISCUSSION

This study is scheduled to be completed in April of 1982, followed by a decision from the Corps of Engineers concerning approval or disapproval of a permit for developing the primary area investigated. The data collected so far appear to be insufficient to evaluate fully the importance of the area to the adjacent marsh/estuarine ecosystem.

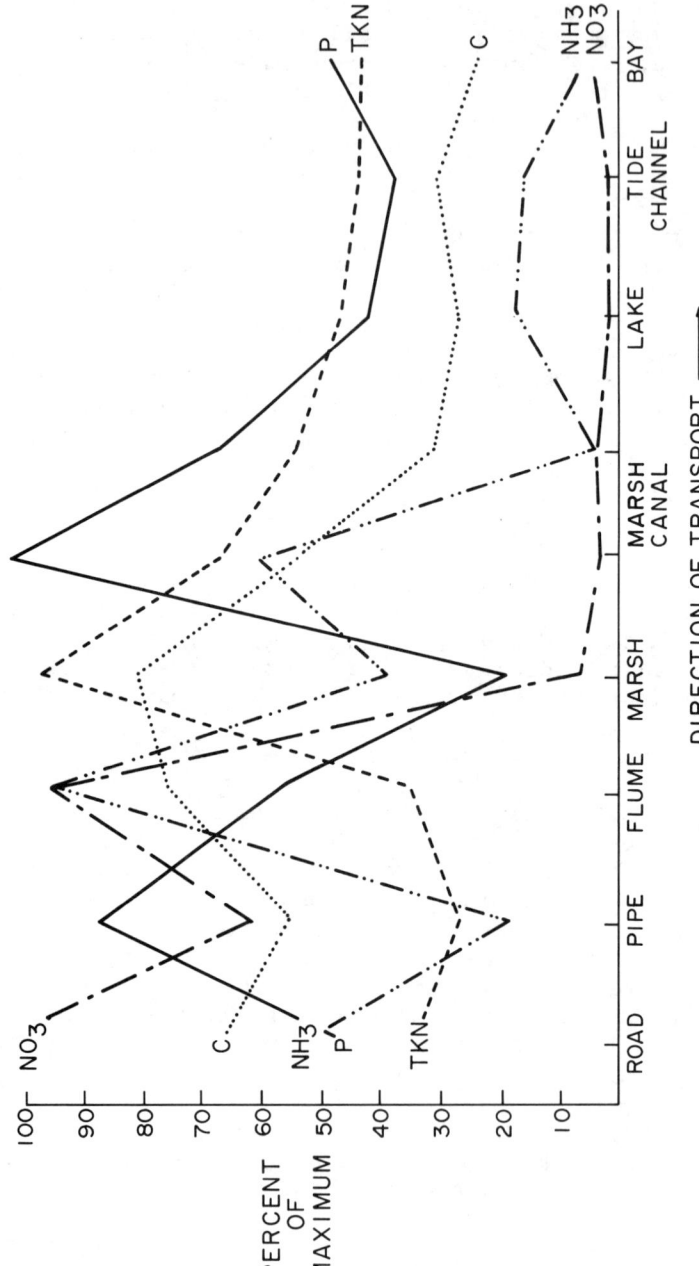

Figure 4. Surface Water Nutrients.

TABLE I. COMPARISON OF NUTRIENTS IN SURFACE AND GROUND WATER SOURCES (CONCENTRATIONS IN MG/L)

		RAINFALL EVENTS			
		1		2	
		SURFACE WATER	GROUND WATER	SURFACE WATER	GROUND WATER
STATION 1	TOC	18	14	NO DATA COLLECTED	
	NO_3	0.16	0.38		
	TKN	0.88	4.0		
	NH_3	0.04	0.10		
	T-P	0.16	0.15		
STATION 2	TOC	21	17	18	19
	NO_3	0.24	0.17	0.21	0.074
	TKN	1.2	4.3	1.6	0.88
	NH_3	0.10	0.20	<0.05	<0.05
	T-P	0.08	<0.01	0.18	0.26

Nevertheless, the design of this investigative procedure and the means by which interpretations may be made should be instructive for other industries in a similar situation.

Characterization of the ecological significance of the study area may be made in several ways. The most direct method can be based on quantifying key compartments of pathways in Figure 2. For example, what is the annual flow of freshwater (J_2), organic matter (J_{20}) and nutrients (J_{24}) to the tidal marsh? Direct evaluation of _in situ_ value may be made by comparison with existing literature. Another method of evaluation will be to compare the export of the study area to the estimated total storage capacity of the entire drainage basin. This estimate will be based on selected field measurements and literature data. Thus, the total quantity of dead organic matter in the marsh and the total annual input (both in terms of marsh productivity and export from all adjacent lands) will be compared to contributions measured as export from the study areas.

The most significant question to be answered from this study is what will be the effect on the estuarine ecosystem from replacement of the existing mosaic wetland-pine forest community with an industrial development? The present study considers an estimate of nutrient resources available relative to the amounts that reach the estuary, as a logical measure of the importance that can be attributed to the study area.

There is obviously a general export of nutrients out of the study area into the estuary, through decomposition of dead vegetation followed by flushing during rainfall events.

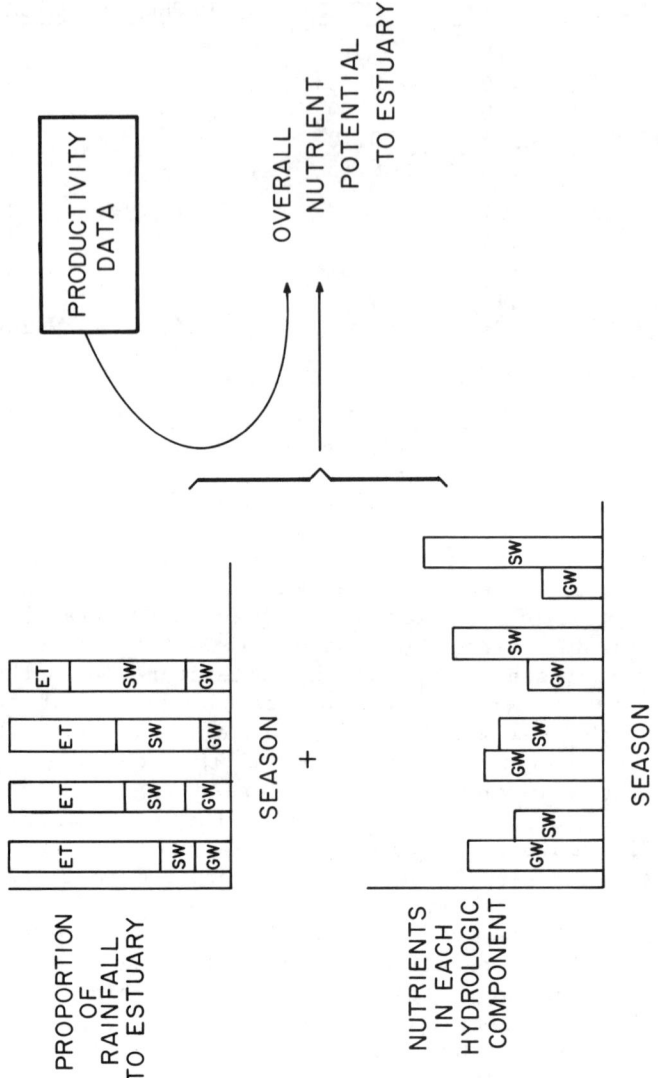

Figure 5. Summary of Decision Inputs.

It is also apparent that both surface and ground water routes are important means of export for these decomposition products. However, no readily accessible data can predict the effects on the estuary by the loss of these specific concentrations of nutrients. A logical approach to this analysis can be made by a comparison of the proportion of nutrient exported into the estuary from the primary study area (Area 15), expressed as a percent of total nutrients exported for all similar drainage systems into the estuary. The final analysis will also need to consider the proportion of nutrients exported (by various hydrologic regimes) versus the nutrient sources available (peak biomass estimates).

As illustrated in Figure 5, the major data sources to be used in the final decision criteria will include:
- o Proportion of seasonal inputs for each nutrient through both surface water (SW) and ground water (GW),
- o Seasonal variations in overall hydrologic regime,
- o Productivity estimates for each area affording these inputs.

LITERATURE CITED

Cottam, Grant and J. T. Curtis, 1956. The use of distance measures in phytosociological sampling. Ecology 37(3): 451-460.

Cox, George W., 1976. Laboratory Manual of General Ecology. Wm. C. Brown Company, Publishers, Dubuque, Iowa. 232 p.

Curtis, J. T., 1959. The Vegetation of Wisconsin. An Ordination of Plant Communities. University of Wisconsin Press, Madison. 657 p.

Daubenmire, Rexford, 1968. Plant Communities. Harper & Row, Publishers, New York. 300 p.

Gosselink, J. G., C. S. Hopkinson, Jr., and R. T. Parrondo, 1977. Common Marsh Plant Species of the Gulf Coast Area. Volume 1: Productivity. U. S. Army Engineer Waterways Experiment Station. Vicksburg, Mississippi.

Mueller-Dombois, D. and Heinz Ellenberg, 1974. Aims and Methods of Vegetation Ecology. John Wiley and Sons, New York. 547 p.

Odum, Howard T. and Elizabeth C. Odum, 1976. Energy Basis for Men and Nature. McGraw-Hill Book Company, New York, 296 p.

Phillips, E. A., 1959. Methods of Vegetation Study. Holt, Rinehart and Winston, Inc., 107 p.

DEVELOPMENT OF AN AQUATIC BIOLOGICAL
IMPACT ASSESSMENT METHODOLOGY FOR ANALYSIS
OF HYPOTHETICAL COAL SLURRY PIPELINE SPILLS[1]

Eugene R. Mancini
  Senior Staff Scientist
  Woodward-Clyde Consultants
  3489 Kurtz Street
  San Diego, California 92110

ABSTRACT

An integral part of the environmental impact statement process for the Energy Transportation Systems Incorporated (ETSI) coal slurry pipeline project was a biological impact assessment of hypothetical coal slurry pipeline spills in a variety of terrestrial and freshwater habitats. For purposes of aquatic biological analyses, various hydraulic and water quality models were used to estimate the post-spill physico-chemical characteristics of affected rivers and streams.

Impact significance criteria were based on (1) water quality and quantity phenomena and (2) coal particle suspension and sedimentation phenomena. Water quality and quantity characteristics of major biological concern were limited to sulfate, total dissolved solids, and dissolved oxygen concentration in addition to water temperature and spill volume, since laboratory studies did not indicate the presence of detectable levels of potentially toxic contaminants in the slurry water. Coal particle phenomena of major biological concern were the suspended solids concentration of particles in the water column and the rates and locations at which these particles could be expected to settle out of the water column. In all phases of impact analysis, emphasis was

---

[1] Funding for the research upon which this manuscript is based was provided by Energy Transportation Systems Incorporated as a result of data requests from the Special Projects Staff of the Bureau of Land Management.

placed upon fishes and macroinvertebrates, and/or sensitive classification species.

Hypothetical spill analysis indicated that, in general, spills in "small" streams or under low streamflow conditions would be expected to cause more severe biological impacts than spills in large rivers or under high streamflow conditions. Worst case impact assessment suggested that aquatic biological impacts could include (1) approximately 85% mortality of fish eggs and larvae in the affected area; (2) smothering of benthic macroinvertebrates, or at least emigration from affected substrate areas, and the elimination of substrate habitat; (3) heat shock and cold shock fish and invertebrate kills; (4) potential bioaccumulation of some toxic metals; (5) low dissolved oxygen tension stress, and/or decreases in invertebrate and fish productivity in the affected area. It is important to note, however, that it is not possible for all of these impacts to occur simultaneously at any single spill location.

INTRODUCTION

The Final Environmental Impact Statement on the Energy Transportation Systems Inc. coal slurry pipeline transportation project discusses the environmental consequences anticipated to result from project implementation [Department of the Interior 1981]. Recent Council on Environmental Quality guidelines, however, have limited the analytical detail included in EIS's by imposing document page limitations. In the case of the ETSI project this limitation resulted in the development of numerous "technical reports" each of which contains extensive analytical and technical data regarding one or more aspects of the proposed project. All of the data presented in this manuscript are discussed in either the final EIS or the technical report entitled "Ruptures and Spills" [Woodward-Clyde Consultants 1980].

A cursory description of the ETSI project would be instructive in order to place this spill analysis in perspective. ETSI proposes to develop a coal slurry transportation project which would transport an approximately 40/60 mixture of powdered coal/water through 1828 miles of pipeline from Wyoming to power plants in Oklahoma, Arkansas, and Louisiana. Approximately 37 MMTA would be transported to dewatering plants where the coal would be prepared for use by the utility stations.

Approximately 650 ephemeral, intermittent, and perennial streams would be crossed by the slurry pipeline. Even though the pipeline would be designed to specifications which would minimize the probability of a product spill, worst case analysis indicates that thousands of barrels of coal slurry could be released as a result of a total line rupture. It should be emphasized, however, that the majority of spill scenarios would be expected to involve significantly less slurry volume and, in fact, the "average" anticipated spill volume would be approximately 565 barrels [Woodward-Clyde Consultants 1980].

The general approach to the aquatic biological impact assessment required two distinct steps. The first involved quantification of physicochemical changes anticipated in the receiving streams as a result of a slurry spill. The second step concerned estimation of the biological impacts of these physicochemical changes on freshwater biota.

METHODS

Due to the length of the pipeline and the diversity of lotic habitats that could be affected by a coal slurry spill, potential impacts were assessed by choosing five spill scenario locations which were believed to be representative of the lotic habitat types located along the pipeline route. The following criteria were used in selecting spill locations.

- To account for the variability in dilution and dispersion of potential spills associated with different flow regimes, sites should include: a small perennial creek, medium-size river, major river, and bayou waterway.

- Because the slurry flow rate varies (decreases) from the beginning to the end of a given route, spill sites should be geographically distributed along the general project route in order to represent the range in potential spill volumes. For a given spill area, the site should be selected at a topographic low point in order to represent the worst case spill rate and volume.

- Site selection should incorporate specially identified sensitive habitats.

- All selected sites should be located in close proximity to a USGS gaging station.

Table I presents a list of the sites selected for simulated slurry spill analysis and indicates the criteria applicable to each site.

Table I. Slurry Spill Simulation Sites and Their Selection Criteria

Waterbody	Flow Type	Criteria Sensitivity	Distribution
1. North Platte River, Nebraska	Major river	Special state concern	Northern
2. Sand Creek, Oklahoma	Small perennial creek	Unique fish habitat	Central
3. Illinois River, Oklahoma	Medium-size river	State scenic river, recreational area, municipal water use	Central
4. Saline River, Arkansas	Medium-size river	Active fishery, special state concern	South central
5. Bayou Cocodrie, Louisiana	Bayou waterway	State scenic river, special state concern	Southern

Computer models were used to estimate maximum spill rates and volumes at each scenario location for both a complete break and orifice-type rupture. The results were used to model the spill's movements through surface waters at the selected river crossing locations. The conclusions of the modelling exercises were then used to quantify the effects of spills on surface water quality. The reader is referred to Woodward-Clyde Consultants [1980] for details regarding the hydraulic and water quality models used for these analyses.

In all phases of aquatic biological impact analysis,

emphasis was placed on fishes, macroinvertebrates, and sensitive classification species. Other components of the freshwater community including phytoplankton, zooplankton, and periphyton were considered only in the few situations where anticipated impacts could significantly affect their populations. In general, the high and rapid reproductive rate of plankton and periphyton populations was anticipated to make them reasonably resistant to short-term spill impacts.

Aquatic biological impacts associated with spills were conveniently divided into (1) physicochemical water quality phenomena and (2) coal particle suspension and sedimentation phenomena. Synergistic effects were addressed only when such effects were anticipated to cause significant biological damage.

In order to judge the severity of anticipated biological impacts, significance criteria were established. Since water quality investigations indicated that no detectable concentrations of toxic "priority pollutants" would be anticipated in the slurry water, water quality characteristics of major biological concern were limited to sulfate, total dissolved solids (TDS), and dissolved oxygen concentrations in addition to water temperature and spill volume. In order to establish significance criteria for each of these concerns the available technical literature was consulted.

One potential consequence of slurry spills which could not be credibly quantified concerned the potential bioaccumulation of toxic metals in invertebrates which may indiscriminately feed on spilled coal particles. Various light and heavy metals have been found in coal samples from the Wyoming mine sites. The extent to which bioaccumulation of these metals would be anticipated in the affected invertebrates and other organisms in the food web which feed on them is dependent on a number of variables which, themselves, are not easily quantifiable. However, this phenomenon has been recently documented for invertebrate populations exposed to a coal ash spill in a South Carolina river [Cherry et al. 1979], and it is anticipated that a similar situation could develop at a slurry spill site.

RESULTS AND DISCUSSION

In order to establish significant impact criteria for water quality alterations, publications like the "red book" [EPA 1976] and the American Fisheries Society response document [Thurston et al. 1979] were consulted and proved to be valuable information sources. Water quality criteria, when suggested in these documents, however, are more specifically designed for long-term point source discharges than for instantaneous spill situations. Criteria established in this manuscript, therefore, may not agree with criteria from either or both of these sources.

The quantity of slurry spilled, in relation to the volume of water in the affected water body would ultimately determine the degree of biological damage. If a slurry spill represented a significant increase in stream discharge (e.g., a doubling of volume, or greater) then the physical biological impacts would be similar to those caused by a severe rainstorm. The impacts on affected biota could be involuntary downstream or out-of-stream displacement and "scouring" of the streambed [Waters 1972, Hynes 1970, and others].

The temperature of the slurry within the pipeline (Table II) could significantly affect ambient stream temperatures at spill sites. A sudden increase or decrease in water temperatures could result in a localized fish and invertebrate kill as a result of "heat shock" [Talmage and Coutant 1978] or "cold shock" [Burton et al. 1979].

Table II. Estimated Coal Slurry Temperatures

State	Temperature	
	°C	°F
Wyoming	7.8	46.0
Nebraska	9.2	48.6
Colorado	7.8	46.0
Kansas	13.1	55.6
Oklahoma	15.6	60.1
Arkansas	15.6	60.1
Louisiana	19.2	66.6

Available data indicate that the rate at which water temperature changes determines the severity of impact [Burton et al. 1979; U.S. EPA 1976]. A slurry spill could cause a relatively rapid temperature change and, therefore, a conservative temperature criterion should be established for these spill analyses. Heat shock and cold shock studies have generally indicated that fish mortality declines as temperature change (ΔT) is limited to approximately 3°C or less. It will be assumed, therefore, that temperature changes of less than 3°C would have no significant impact on aquatic communities. It should be noted that water temperature alterations due to slurry spills would be expected to last for fewer than 24 hours.

The U. S. Environmental Protection Agency [1976] has established a general dissolved oxygen minimum concentration criterion of 5.0 mg/l. As Thurston et al. [1979] have pointed out, this criterion is not necessarily justified under all conditions in all water bodies. Nevertheless, for the purposes of these spill analyses, this criterion will be used to establish degrees of biological impact.

Maximum TDS and sulfate concentrations in the slurry water are anticipated to be approximately 1900 and 1100 mg/l, respectively. While these levels are in excess of the suggested domestic water supply concentrations of 250 mg/l [U. S. EPA 1976], it has been demonstrated that common freshwater fishes survived concentrations as high as 10,000 and 15,000 mg/l [U. S. EPA 1976]. It is anticipated, therefore, that short-term exposure to 1900 and 1100 mg/l TDS and sulfate concentrations (undiluted) would not have a significant impact on affected aquatic communities.

The approximate suspended solids concentration of the coal slurry would be 400,000 mg/l and, for these analyses, it is assumed that essentially all of the suspended material would be coarse, medium and fine particulate coal. The potential physical biological damage associated with such a dense slurry mixture is, undoubtedly, the major aquatic biological spill issue.

Although general turbidity, suspended solids, and sedimentation data are abundant in the available literature, few published data are available specifically regarding the impacts of coal particles on aquatic biota [Scullion and Edwards 1980]. Impact analysis, therefore, will rely on the application of generalized sedimentation and turbidity data. Cordone and Kelley [1961] and Stern and Stickle [1978] have completed extensive reviews of the literature regarding the biological effects of increased sedimentation and turbidity. It has been found that elevated turbidity levels can affect productivity throughout all trophic levels [Karr and Schlosser 1978; Stern and Stickle 1978; Peters 1967; Cordone and Kelley 1961; Gangmark and Bakkala 1960; and many others].

Although sustained periods of exposure to high suspended solids under laboratory conditions have been shown to cause adult and juvenile fish mortality [Herbert and Merkens 1961; Herbert and Richards 1963; and several studies reported in Stern and Stickle 1978], increases in ventilation [Horkel and Pearson 1966], physical damage to gills and other exposed tissues [Herbert and Merkens 1961; Ellis 1944 in Cordone and Kelley 1961], plus other behavioral effects, it has been shown that under natural conditions [Peters 1967; Herbert et al. 1961; Burnside 1967] fish do not remain in areas of high turbidity. These data suggest that the likely response of affected adult and juvenile fishes to increased turbidity would be temporary emigration from the affected area and this sort of response has been documented in the literature [Gammon 1970].

The U. S. Environmental Protection Agency [1976] suggested a suspended solids criterion which ". . . should not reduce the depth of the compensation point for photosynthetic activity by more than 10% from the seasonally established norm for aquatic life." Clearly, a site-specific criterion of this sort has more realistic applicability to continuous discharges than to transient (less than 24-hour) spill conditions.

A more realistic criterion for the purposes of coal slurry spill analyses should be related to potential fish and invertebrate mortality. Data cited or published by Wallen [1958] and Hocutt et al. [1976] indicate that there is no evidence of gill clogging or damage in fishes exposed to short-term suspended solids concentrations of approximately 70,000 mg/l. Behavioral effects on fishes (primarily avoidance) have been reported as a result of concentrations as low as 20,000 mg/l. For the purposes of these analyses, avoidance behavior will not be considered a significant fisheries impact since repopulation of the affected habitat would be anticipated after pipeline repair. Avoidance behavior is, in fact, a desirable response under spill conditions.

The suspended solids concentration below which adult and juvenile fish mortality would not be expected to occur is considered to be 70,000 mg/l. Significant impacts on the indigenous invertebrate population, however, would be anticipated to occur at much lower concentrations of approximately 400 mg/l [U. S. EPA 1976] and these will be discussed below.

The most notable fisheries impact anticipated to be associated with spill-induced turbidity would be a potential reduction in reproductive success. When fine sediment settles on coarse unconsolidated substrates the permeability of those substrates is decreased. When eggs are laid in affected areas water may not flow freely over the eggs which can result in a decrease in hatching success [Meehan and Swanston 1977; Auld and Schubel 1978]. Auld and Schubel [1978] found that varying amounts of suspended sediments affected the hatching success of species of fish differently. Less than

1000 mg/l did not affect the hatching success of yellow perch, blueback herring, alewife or American shad eggs, but 1000 mg/l significantly reduced the hatching success of white perch and striped bass.

It is also believed that sediment affects the flow of water through gravel, preventing the removal of metabolic waste and entrance of oxygen [Cooper 1965; Sheridan and McNeil 1968; Meehan and Swanston 1977; and others]. Shelton and Pollock [1966] found that if 15 to 30 percent of the interstices in gravel were filled with sediment, there resulted an 85 percent mortality of salmon eggs. The sediment may act as a physical barrier to the fry, even if they do hatch successfully. Sedimentation can also disrupt reproduction by covering spawning grounds [Karr and Schlosser 1978], making them unavailable for reproduction. If a spill were to occur during the spawning period in a nursery area, locally significant impacts on eggs and larval fishes would be anticipated.

If a spill were to coincide with fish migration periods there is a possibility that a spill would interfere with pre- or post-reproductive migration. Such interference has been reported in the literature [U. S. EPA 1976] and the severity of the impact would depend upon the spawning behavior of the species involved, the suspended solids increase anticipated, and the delineation of the downstream area to be affected.

The primary impacts of stream siltation on aquatic invertebrates would be gill membrane abrasion, smothering, and/or loss of acceptable substrate habitat as a result of substrate in-filling. Casey [1959 as reported in Cordone and Kelley 1961] found that siltation for about one-fourth mile downstream from a dredging operation eliminated macroinvertebrates. There was a 50 percent reduction in numerical abundance one mile below the dredge. Not only do macroinvertebrate populations decrease when sedimentation increases [Tebo 1955], low level sedimentation can alter species composition [Conlan and Ellis 1979; Rosenberg and Wiens 1978]. Conlan and Ellis [1979] found that a one-cm layer of wood waste resulted in a reduction in biomass and the loss of the majority of suspension feeders with the affected area becoming dominated by deposit feeders. Rosenberg and Wiens [1978] noticed that sediment additions resulted in different drift rates for various invertebrates. White and Gammon [1977] reported that increases in suspended solids to concentrations of less than 100 mg/l resulted in increased drift rates to more than double the normal rate. Since an increase in sedimentation results in a decrease in habitat diversity by filling in the substrate interstices, macroinvertebrate standing crop has been found to decrease [Williams and Mundie 1978; Allan 1975; Barber and Kevern 1973].

Mollusks, in addition to insects, may also be adversely affected by stream siltation. Ellis [1936] reported a general increase in mortality of mussels affected by silt, and many species of snails and mussels specifically avoid silt-substrate areas [Pennak 1978].

While suspended solids concentrations as low as approximately 100 mg/l would be anticipated to stimulate some aquatic insect emigration from the affected area in the form of invertebrate drift [White and Gammon 1977], deposition of as little as approximately 0.3 inches of coal particles would be expected to significantly reduce invertebrate production in the affected area.

Data summarized from Neves [1979], Bane and Lind [1978], and Ragland [1974] suggest that "typical" freshwater benthic invertebrate dry weight estimates range from 0.02 to 0.60 oz/yd^2, when appropriately converted. For assessment purposes a mean dry weight biomass of 0.50 oz/yd^2 has been assumed for all potentially affected macroinvertebrate populations. These estimates will be used to determine the invertebrate productivity loss for each of the spill scenarios where coal particles would be expected to settle on the stream bottom to a depth of approximately 0.3 inches, or greater.

The anticipated secondary impacts of this invertebrate productivity reduction on affected fishes would be a similar decrease in fish food availability. If a fish is assumed to be approximately 15 percent efficient in converting its food to flesh [Russell-Hunter 1970] then, for every yd^2 of substrate covered, approximately 0.08 oz (dry weight) of the fish flesh would be "lost." It should be noted that these productivity estimates are very liberal in order to present the "worst-case" data in a quantitative fashion. Potential invertebrate and fish biomass loss estimates are provided in Table III.

Table IV is an impact matrix which summarizes the aquatic biological impacts anticipated at the various spill sites using the impact criteria discussed above. As indicated in Table IV, the anticipated physical effects and/or biological impacts are identified for both complete and orifice ruptures during both high and low streamflow conditions.

These specific spill scenario data tend to substantiate the theory that, in general, spills in "small" streams or under low streamflow conditions would be anticipated to cause more severe biological impacts than spills in large rivers or under high streamflow conditions. Depending on the spill location, seasons, and sensitivity of the affected biota, significant aquatic biological impacts could include (1) approximately 85 percent mortality of fish eggs and larvae in the affected river area; (2) smothering of benthic macroinvertebrates and substrate habitat, or at least emigration of invertebrates and fishes from affected substrate areas; (3) heat shock or cold shock fish and invertebrate kills; (4) bioaccumulation of some toxic metals; (5) low dissolved oxygen tension stress; and/or (6) decreases in invertebrate and fish productivity in the affected areas. It is important to note, however, that these are "worst-case" analyses and it is impossible for all of these impacts to occur at any single spill location.

Table III. Estimates of Macroinvertebrate and Equivalent Fish Flesh Biomass that May Be Lost as a Result of Particulate Coal Deposition at Coal Slurry Pipeline Spill Sites

River Substrate Area Covered (yd^2)	Macroinvertebrate Biomass (lbs, dry weight)	Fish Biomass (lbs, dry weight)
10	0.3	0.05
20	0.6	0.1
50	1.6	0.2
100	3.1	0.5
500	15.6	2.5
1,000	31.2	5.0
2,000	62.4	10.0
5,000	156.0	25.0
10,000	312.0	50.0

Table IV. Quantitative Comparison of Anticipated Aquatic Biological Impacts at the Various Spill Scenario Sites

	Site 1 (N. Platte River)			
	High Flow 11,450 cfs		Low Flow 266 cfs	
Impact Category	CR	OR	CR	OR
Maximum temperature change expected (°C)	0.0	0.0	6.1	1.6
Heat or cold shock (at least 3°C change)	No	No	Cold	No
Dissolved oxygen stress anticipated (DO below 5 mg/l)	No	No	No	No
Suspended solids emigration (ft)	>10,000	100	>10,000	>10,000
Time to emigrate (hr)	1	5	1	1
Sedimentation mortality (lbs, dry weight) Fish Invertebrates	No	No	No	No

CR = Complete rupture.
OR = Orifice rupture.

Table IV. Continued

	Site 2 (Sand Creek)			
	High Flow 317 cfs		Low Flow 211 cfs	
Impact Category	CR	OR	CR	OR
Maximum temperature change expected (°C)	6.1 4.1	0.4	12.4 13.8	10.2 11.3
Heat or cold shock (at least 3°C change)	Heat or Cold	No	Heat or Cold	Heat or Cold
Dissolved oxygen stress anticipated (DO below 5 mg/l)	No	No	Yes (0.7)	Yes (2.0)
Suspended solids emigration (ft)	>10,000	50	>43,000	>36,000
Time to emigrate (hr)	1	1	6	100
Sedimentation mortality (lbs, dry weight) Fish	No	No	No	0.3
Sedimentation mortality (lbs, dry weight) Invertebrates				1.7

CR = Complete rupture.
OR = Orifice rupture.

Table IV. Continued

	Site 3 (Illinois River)			
	High Flow 16,300 cfs		Low Flow 152 cfs	
Impact Category	CR	OR	CR	OR
Maximum temperature change expected (°C)	0.1	0.1	7.0	0.8
Heat or cold shock (at least 3°C change)	No	No	Cold	No
Dissolved oxygen stress anticipated (DO below 5 mg/l)	No	No	No	No
Suspended solids emigration (ft)	50	50	>4,000	>400
Time to emigrate (hr)	1	1	1	1
Sedimentation mortality (lbs, dry weight) Fish	No	No	4.7	0.07
Sedimentation mortality (lbs, dry weight) Invertebrates	No	No	13.2	0.4

CR = Complete rupture.
OR = Orifice rupture.

Table IV. Continued

Impact Category	Site 4 (Saline River)			
	High Flow 11,400 cfs		Low Flow 246 cfs	
	CR	OR	CR	OR
Maximum temperature change expected (°C)	0.1	0.1	0.7	0.1
Heat or cold shock (at least 3°C change)	No	No	No	No
Dissolved oxygen stress anticipated (DO below 5 mg/l)	No	No	No	No
Suspended solids emigration (ft)	500	50	400	50
Time to emigrate (hr)	0.5	1	0.5	1
Sedimentation mortality (lbs, dry weight) Fish	0.7	1.0	No	No
Sedimentation mortality (lbs, dry weight) Invertebrates	4.9	6.4	No	No

CR = Complete rupture.
OR = Orifice rupture.

Table IV. Concluded.

		Site 5 (Bayou Cocodrie)		
		High Flow 1,900 cfs		Low Flow 104 cfs	
Impact Category		CR	OR	CR	OR
Maximum temperature change expected (°C)		0.1	0.1	0.3	0.1
Heat or cold shock (at least 3°C change)		No	No	No	No
Dissolved oxygen stress anticipated (DO below 5 mg/l)		Yes (3.0)	Yes (3.0)	Yes (4.9)	Yes (4.9)
Suspended solids emigration (ft)		50	50	50	50
Time to emigrate (hr)		0.3	1	0.3	1
Sedimentation mortality (lbs, dry weight)	Fish	3.2	2.4	>0.2	0.2
	Invertebrates	21.3	16.1	>1.3	1.3

CR = Complete rupture.
OR = Orifice rupture.

ACKNOWLEDGEMENTS

Physicochemical spill characteristics, upon which all biological analyses were based, were derived from various hydraulic and water quality models used by Patrick Ritter and Peter Mangarella of Woodward-Clyde Consultants. Funding for preparation of this manuscript was provided by the Environmental Systems Division of Woodward-Clyde Consultants. I would like to sincerely thank Gerry Likens and Kathy Trimble for typing this manuscript, and Betty Dehoney for considerable technical assistance in preparing it.

LITERATURE CITED

Allan, J. D. 1975. The distributional ecology and diversity of benthic insects in Cement Creek, Colorado. Ecology 56:104-153.

Auld, A. H. and J. R. Schubel. 1978. Effects of suspended sediment on fish eggs and larvae: A laboratory assessment. Estuary and Coast Marine Science 6:153-164.

Bane, C. A. and O. T. Lind. 1978. The benthic invertebrate standing crop and diversity of a small desert stream in the Big Bend National Park, Texas. Southwestern Naturalist 23:215-226.

Barber, W. E. and N. R. Kevern. 1973. Ecological factors influencing macroinvertebrate standing crop distribution. Hydrobiologia 43:53-75.

Burnside, K. R. 1967. The effects of channelization on fish populations in Boeuf River in northeast Louisiana. Louisiana Wildlife and Fisheries Commission.

Burton, D. T., P. R. Abell and T. P. Capizzi. 1979. Cold shock: Effect of rate of thermal decrease on Atlantic menhaden. Marine Pollution Bulletin 10:347-349.

Cherry, D. S., R. K. Guthrie, F. F. Sherberger and S. R. Larrick. 1979. The influence of coal ash and thermal discharges upon the distribution and bioaccumulation of aquatic invertebrates. Hydrobiologia 6:257-267.

Conlan, K. E. and D. V. Ellis. 1979. Effects of wood waste on sand-bed benthos. Marine Pollution Bulletin 10:262-267.

Cooper, H. G. 1965. The effects of transported stream sediment on the survival of sockeye and pink salmon eggs and alevin. International Pacific Salmon Fishery Commission Bulletin XVIII.

Cordone, A. J. and D. W. Kelley. 1961. The influences of inorganic sediment on the aquatic life of a stream. California Fish and Game 47:189-228.

Department of the Interior. 1981. Final environmental impact statement on the Energy Transportation Systems Inc. coal slurry pipeline transportation project. Bureau of Land Management and Woodward-Clyde Consultants.

Ellis, M. 1936. Erosion silt as a factor in aquatic environments. Ecology 17:29-42.

Gammon, J. R. 1970. The effect of inorganic sediment on stream biota. Water pollution control research series, 18050 DWC 12/70. U. S. Environmental Protection Agency.

Gangmark, H. A. and R. G. Bakkala. 1960. Comparative study of unstable and stable (artificial channel) spawning streams for incubating king salmon at Mill Creek. California Fish and Game 46:151-164.

Herbert, D. W. M. and J. C. Merkens. 1961. The effect of suspended mineral solids on the survival of trout. Journal of Air and Water Pollution 5:46-55.

Herbert, D. W. M. and J. M. Richards. 1963. The growth and survival of fish in suspensions of solids of industrial origin. International Journal of Air and Water Pollution 7:297-302.

Herbert, D. W. M. et al. 1961. The effect of china-clay wastes on trout streams. International Journal of Air and Water Pollution 5:56-74.

Hocutt, C. H., K. L. Dickson, M. T. Masnik and J. R. Stauffer. 1976. Observations of an ash lagoon spill on the New River, Virginia. Hydrobiologia 48:241-245.

Horkel, J. D. and W. D. Pearson. 1976. Effects of turbidity on ventilation rates and oxygen consumption of green sunfish, *Lepomis cyanellus*. Transactions of the American Fisheries Society 105(1):107-113.

Hynes, H. B. N. 1970. The ecology of running waters. University of Toronto Press, Toronto, Ontario. 555 pp.

Karr, J. R. and I. J. Schlosser. 1978. Water resources and the land water interface. Science 201:229-234.

Meehan, W. R. and D. N. Swanston. 1977. Effects of gravel morphology on fine sediment accumulation and survival of incubating salmon eggs. Research paper PNW220. USDA Forest Service.

Neves, R. J. 1979. Secondary production of epilithic fauna in a woodland stream. American Midland Naturalist 102: 209-224.

Pennak, R. W. 1978. Fresh-water invertebrates of the U. S. 2nd ed. New York, Ronald Press Company.

Peters, J. C. 1967. Effects on a trout stream of sediment from agriculture practices. Journal of Wildlife Management 31:805-812.

Ragland, D. V. 1974. Evaluation of three side channels and the main channel border of the middle Mississippi River as fish habitat. U. S. Army Engineer District, St. Louis.

Rosenberg and Wiens. 1978. Effects of sediment addition on macrobenthic invertebrates in a northern Canadian river. Water Research 42:753-763.

Russell-Hunter, W. D. 1970. Aquatic productivity: An introduction to some basic aspects of biological oceanography and limnology. New York, MacMillian Company.

Scullion, J. and R. W. Edwards. 1980. The effects of coal industry pollutants on the macroinvertebrate fauna of a small river in the South Wales coalfield. Freshwater Biology 10:141-161.

Shelton, J. M. and R. D. Pollock. 1966. Siltation and egg survival in incubation channels. Transactions of the American Fisheries Society 95:183-187.

Sheridan, W. L. and W. J. McNeil. 1968. Some effects on logging on two salmon streams in Alaska. Journal of Forestry 66:128-133.

Stern, E. M. and W. B. Stickle. 1978. Effects of turbidity and suspended material in aquatic environments. Technical report D-78-21. U. S. Army Engineer Waterways Experimental Station, Vicksburg, Mississippi.

Talmage, S. S. and C. C. Coutant. 1978. Thermal effects. Journal of Water Pollution Control Federation:1514-1553.

Tebo, L. B., Jr. 1955. Effects of siltation resulting from improper logging on the bottom fauna of a small trout stream in the southern Appalachians. Progressive Fish Culturist 17:64-70.

Thurston, R. V., R. C. Russo, C. M. Fetterolf, Jr., T. A. Edsall and Y. M. Barber, Jr. (Eds.). 1979. A review of the EPA Red Book: Quality criteria for water. Water Quality Section, American Fisheries Society, Bethesda, Maryland, 313 pp.

U. S. Environmental Protection Agency. 1976. Quality criteria for water. Washington, D. C. 256 pp.

Wallen, G. H. 1958. Fishes of the Verdigris River in Oklahoma. M.S. Thesis. Oklahoma State University, Stillwater.

Waters, T. F. 1972. The drift of stream insects. Annual Review of Entomology 17:253-272.

White, D. S. and J. R. Gammon. 1977. The effect of suspended solids on macroinvertebrate drift in an Indiana Creek. Proceedings of the Indiana Academy of Science 86:182-188.

Williams, D. D. and J. H. Mundie. 1978. Substrate size selection by stream invertebrates and the influence of sand. Limnology and Oceanography 23:1030-1033.

Woodward-Clyde Consultants. 1980. Ruptures and spills: Technical report. U. S. Department of the Interior.

APPENDIX

THE INTERNATIONAL SOCIETY OF PETROLEUM INDUSTRY BIOLOGISTS

The International Society of Petroleum Industry Biologists, sponsor of this Symposium, is an organization of professional biologists associated with the petroleum industry. The Society was established by several biologists employed by Atlantic Richfield, British Petroleum, Shell and Union Oil Companies who met often at oil spill and other petroleum-related conferences where, in impromptu meetings, they shared environmental and biological information.

Recognizing the professional value to their companies and to one another of these informal meetings, they formalized the association in 1976, thereby opening the discussions to a wider audience and a broader array of subject matters. The Fourth Annual Meeting of the Society in Denver in September, 1981, of which this volume is the outcome, was directed specifically to freshwater and terrestrial topics.

The Articles of Association of the Society summarize well the purpose of the organization:

    a. "To promote and encourage the application of sound biological science to environmental issues, to increase awareness and appreciation of natural ecosystems, and to promote the sound stewardship of living resources and the environments upon which they depend;

    b. "To gather and disseminate technical and other information on biological communities associated with energy exploration, production, processing and transportation for the purpose of recommending sound measures to mitigate any adverse impact of these operations on those communities;

    c. "To provide a forum for the discussion of environmental issues relevant to energy development activities;

d. "To promote understanding and cooperation among the scientific community, public environmental organizations, and energy-producing industries."

An integral part of the Articles of Association is the Code of Ethics to which all members subscribe:
"All members of the Society assume the responsibility to increase awareness and understanding of man's proper relationship with biological systems in connection with the exploration for and production of energy. The Society's members will be held to the highest standard of professional integrity and conduct at all times. Since each member recognizes that the conservation of living resources is the primary objective of his/her profession, all members agree to encourage and support research to maintain and improve conditions for those ecosystems that are associated with energy recovery. All members agree to encourage the exchange of information among members of the biological sciences profession, the energy industry, and any and all members of the interested general public. All members will strive to achieve public and industry understanding of the intricate complexities surrounding the relationship between the conservation of living resources and the activities of a highly industrialized society. All members agree to increase their knowledge and to utilize their respective personal skills to advance the practice of resource conservation and education in connection with energy-related activities. All members will execute their professional responsibilities in an objective manner and will base any decisions reached upon fundamental ecological principles."

Membership in the Society today approaches 200 members who reside in Canada, England, Scotland, France, and Holland as well as in the United States. When asked to describe their areas of training or expertise, members' replies range, literally, from "air pollution" to "zoology" and reflect subject matters and jargon currently fashionable as well as more classic fields. In the following list of specialities, members' own terminology is retained, hence descriptors are not mutually exclusive.

BIOLOGY: botany, zoology;applied biology; field biology; biochemistry; microbiology, virology, environmental microbiology, soil microbiology; entomology, ornithology, parasitology, mammalogy; marine biology/ecology, coral reef ecology, intertidal zone biology, marine zooplankton, marine invertebrates, benthic ecosystems, biological oceanography, marine mammals, marine phycology, estuarine ecology; ichthyology, fisheries, fisheries biology/ecology, fisheries management, zoogeography of fishes, freshwater fishes; freshwater ecology/

biology, limnology, aquatic toxicology, aquatic macroinvertebrate distribution and ecology; wildlife, wildlife management; environmental physiology; ecology, ecosystem energetics, plant ecology, biogeography, ethology, arctic biology/ecology, terrestrial ecology/biology, wildland ecology, vegetation;

HEALTH SCIENCES: medicine, veterinary medicine, health and safety, pharmacology, toxicology, human physiology and nutrition;

ENVIRONMENTAL SCIENCES: air pollution effects on vegetation; heavy metals effects; environmental impact, environmental audit, environmental science, mitigation of environmental disturbances, resource assessment; water quality; oil spill response, assessment of oil spill damage, oil spill control and recovery, oil pollution, reclamation of spills; botanical aspects of land disposal of oily wastes, biological oxidation processes, biodegradation; hazardous material disposal, hazardous waste; land reclamation, soil reclamation, agriculture, revegetation research, reclamation of surface mines; wildlife and pipeline interactions; habitat management, coastal zone management, impact assessment of dredging, perturbations in wetlands, environmental management, ecological management, ecology of land management; regulations, federal liaison;

OTHER: bioassays, ocean systems, oceanography, marine sediments.

Current officers of the Society from whom further information can be obtained are:

    President: James P. Ray, Shell Oil Co., Houston, Texas
    Vice Pres: Bernard Tramier, Soc. Nat. Elf Aquitaine, Pau, France
    Secretary: Mary Anne Sharpe, Gulf Canada Resources, Inc., Calgary, Alberta, Canada
    Treasurer: Dilworth W. Chamberlain, Atlantic Richfield Co., Los Angeles, California

INDEX

Abies balsamea 340
Abies lasiocarpa 340
Accipiter cooperii 62
Accipiter striatus 62
adaptive environmental assessment and management 1
adenosine triphosphatase 281,282
aerial photography 165,405
Agropyron
  cristatum 109,115
  dasystachyum 109,129
  elongatum 341
  intermedium 108
  riparium 109,129
  smithii 109,129
  spicatum 109
  trachycaulum 109,129
air quality 38,46
Alaska
  Bering Sea 10
  mapping 165
  National Petroleum Reserve 403
  North Slope 165,403
  Outer Continental Shelf leasing 10
  seismic exploration 403
  tundra 403
Alberta 185,339
Alces alces 406
American Petroleum Institute 321
ammonia 319
Amoco Production Co. 365
amphibians and reptiles 74
Anaconda Copper Co. 365
animal
  censusing methods 59
  population cycles 19,65,224
  See also birds; fish; invertebrates; mammals; specific kinds
antelope, pronghorn 221
Antilocapra americana 221
apparatus for oil solubilization 179
aquatic biology 74
  community species diversity 169
  effects of ammonia 319
  effects of coal slurry pipeline spills 433
  effects of pipeline stream crossings 385
  food web 297
  plants 213
  plants and pipeline construction 395
  plants in wetlands 164
  See also invertebrates
Aquila chrysaetos 66
ARCO Coal Co. 222
ARCO Transportation Co. 234
arctic mapping 165
arctic tundra 165,403,405
Arctostaphylos rubra 412
Arkansas 433
Army Corps of Engineers
  See Engineers, Corps of
Artemisia tridentata 221
aspen 340
Atlantic Richfield Pier E Tanker Terminal Relocation Project 234
automated data processing systems 75
Avicennia spp. 304

bacteria 190
  marine, bioassay 204,206
bald eagle 153,154

baseline data integration and planning 35,58,77,86,99
bass, largemouth 324
Bayou Cocodrie, LA 436
bearberry 412
bears, grizzly 153,155
bears, polar 263
benthos, freshwater 358,385, 392,435
benthos, marine 13,174,246
benz(a)anthracene 285
Bering Sea 10
Betula nana 405
Betula papyrifera 340
bioaccumulation
  methods 246
  of cadmium 237
  of heavy metals 233,437
bioassay
  apparatus 180
  clams (Macoma) 236
  Daphnia 190,203
  fish 190,202,319
  marine bacteria 190,204,206
  methods 180,246
  native plants 343
  polychaete worms 246
  solubilizer 180
  standards 253
biological assessment 59,73, 86,151,169,222,323,340, 355,365,386,405,419
biological damage
  See environmental impact assessment
biological impact assessment
  See environmental impact assessment
biomonitoring 59,343
birch, dwarf 405
birch, white 340
birds 38,40,62,66,74,153,155, 222,257,406
  treatment of oiled 257
  See also raptors, songbirds, specific kinds; waterfowl
bitterbrush, antelope 66
bitumen extraction 186
black-footed ferret 153
BLM
  See Bureau of Land Management
blueberry 344
bluegills 324,330
bluegrass, Kentucky 108,129
bobcat 380
bog 420
boreal forest 340
brine effects on plants 339
brine spill reclamation 339
Broad Run, PA 389
bromegrass, smooth 129
Bromus inermis 129
Brooks Range, AK 406
browse plants 66,222
Bunker C oil effects on mangroves 306
Bureau of Land Management
  Alaska 403
  coal leasing program 139
  Colorado 75
  crude oil pipelines 103,152
  Outer Continental Shelf leasing programs 10,179
  Technical Investigations 139
  visual resources classification 51
  Wyoming 122,125
Bushkill Creek, PA 391
Buteo jamaicensis 66

cadmium in clams 233
cadmium in dredge spoil 241
calcium in soil 344
California 212,233
  Coastal Commission 233
canarygrass, reed 108,341
Canis latrans 62,66
Carex aquatilis 405
Caribbean Sea 305
caribou 406
carnivores
  See specific kinds
carp 326
catfish, channel 324,325
census techniques for wildlife 424
Centrocercus urophasianus 37, 40,66,222
Cervus elaphas 364
Charter Company 58
Chautauqa Creek, NY 387

chemical treatment of tailings
   pond waters   204
chickadee, mountain   64
chipmunks   63
chloride   339
<u>Chlorura chlorura</u>   62
cichlids   326
clams   233,236,246,249
classification of wetlands   157
climate, effect on revegetation planning   99
cloudberry   414
clover
   alsike   341
   subterranean   108
   white   108
coal
   conversion effluents   320
   leasing program   139
   mining   35,87
   slurry pipeline   433
Coal Management Program   139
Coal Program Technical Investigations, BLM   140
COE
   <u>See</u> Engineers, Corps of
Cold Regions Research and
   Engineering Laboratory
   (CRREL)   165
colonization of disturbed
   areas by plants   412
Colorado
   Division of Wildlife   74
   energy development and
      wildlife   71
   Piceance Basin ecology   58
   State University   74
   University of   74
   uranium disposal reclamation
      111
Columbia Gas System   385
computer-assisted modeling
   <u>See</u> modeling
conflict resolution   8,155,
   230,233
conifer forest, Canada   339
conifer forest, Wyoming   125
<u>Conocarpus</u>   314
conservation biology   82,142
consultation process, endan-

   gered species   143
Cook Inlet crude oil   181
cordgrass, salt meadow   419
corridors
   <u>See</u> rights-of-way
cottongrass   405
coyotes   62,66
crabs   11,14
crane, whooping   153,155
critical habitat, Endangered
   Species Act   143
cross-country travel effects
   404
crude oil
   <u>See</u> oil, crude; <u>specific</u>
      <u>kinds</u>
cumulative wildlife impacts   71

damage by livestock   137
damage by wildlife   226
Dames and Moore   111
<u>Daphnia</u>
   <u>magna</u>   190,206
   <u>pulex</u>   297,301
   spp.   203
decision-making techniques   1,
   23,35,57,71,85,95,111,141,
   165,169,221,233,419,433
deer, mule   38,60
Delaware River-Bay oil spills
   257
depuration
   methods   249
   of cadmium by clams and worms
      249
   of hydrocarbons by <u>Daphnia</u>
      301
detoxification of tailings pond
   waters   202
diesel engine emissions   46
diesel fuel effect on mangroves
   307
disposal sites   242
diversity
   estimates   169
   indices   169
   measurements   169,392
dredge-and-fill permits   236
dredge sediment analysis   237
dredging effects

on freshwater biota 386
on marine biota 240
on marine sediments 240
dredging spoils disposal 240
drilling fluid pits 127,353
drilling operations and wildlife 370
drill site reclamation 122
Dryas integrifolia 405
ducks
   See waterfowl
dumpsites, dredge sediments 242
dust 46

eagles, golden 66
ecological diversity estimates 169
economic feasibility analyses 150
economics of revegetation 133
economics of wetland enhancement 214
ecosystem energetics 13,58,419
effects
   of ammonia on aquatic biota 319
   of brine spills on vegetation 339
   of coal mining on the environment 37
   of coal slurry pipeline on aquatics 437
   of development on wildlife 57,71
   of fertilizer on arctic tundra 412
   of haul roads on wildlife 38
   of pipeline stream crossings on benthos 385
   of railroads on wildlife 38
   of seismic exploration on vegetation 403
effects of crude oil
   on birds 257
   on crabs 11
   on Daphnia pulex 297
   on fish 179,281
   on mangroves 306
   on polar bears 263
   on shellfish 179

effects of hydrocarbon development
   on elk 364
   on marine fisheries 11
   on recreation 47
effluents used to create wetlands 211
EIS
   See environmental impact assessment
elk 364
EMRIA
   See Energy Minerals Rehabilitation Inventory and Analysis
emulsification of crude oil 180
Endangered Species Act 142
Energy Minerals Rehabilitation Inventory and Analysis (EMRIA) 139
energy resource development 35,72,211,221
Energy Transportation Systems Inc. (ETSI) 433
engineering analysis 51
Engineers, Corps of, Alaska 165
Engineers, Corps of, permits
   for harbor improvement 26
   for pipeline river crossings 103
   for tanker terminal 233
   for wetlands use 419
environmental hazard of tailings pond waters 201
environmental impact assessment 1,23,35,57,71,85,95,111,121,139,141,157,165,169,185,211,221,233,303,319,339,353,363,385,403,419,433
   for mined land reclamation 57,85,111,139,221
   for wildlife management 35,57,71,85,141,221,363
   of conveyance routes 35
   of energy resource development 1,35,71,85,139,157,165,221,403
   of harbor improvements 26,233

460

of pipelines 95,141,385,433
of seismic exploration 403
of wetlands 157,211,303,419
environmental management
  methods 1,57,71,85,95,111,
    121,139,141,165,233
  of endangered species 141
  of water resources 23,157,
    185,211,221,385,419,433
  of wetlands 164
environmental planning
  See planning
Environmental Protection
  Agency 212,235,236
Environmental Research and
  Technology, Inc. (ERT) 152
enzymes, effect of crude oil
  on 281
EPA
  See Environmental Protection
    Agency
Eriophorum vaginatum 405
erosion control 117,132
estimation of species diversity 169
estuaries 419,423
ETSI
  See Energy Transportation
    Systems Inc.
Eutamius minimus 63
Eutamius quadrivittatus 63
exploration drilling 127
exploration for petroleum
  See oil and gas exploration

fatal flaw analysis 150
feasibility study 212
fertilizer treatments 131,341
fescues 108
fir, alpine 340
fir, balsam 340
fish 179,292,329,357,386,438,
  440,442-443
  bioassay with 190,202,319
  See also specific kinds
Fish and Wildlife Service 74,
  103,146,152,155,157,211
fisheries 11,49,319,385,433
Florida 303,315
forage 223
Forest Service 75,103

fouling
  See petroleum hydrocarbon
    effects
foxtail, meadow 108
France 353
freshwater benthic community 385
freshwater habitat analysis
  385,433

Gambusia affinis 330
game animals
  See wildlife
Geological Survey, U.S.
  Alaska 403
  Colorado 58,75
  Water Resources Division 140
geothermal effluents 211
geothermal energy resources 211
geothermal springs 213
gill enzymes, effect of crude
  oil on 285
grasses, response to salts 350
grasslands 222
grazing 103,137
Great Divide Basin, WY 124
Green Mountain, WY, Uranium
  Exploration Project 121
grizzly bear 153,155
groundwater analysis 427
groundwater hydrology 38,190,
  218
grouse, sage 38,40,62,66,222

habitat, characterization for
  wildlife use 90
habitat recovery from disturbance 385,408
harbor improvement 26
harbor sediment analysis 237
hardwood swamp 389
harvest of game animals 224
haul road effects 35
hawks
  Cooper's 62
  red-tailed 66
  sharp-shinned 62
  See also raptors
Haystack Mountains, WY 35
heavy metals
  See metals, heavy
hemorrhagic enteritis 260

horses, feral 59,66
Hudson Bay 264
Hungary 26
hunting 49,224
hydraulic model 436
hydrocarbon
  See petroleum hydrocarbons
hydrography, Bering Sea 11
hydrologic budget 421
hydrologic cycle 421
hydrology
  groundwater 218
  impacts of surface mining on 139
  of wetlands 164
  surface water 47
hydrophytes
  See aquatic plants; specific kinds
Hylocomium splendens 344

Ictalurus punctatus 324
Idaho 212
Illinois River, OK 436
impact analysis
  See environmental impact assessment
impact matrix 152
impacts
  See effects
Incline Village General Improvement District (IVGID) Project 214
indices of diversity 170
Indonesia 305
industrial development, effects on wetlands 419
insects 393,442
Institute of Animal Resource Ecology, University of British Columbia 5
Institute of Applied Systems Analysis, Vienna, Austria 5
International Environmental Consultants, Inc. 37
invertebrates 11,14,74,179, 190,203,206,233,236,246, 249,297,301,319,325,356, 358,385,393,437,439,442
  See also specific kinds

ion regulation in trout 281

Juncus roemerianus 419

Kuwait crude oil 306

Labrador tea 344,405
Lagopus lagopus 406
Lagurus curtatus 63
land use management 116
land use planning 23,35,57,71, 85,95,111
Larix laricina 340
leasing stipulations 10,58,125, 140
Ledum groenlandicum 344
Ledum palustre 405
leks, sage grouse 40,66
lemmings, brown 405
Lepomis macrochirus 324,330
LGL Ecological Research Associates 58
lichens 405
lingonberry 405
livestock 137,223
lodgepole pine 340
Long Beach, CA, Harbor 233
lotic habitat analysis 385,435
Louisiana 433
  crude oil 307
low-ground-pressure vehicles
  See off-road vehicles
Lupinus arcticus 405

Macoma carlottensis 246,249
Macoma sp. 233,236
Macrobrachium rosenbergii 325
macroinvertebrates
  See invertebrates
mammals 19,38,59,74
  See also specific kinds
mangroves
  black 305
  buttonwood 314
  red 305
marine fish
  See specific kinds
marine mammals
  See specific kinds
marine tanker terminal 234
marsh development 214

Maryland 387
Massachusetts 26
MBC Applied Environmental Sciences 179
metabolism, polar bears 266
metabolism, trout 281
metals, heavy
  bioaccumulation in clams 233
  bioaccumulation in freshwater invertebrates 437
  in wetlands 214
metals, toxicity of 214
metals, trace, in tailings pond waters 198
Mexico, Gulf of 305
mice, deer 63
Michigan 363
Micropterus salmoides 324
Midvale crude oil 266
milkvetch, cicer 116
mined land reclamation 85,111,221
mine haul roads 35
mine site environmental assessment 86
mining
  coal 35,87
  oil sands 186
  sodium 58
  uranium 111
minnows, fathead 190,202,326,328
mitigation
  for endangered species 155
  of energy development effects on wildlife 71
  of geothermal energy development 211
  to minimize effects of coal mining 222
  See also reclamation; revegetation
modeling
  coal slurry pipeline effects 435
  computer-assisted 3,24,436
  ecological diversity 169
  ecological interrelationships 59,75,419
  environmental management strategies 1,23
  effects of energy resource development on the environment 3,13,37,73,434
  mangrove forest response to stress 303
  multiple attribute decision 23
  nutrient export to an estuary 421
  rights-of-way effects 37
  worst-case analysis 435
mollusks 233,236,246,249,442
monitors of impact 59
Montana 100,106,212
moose 406
Morone americana 325,327
mosquito fish 330
mosses 344,405
Mountain View Sanitation District, CA 217
mousse, formation and removal 181
mulching 117,129
mule deer, Rocky Mountain 38,59
Multi Mineral Corp. 58
multiple attribute-information theoretic decision model (MAIT) 23
multiple-use land management 221

National Marine Fisheries Service, Tiburon Laboratory 181
National Oceanic and Atmospheric Administration (NOAA) 146
  Alaska Office of Marine Pollution Assessment 11
National Petroleum Reserve—Alaska 403
national wetlands inventory 157
natural gas pipelines 385
natural resource management 73,86
Neanthes arenaceodentata 246
Nebraska 433
needle and thread grass 109
needlegrass, green 109,129
nesting habitats 41

Nevada 212
New Mexico 111,212
New York 387
Nigeria 305
nitrate concentration in soil 116,427
nitrate concentration in water 321
nitrification in surface waters 321
Nitrobacter 323
Nitrosomonas 323
NOAA
　See National Oceanic and Atmospheric Administration
Norman Wells crude oil 281,298
North Dakota 154
Northern Colorado, University of 74
Northern Tier Pipeline 101,142
North Platte River 47,436
North Slope
　See Alaska
Northwest Colorado Wildlife Consortium 72
Notropis lutrensis 327
Number 2 oil 306
nuthatch, white-breasted 64
nutrient transport in wetlands 419

oceanography, Bering Sea 11
OCS
　See Outer Continental Shelf
Odocoileus hemionus 38,59
off-road vehicle effects on tundra 403
oil
　pipelines 152
　production, Bering Sea 19
　production, Michigan 370
　products, Bunker C 306
　products, Number 2 306
　sands 186
oil and gas exploration
　Alaska Outer Continental Shelf 10
　effects on Alaska tundra 403
　effects on Michigan wildlife 363

oil and gas production effects
　on Bering Sea crab population 19
　on wildlife 370
oil, crude
　Cook Inlet 181
　Kuwait 306
　Louisiana 307
　Midvale 266
　Norman Wells 281,298
　Santa Barbara 179
　toxicity to fish and shellfish 179
　Venezuelan 281,298
oiled bird rehabilitation 257
oil spills
　modeling marine impacts 13
　response teams 258
　See also petroleum hydrocarbon effects
Oklahoma 433
Oregon 212
outdoor recreation, effects on animal populations 49,371
Outer Continental Shelf
　Bering Sea 11
　Environmental Assessment Program (OCSEAP) 11
　Pacific Office, BLM 179

Pacific Outer Continental Shelf Office, BLM 179
pelagic food web 12
pelican, brown 153
Pennsylvania 389
perch, white 325,327
peregrine falcon 153
periphyton, freshwater 437
permitting procedures 26,58, 91,103,125,142,254
permitting process, negotiation 156,234
Peromyscus maniculatus 63
Peromyscus truei 63
petroleum exploration
　See oil and gas exploration
petroleum hydrocarbons
　in tailings ponds 195
　See also oil and gas production

petroleum hydrocarbon effects
　on birds　257
　on crabs　11
　on *Daphnia*, of aromatic
　　fractions　297
　on *Daphnia pulex*　297
　on fish　11,179,281
　on mangroves　303
　on polar bears　263
　on seals　263
　on shellfish　179
Phleum pratense　129
Phoca hispida　264
Phoxinus phoxinus　326
phytoplankton, freshwater　437
Picea glauca　340
Picea mariana　340
Piceance Basin, Colorado　58
pier relocation, permitting
　234
Pigeon River County State
　Forest, MI　363
Pimephales promelas　190,202,
　206,328
pine
　forest　126,419
　limber　126
　lodgepole　126,131,340
　slash　419
Pinus contorta　340
Pinus elliotti　419
pinyon-juniper woodland　62
Pioneer Nuclear, Inc.　121
pipeline construction
　effects on freshwater biota
　　386
　planning process　142,152
　revegetation requirements　96
　river crossings　385,436
　tanker off-loading　234
pipelines
　coal slurry　433
　crude oil　96,142
　natural gas　385
planning　23,35,57,71,85,95,
　111,121,139,141,165,221,
　433
plant communities
　See specific kinds; vegeta-
　　tion types
plant cover

　See vegetation
Pleurozium schreberi　344
Poa pratensis　129
polar bears, effects of crude
　oil on　263
pollock, walleye　11
pollution effects　111,169,185,
　211,233,257,263,281,297,
　303,319,339,353,385,433
polychaete worms　246,249
polycyclic aromatic hydrocar-
　bons
　See petroleum hydrocarbons
ponds, tailings, physical and
　chemical properties　190
poplar, balsam　340
Populus balsamifera　340
Populus tremuloides　340
Port of Long Beach, tanker
　terminal　233
Powder River Basin, WY　222
prairie dogs　154
predation　63
primary productivity
　See productivity
probability models　24
problem solving
　See decision-making
Processes and Resources of the
　Bering Sea Program (PROBES)
　11
productivity, plant　45,59,424
project planning
　See planning
pronghorn　38,221,224
ptarmigan, willow　406
public policy decisions　20
Puerto Rico　303
Purshia tridentata　66

rabbits　59
railroad routing　35
ranching　225
range productivity　45
raptors　38,59,62,66,153,154
rare species　142
reclamation
　and livestock use　137
　and wildlife　76
　objectives　87,111,221
　of brine spills　339

of drilling fluid reserve
   pits 359
of exploration drill sites
   121
of haul roads 44
of mined land, Wyoming 221
of railroad spur rights-
   of-way 44
of rights-of-way 44
of soils 112
of surface-mined lands 139
of tailings disposal areas
   190
of tailings pond waters 202
of uranium mined lands 111
plans 87
Wyoming 37,221
recovery of arctic vegetation
   from disturbance 412
recovery of vegetation from
   petroleum drilling 359
recreation 47,122,221
Red Desert, WY 123
regulatory procedures,
   negotiation 234
rehabilitation
   See reclamation; revegeta-
      tion
remote sensing for animal
   population inventory 368
remote sensing for vegetation
   mapping 165,423
reptiles and amphibians 74
reseeding
   See revegetation
reserve pit reclamation 354
resource
   development and wildlife
      35,57,71,86
   management 1,221
   planning 23
restoration of wildlife habitat 87,223
revegetation
   and livestock grazing 137
   costs 133
   effects on animal popula-
      tions 88,371
   equipment 133
   for wildlife habitat res-
      toration 85
   high elevation 122
   methods 96,128
   of mine sites 85
   of pipelines 95
   of uranium exploration 122
   planning 85,95
   plant species selection 114
   research plots 137,341
   See also reclamation
Rhizophora 305
Rhodeus sericeus amarus 326
ricegrass, Indian 109
rights-of-way, coal mine conveyance route 35
rights-of-way, pipeline 152
ringed seal 264
riparian willow community
   126,406
rivers
   See streams
roads 35
rodents 59,61,63
rooting depths 114,115
route selection 35
Rubus chamaemorus 414
rush, black 419
ryegrass, perennial 108

sacaton, alkali 109
sagebrush
   reestablishment 222
   removal 222
   steppe 63,125,222
sage grouse 38,40,62,66,222
Saline River, Arkansas 436
Salix alaxensis 406
Salmo
   clarki 324
   gairdneri 190,202,206,282,
      324,325,326,327,328,387
   trutta 391
salmon
   Atlantic 326
   coho 325
   pink 12
salt marsh
   See estuaries
salts, effect on boreal forest 339

Sand Creek, OK   436
sandreed, prairie   109
Santa Barbara crude oil   179
Saudi Arabia   305
scenario, coal slurry pipeline spill   433
scenario, energy development and wildlife   78
scenic resources   50
seal, northern fur   12
seal, ringed   264
sea turtles   153
sedge meadows, arctic   405
sedges, arctic   405
sedimentation, effects on freshwater biota   385,437
sediments, harbor, analysis   242
seeding, experimental plots   341
seed mixtures for revegetation   108,109,129,341
seismic exploration   403
sensitive areas   101
sewage effluent   214
sheep   137
shellfish   179
Shell Oil Co.   234
shiner, red   327
siltation, effects on freshwater biota   386,437
simulation modeling   2,13
site inventory   86
site-selection process methods   23,37,55,75,165
skin, effects of crude oil on   269
slurry pipelines   433
snow   223
socioeconomic analyses   72, 218
sociopolitical conflicts   221
sodium concentrations in soil   344
Sohio Alaska Petroleum Co.   165
soil
   analysis   101,113,124,341, 425
   electrical conductivity   344
   erosion control   117,125
   in wetlands   164
   nutrients   116,425
   physical properties   117
   resource management   112
   rooting depths   114
   stabilization   117,132
   survey   112
   topsoil   112
Soil Conservation Service   103
   range site productivity values   45
solubilizer   179,181
songbirds   61,62,64
Southern California, University of   236
sparrow
   Brewer's   62
   chipping   64
   vesper   64
Spartina patens   419
species abundance   173
species diversity   115
spills
   analysis   433
   brine   339
   coal slurry pipeline   433
   oil   303
Spizella breweri   62
spruce   340
squirrel, golden-mantled ground   65
Stipa viridula   129
streams
   benthos sampling methods   392
   crossing impacts   47
   habitat enhancement   385
   physico-chemical characteristics   433
   siltation, effect on biota   385,439
surface mining and wildlife   86
surface water hydrology   47
surface water quality   321,423, 433
swamp, mixed deciduous   420
Syncrude Canada Ltd.   186
synthetic fuels and ammonia production   320
synthetic fuels production from oil sands   186

tailings ponds 185
  physical properties 198,205
  water quality 189
tamarack 340
tanker terminal permitting 233
Technical Investigations in Support of the Coal Program, BLM 139
techniques
  See specific kinds
Texas Eastern, Inc. 121
thermoregulation, polar bears 269
Tiburon solubilizer 181
tidal marsh
  See estuaries
Tilapia aurea 326
timothygrass 129
Tisza River basin, Hungary 26
topsoil
  See soil
towhee, green-tailed 62
toxicity of crude oil to fish and shellfish 179
tractor trains, effect on tundra 403
trails, effects on plants 403
treatment of tailings pond waters 202
tree plantation 360
tree planting 131
trefoil, birdsfoot 108
Tri-State Bird Rescue 257
trophic dynamics 60,297
trout
  brown 391
  cutthroat 324
  rainbow 190,202,206,282, 324,325,326,327,328,387
tundra, effects of cross-country travel on 403
tundra types 405
turtles, sea 153

uncertainty in environmental management 2
upland hardwood forest, MI 370
uranium exploration 121

uranium mining and milling 111
Ursus maritimus 263
Utah 111,212

Vaccinium myrtilloides 344
Vaccinium vitisidaea 405
vegetation
  analysis 45,369
  and wildlife 35,57,71,85,369
  classification 157
  impact assessment 37,87
  mapping 165,370,423
  productivity 45,59
  sampling 341,407,423
  structure and bird use 62
  succession 82
vegetation recovery
  from brine spills 341
  from drilling operations 358
  from vehicular disturbance 412
vegetation types
  forest, aspen 340,373
  forest, conifer 62,125,340, 373,419
  forest, deciduous 373
  forest, mangrove 303
  forest, wetlands mosaic 420
  grasslands 44,222
  pinyon-juniper woodland 63
  riparian 126,405
  sagebrush steppe 44,63,125, 222
  sedge meadows 405
  tundra 165,405
  wetlands 157,164,213,389,420
  See also specific kinds
Venezuelan crude oil 281,298
vireo, solitary 64
visual impacts 50
visual resources classification, BLM 51
vole, sagebrush 63

walleye pollock 11
warbler, black-throated gray 64
Washington state 101,105,110, 154
water analysis 190,213,321,

355,423,437
waterfowl
  effect of oil on   257,260
  habitats   211
  toxicity information   212
water quality
  and ammonia   321
  and coal slurry pipeline
    spills   433
  and drilling fluid reserve
    pits   355
  and geothermal effluents   211
  and oil sands tailings
    ponds   189
  model   436
water resources   419
  allocation   26
  planning   26
water, surface, hydrology   47
well pad construction effects
  on environment   403
wetlands
  classification   157
  construction costs   216
  creation from geothermal
    effluents   211
  definition   164
  environmental assessment
    of   419
  hardwood swamp   389
  operation and maintenance
    costs   216
  role in purification of
    effluents   214
whales   153
wheatgrass   109
  crested   115
  intermediate   108
  orbital tall   341
  slender   129
  streambank   129
  thickspike   129
  western   129
whooping crane   153,155
wilderness, arctic scenic
    values   416
wildlife
  and development in Colorado
    71
  damage   226
  foods   60,66,89,103,222

  harvest   224
  inventory   72,365,424
  population dynamics   224,364
  use of reclaimed mine land   88
wildlife habitat
  effects of coal conveyance
    route on   38
  effects of winter cross-
    country travel on arctic
    416
  enhancement   211
  in Wyoming coal leases   221
  restoration   85,89
wildlife management
  legal/political aspects   76
  of elk, MI   373
  of game, WY   221
  of pronghorn, WY   224
  policy decisions   71
willow ptarmigan   406
willows, arctic   405
winter cross-country travel   403
workshop on simulation model-
    ing   5
worm, polychaete   246,249
worst-case impact assessment
    15,433
Wyoming
  coal leases   221
  coal mine conveyance route
    35
  Department of Environmental
    Quality   87,112,125,223,228
  Game and Fish Department   40,
    221
  Green Mountain   123
  Haystack Mountains   35
  Powder River Basin coal
    fields   222
  Red Desert   123
  uranium disposal areas   111
  uranium exploration   121

Zekiah Swamp Creek, MD   387
zooplankton
  effects of hydrocarbons on
    297
  freshwater, effect of coal
    slurry pipeline spill on
    437
  in drilling fluid reserve
    pits   358

DATE DUE

C			
C APR 2 '85			
C MAY			
C AUG 5 '85			

DEMCO 38-297